A mi amigo Juan, como muestra de mi formalidad informal en tanto que físico.

Luis

ELEMENTS OF NON-RELATIVISTIC QUANTUM MECHANICS

ELEMENTS OF NON-RELATIVISTIC QUANTUM MECHANICS

Luis Sobrino

University of British Columbia, Canada

World Scientific

Singapore • New Jersey • London • Hong Kong

Published by

World Scientific Publishing Co Pte Ltd

P O Box 128, Farrer Road, Singapore 912805

USA office: Suite 1B, 1060 Main Street, River Edge, NJ 07661

UK office: 57 Shelton Street, Covent Garden, London WC2H 9HE

Library of Congress Cataloging-in-Publication Data
Sobrino, Luis.
Elements of non-relativistic quantum mechanics / Luis Sobrino.
p. cm.
Includes bibliographical references and index.
ISBN 9810223862
1. Nonrelativistic quantum mechanics. I. Title.
QC174.24.N64S57 1996
530.1'2--dc20 96-3188
CIP

British Library Cataloguing-in-Publication Data
A catalogue record for this book is available from the British Library.

Printed in Singapore.

PREFACE

It is almost a trivial observation that many collective endeavours require the accomplishment of numerous complicated tasks, and that the different individuals that perform them may lose sight of the overall design. Like other creative endeavours, the scientific quest is particularly prone to such a state of affairs, for not only does the difficulty of the tasks absorb the mind but the beauty of what is learnt in performing them engages the soul. Like a diver, who may be so charmed by the nearby coral so as to forget that its beauty is but a minute reflection of the majestic beauty of the ocean, a spectroscopist may be so taken by the beautiful symmetries that the molecules exhibit so as to loose sight of the wonderful structure of quantum mechanics that informs them.

* * *

For many years, I have taught a graduate course on non-relativistic quantum mechanics at the University of British Columbia. As a first graduate course in the subject, it was to provide a good understanding of the principles as well as the necessary basic skills for the application of quantum mechanics to a diversity of fields. The students, coming, as the did, from universities all over the world, had very diverse backgrounds and I could not assume that every one of them had a good grasp of the basic principles of quantum mechanics, all I could assume was that they all had studied the standard elementary applications. Because the course was required of all graduate students in Physics, the prospective fields of its audience ranged widely – from quantum field theory to physical oceanography –, which meant that I could not maintain their interest by centring the lectures on any particular field of application. Faced with these constraints, I felt that the best way of achieving the aims of the course was to present the theory of quantum mechanics taking care to exhibit its coherence and beauty, hoping to maintain the students' interest by appealing to their aesthetic sense. Now that I am nearing the end of my teaching career, I feel that the approach taken was successful, and I have the satisfaction of knowing that this feeling is shared by many of my former students.

The present book was born of the experience gained in teaching the course just described. It is not meant to be a textbook for such a course. It covers less material than the course did, but treated in more depth. As a first approximation, the book might be described as a coherent presentation of the core of the course augmented by the answers to questions that many of my students and some of my colleagues have put to me in the course of the years. Many of these questions I could not answer by reference to the literature, for the available answers were presented at a level not accessible to those that formulated them.

In spite of the suggestions that I received from students and colleagues I would not have undertaken the task of writing this book had it not been for my own desire to have a comprehensive overview of quantum mechanical theory and gain a better appreciation of its beauty. I feel now that the exercise was worth the effort and I hope that the reader would gain as much aesthetic satisfaction from reading this book as the author gained from writing it.

To limit the size of the book, I had to decide what not to include. It was clear to me that those problems which are solved in detail in every textbook (e.g., the hydrogen atom) could be left out, unless they were necessary for the development of the presentation (such as the harmonic oscillator) or particularly useful as illustrations of the theory. More difficult was the decision not to include approximate methods and scattering theory because, although these two subjects have in some sense an ancillary status within the general theory, not only are they of essential importance in the applications but they also have had an enormous influence in the way we think about the physical world. Even more difficult was the decision not to treat the quantization of fields more extensively and just to touch it very briefly in connection with the study of an assembly of particles.

I had no difficulty in deciding not to include a discussion of the interpretations of quantum mechanics nor of the closely related problem of measurement. These extremely interesting questions lie in a "metalevel" above quantum theory and, in my opinion, would require a full volume to be properly discussed.

What is included constitutes the core of non-relativistic quantum-mechanical theory, the *elements* needed to properly understand the developments of the subject and its applications.

As can be gathered from the introductory paragraphs of this preface, these elements are presented with a different emphasis than is common in the standard textbooks, and at a higher level, but without requiring from the reader more than a working knowledge of quantum mechanics on the physical side, and the mathematical background in algebra and analysis normally acquired at the undergraduate level. This book occupies a largely empty niche between the usual textbooks and the more mathematically advanced writings on the subject. I expect it would be of interest both to graduate students and to many practising scientists who would like to delve a little into the structure of quantum mechanics and into the mathematical notions that underlie it.

Quantum systems cannot be described in terms of our sensory experience, to understand them it is necessary to use the abstract language provided by the algebra of linear operators on a Hilbert space. For this reason, three mathematical appendices have been included. For ease of reference, their paragraphs are numbered and are often referred to in the main text. Appendix A reviews finite-dimensional vector spaces; the necessary results in the theory of Hilbert spaces are summarized in Appendix B, while Appendix C presents some elements of the theory of distributions leading to an explanation of the mathematical basis of Dirac's formalism. The first four chapters of the book make use of the material in sections 1 to 3 of Appendix A which are mostly intended to serve as a reminder. Before embarking in a thorough reading of the rest of the book, the reader is invited to peruse the final sections of Appendix A as well as appendices B and C. A fourth appendix, D, includes some pertinent results in classical mechanics.

Group theoretical considerations pertaining mostly to the Galilei and to the symmetric groups play an important part in the exposition. It has been possible to develop them without assuming on the part of the reader much more than a knowledge of the definition of group. They are not included in the appendices but in appropriate places in the main body of the text.

A remark on the mathematical aspects of the presentation is in order here. Our experimental instruments are finite, in size, in energy, and in the number of configurations that they can exhibit. This implies that, in principle, only the language of finite-dimensional vector spaces is needed to explain experimental results and to understand the structure of quantum mechanics. However, it is extremely fruitful – and necessary if we want to embed the theory in a space and time continuum – to consider idealized instruments capable of an infinite number of configurations, which require a description cast in the language of infinite-dimensional spaces. Because these infinite instruments are approximations of the actual finite ones, physicists have all reason to ignore those properties of the infinite-dimensional Hilbert space that cannot be obtained from the properties of finite-dimensional ones by some, not necessarily unique, *physically based* limiting procedure. A working knowledge of the mathematical description that results from the adopted limiting procedure is necessary to understand many of the developments of quantum mechanics. The mathematical presentation reflects these considerations: the results pertaining to finite-dimensional spaces, necessary for the understanding of the structure of quantum mechanics, are presented with thoroughness; their generalizations to infinite-dimensional spaces are discussed in less detail, although references to the mathematical literature are given for the benefit of the interested reader.

After an introductory chapter that sets the point of view adopted in the exposition, the main body of the book can be thought of as being divided into two parts. In the first one, Chapters II to IV, are presented and elaborated the basic rules (I prefer this word to "postulates" or "axioms" because it does not imply a mathematical rigour that the exposition does not have) as they apply to model systems describable in the language of finite-dimensional spaces. In this way one can exhibit the basic structure of the theory unencumbered by the mathematical complications that arise in infinite-dimensional spaces.

The second part deals with the generalization of the theory and its applications to different types of systems. The general theory is presented in Chapter V, in which the concept of kinematical symmetries is also introduced and applied to those symmetries that arise from galilean invariance. Chapters VI and VII which discuss the particle without internal degrees of freedom are followed by the study in Chapter VIII of the dynamical symmetries of quantum mechanics; at this point the reader may find it helpful to review the classical results of Appendix D. The particle with spin is discussed in Chapter IX, and composite systems in the remaining two chapters. The connection between galilean invariance and the fundamental observables of a system constitutes the unifying thread that runs through this part of the book.

Notes, mainly but not exclusively bibliographical, to most sections can be found after the appendices, followed by a selected bibliography.

A standard nomenclature has been adopted; minor exceptions are the use of "density operator" in preference to the more generally used "density matrix", of "commensurable observables" that seems more appropriate than "compatible" or "simultaneously measurable observables" and of "maximal observable" instead of the lengthier "complete set of commuting observables".

Dirac's formalism and notation is used throughout the main body of the book. From the mathematical point of view, however, this notation has some disadvantages; for this reason and because the use of Dirac's notation is not universal, the main results of the mathematical appendices are given in ordinary notation. Each section of these appendices ends with an explanation of the connection between the two notations.

For reference, mathematical expressions within a chapter are numbered in the common decimal system. Within each chapter, they are referred to by their number [e.g., (3.5)]. In the references across chapters the expression number is preceded by the chapter number [e.g, (II.3.5)]. Paragraphs in the appendices are also numbered in the decimal system (but not enclosed in parentheses) and referred to in a similar manner (e.g., paragraph 3.2.2 of Appendix B will be referred to as 3.2.2 from within the appendix and as B.3.2.2 from without). Mathematical expressions in the appendices are not numbered.

* * *

I owe a debt of gratitude to all those that made possible the writing of this book. I am particularly indebted to the many former students whose comments and questions contributed so much to my understanding of quantum mechanics. Among them, I would dare to name Ulrike Närger, Steve Patitsas, Mark Shegelski and Mike Sofer, who collaborated with me in the teaching of the subject. My thanks are also due to all those colleagues with whom I have discussed many of the topics included in this book and, especially, to F.W. Dalby whose provocative questions always demanded an answer. My readings, of course, influenced the shaping of my course and are reflected in my writing. Two books I should mention in this connection: Messiah's *Quantum Mechanics* and Jauch's *Foundations of Quantum Mechanics*. To their authors I express my gratitude.

In the latter stages of the preparation of this book, I received valuable comments from Mark MacLean, who kindly agreed to read the mathematical appendices, and essential technical help from Janet Clark, who typeset the manuscript in TEX. To both of them, my sincere thanks.

In concluding this preface, I must ask the reader to forgive the dryness of my style, for the language in which I am writing is not the language in which I learned to laugh and to cry.

Vancouver, April 1995.

CONTENTS

I INTRODUCTION

The subject matter of this book is Non-relativistic Quantum Mechanics, a theory which, needless to say, explains an enormous wealth of observations. This fact by itself would constitute a sufficient reason to study the theory in some depth. But there are also other reasons. Non-relativistic Quantum Mechanics, which we shall simply call Quantum Mechanics, rests on a sound mathematical basis which means that the basic concepts of the theory can be understood precisely. This precise understanding is essential to understand and to contribute to the development of two broad fields at the limits of our present knowledge. One is the relativistic theory of quantum fields which constitutes today one of the frontiers of theoretical physics. The other, that lies at the boundary between physics and philosophy, is the understanding of the relation between ourselves and the world that we study.

While the physical observations are non-mathematical entities ultimately accessible to our senses, the theory introduces some fundamental concepts that are mathematical and cannot be explained by appealing to our sensory experience. As a consequence it is not possible to understand Quantum Mechanics without understanding the mathematical theory in which those concepts arise.

It is the purpose of this chapter to explain in some detail the brief remarks of the previous paragraph.

1. Description of experiments

There are of course many reasons that may lead physicists to conceive of an experiment. One may want to test a prediction of a more or less well developed theory. Another may be seeking some orientation in the development of a tentative theoretical formulation. Still another may want to determine the properties of some material with an eye to possible applications. Whatever the motivation, in most if not in all cases, the experimenter has in mind some more or less elaborated theoretical framework to guide him in the design of the experiment.

Once the design has been completed the experimenter decides what instruments are needed, proceeds to draft them in detail, and has them built in the technical shop.[1] The important point to be made here is that the detailed plans of the instruments are completely understandable to the shop staff even if they know nothing about the theoretical ideas that guided the design.

After the experimental instruments have been built and assembled, the experiment, which may consist of one or more experimental runs, is performed.

At the start of an experimental run some of the instruments are adjusted by manipulating knobs or other devices until their settings, as verified by the readings of appropriate dials, are the desired ones. Some time later the run is completed by noting another set of dial readings which constitutes the experimental results of the run.

When the predictions of a theory are statistical, as is the case in quantum mechanics, an experiment consists usually of many runs with identical initial settings of the instruments. Although normally these runs are performed in succession, we may think of them as starting simultaneously and using identical copies of the experimental setup.

A proper account of the experiment can be phrased in plain language in the sense that it need not contain any reference to the theoretical ideas that entered in its conception; all that is required is a description of the experimental instruments, the instrumental settings at the start, and the dial readings at the completion of each run. As already remarked, the instruments can be built, and therefore described, without any use of those theoretical ideas, while the readings of the different instruments consist merely of a set of numbers that refer to some variable characteristics incorporated in the construction of those instruments. Of course, in reporting the experiment, the experimenter usually discusses the theoretical ideas that led to its conception and the significance of the results in relation to those ideas but this part of the report constitutes an *interpretation* of the experiment, not an *account* of it.

In practice it is generally true that the instructions for the construction of the experimental instruments are phrased in plain language. On the other hand, in an article intended to describe an experiment to other workers in the field one often finds sentences such as "the proton beam was focused on the target" which

[1] Some instruments may, of course, have been designed by somebody else and be available from the shelf but this and other such details are clearly irrelevant for our purposes.

is certainly not expressed in plain language, for it contains an obvious reference to the theoretical ideas used in designing the experiment. That sentence, however, could have been replaced by something like "the current in the coils labelled A in the diagram was adjusted until a maximum signal was obtained in the detector". Another experimenter will know that that was precisely what was done and what he would have to do if he wanted to repeat the experiment. The original sentence is expressed in a sort of shorthand which is based on shared theoretical ideas. Such a shorthand is extremely useful and even essential, for without it communication among physicists would be unbearably slow, but it should be used with care if one wants to avoid ambiguities. Indeed the history of the development of Quantum Mechanics shows that its uncritical use can lead to apparent paradoxes and conceptual errors.

2. The concept of physical system

In analyzing an experiment which can be satisfactorily explained by classical physics, it is not the experimental instruments which are stressed in the analysis but rather the *system* on which the experiment is done. This is so because the concept of system as some physical object separate from the other instruments used in the experimental setup is easy to define precisely.

To illustrate the last remark let us look at a simple example from classical mechanics. Imagine an experiment designed to determine the range of an artillery piece. In each run one measures the velocity of the shell when it leaves the nozzle as well as the horizontal distance that it travels. To analyze the experiment one may choose the shell as "the system". The *state of the system* at any time is given by the values at that time of an appropriate set of coordinates and velocities of the shell. We say that the theory of classical mechanics (including in this case aerodynamics) explains the experiment because it permits us to write a closed system of equations that describes the time evolution of the state of the shell in accordance with the experimental results.

There are three important properties of a classical mechanical system that need to be stressed here.

(i) The system (in our example, the shell) is a part of the experimental setup, constructed as a piece separate from the other experimental instruments, and can therefore be described precisely in plain language.

(ii) The state of the system can be determined, at least in principle, at any time (for example by stroboscopic photography) without altering the predicted states at future times.

(iii) The state uniquely determines the value of every mechanical property (that is, every dynamical variable) of the system.

The precise definition of a classical mechanical system requires, first of all, that a complete set of coordinates be specified, in the sense that prescriptions must be given on how to measure them. It is, of course, possible to define the complete set of coordinates in different ways but the sets so defined are all equivalent because their coordinates are functions of the coordinates of any particular set. The characterization of the system is completed by giving certain intrinsic (that is, pertaining to the system, not to the rest of the experimental setup) parameters, such as masses or moments of inertia.

Systems that share the same complete set of coordinates will be said to belong to the same *type*. Thus the shell of the above example belongs to the type of rigid bodies. But for two shells to constitute identical systems they must have the same moment of inertia and geometrical shape (because the aerodynamic forces depend on the latter). In general we call two systems identical if they belong to the same type and have the same intrinsic parameters.

A parenthetical remark about nomenclature is in order here. In the above description we have implicitly assumed that one characteristic of a system is that its state at any given time uniquely determines, through a *closed* system of equations of motion, the state at any future time. A part of a classical mechanical system can also be assigned a state which is a subset of the set of coordinates and velocities that constitute the state of the system. Although such a part should properly be called a *subsystem*, the prefix is often omitted and the word "system" is loosely applied to it. As we know, the state of a subsystem at a given time does not, in general, determine the state at future times, for the motion may depend on the values of the remaining coordinates and velocities.

There are theories in classical physics in which a system can be defined which does not have property (i) above. An example is classical electromagnetism. In many cases it is convenient in this theory to regard the electromagnetic field as being "the system", although it is obviously not a piece of equipment that can be constructed in the shop separately from the rest of the experimental instruments. The field can be thought of as a physical agent which is produced by some of the experimental instruments and causes the observed correlations between measurements. However, within the domain of applicability of classical physics, it is possible, even in this case, to regard the state of the system, i.e., the field, as given by a set, albeit infinite, of generalized coordinates and velocities such that properties (ii) and (iii) still hold.[2] Thus, the field can be treated as a kind of mechanical

[2] Admittedly property (ii) requires the introduction of idealized test bodies with vanishingly small charges.

system with an infinite number of degrees of freedom which is again distinct from the rest of the experimental setup.

The concept of *system* in quantum mechanics is much more difficult to define than in classical mechanics. As in classical electromagnetism, property (i) above is not satisfied: it is in general not possible to identify a particular piece of equipment constructed in the shop as being "the system". Moreover, contrary to the situation in classical electromagnetism, experience tells us that one can not define a system and its state in such a way that properties (ii) and (iii) are obeyed. As a result, the concept of system in quantum mechanics cannot be introduced as a straightforward generalization of the corresponding classical concept.

Rather than basing the presentation in this book on the difficult concept of quantum mechanical system, I have found it preferable to take the concepts of preparation and measurement, which will be discussed Section 3, as the primary ones. Nevertheless the idea of system as the physical agent produced at the start of an experiment that causes the readings of the measuring devices, is extremely useful in quantum mechanics. Because we are taking the concept of system as secondary, a precise definition is not needed here and shall be given in due time. For the moment it is sufficient to state that the definition is somewhat analogous to the one used in classical mechanics: the type to which a system belongs is given by specifying a particular complete set of physical attributes which can be regarded as taking the place of the classical coordinates. But, while in classical mechanics different sets of coordinates are equivalent, in quantum mechanics there are many different complete sets of attributes that are not. As in classical mechanics, a particular system within a type is defined by appropriate intrinsic parameters, and to every experimental arrangement there corresponds a hamiltonian which is a function of appropriately chosen inequivalent complete sets of physical attributes that take the place of the coordinates and momenta.

If one were to conform rigidly to the approach used in this book, one should refrain from referring to a "system" until the concept is defined and describe, instead, the appropriate preparation procedure. This would result in a rather lengthy exposition which we can fortunately forego since the reader is assumed to be already acquainted with quantum mechanics. For example the sentence, "a measurement is performed in an experiment in which the preparing instrument consists of an oven filled with helium and equipped with a collimator" will be shortened to "a measurement is performed on a beam of helium atoms". This is an example of the shorthand referred to in Section 1 and we should be careful to treat it as such.

3. Preparation and measurement

As hinted in the previous sections, the point of view taken here is that an experiment is defined by the experimental instruments. Whatever these instruments might be, it will be assumed that they include a clock, a switch, or other mechanism that the experimenter activates in order to start a run, and some recording devices that, for brevity, we shall call dials, whose readings give the result of the experiment.

Very often in practice the instruments have adjustments that are set before the start of a run. Runs with different settings of the adjustments belong to different experiments. One can regard the corresponding experimental setups as being different but it is often convenient not to do so and speak instead of experiments that share the same experimental setup with different settings of the adjustments.

To analyze an experimental run we shall regard it as divided into two phases. In the first phase, called the *preparation*, the experimental instruments are put in place, adjusted if necessary, and the run is started. The second phase, called the *measurement*, follows the first and includes the recording of the appropriate readings of the instrument dials.

In many experiments the instruments that are activated at the start may have been constructed as distinct from those whose dials are read during the measurement phase. But in other experiments this may not be the case. In general, therefore, it is difficult or even impossible to define the two phases by referring to the instruments used in them. Instead the two phases can be defined by selecting an appropriate *reference time* τ, read by the laboratory clock, to separate them. The preparation phase is the part of the experimental run that precedes τ and the measurement time the one that follows it.

The reference time τ is restricted to an interval but otherwise arbitrary. Clearly τ must be a time not before the starting switch is closed and not after the dials are read. Although the precise limits of the interval may be in principle very difficult to define, this should not concern us here. In practice, a *preparation time* t_0 and a *measurement time* t can be adequately, though to some extent arbitrarily, defined by an experimenter with full knowledge of the operation of the experimental instruments. The reference time τ may then be taken to lie in the interval (t_0, t). When thinking of the collection of runs that constitute an experiment as starting simultaneously, the time t_0 will be the same in all runs but the time t, not under the experimenter's control, may be different in different runs. We can select the same reference time for all runs by choosing it sufficiently close to t_0.

To the two phases, preparation and measurement, in which we imagine the experiment divided, we make correspond two apparatus.

The *preparation apparatus* is defined by the experimental setup and the interval (t_0, τ). Its *configuration* is given by the setting of the adjustments.

The *measurement apparatus* is defined by the experimental setup and the interval (τ, t). Its *configuration* is given by the readings of the measurement dials.

One should be careful not to interpret the word "apparatus" as used in these definitions as merely synonymous of "instrument" or "collection of instruments". The word once meant not only the equipment necessary to accomplish an end but also the processes leading to it. As used here, it refers not only to the collection of experimental instruments but also to the processes they cause during the corresponding time interval. In some cases, as we shall see in the example that follows, it is possible to define each apparatus with reference to distinct experimental instruments, which shall be referred to as preparing or measuring instruments, but this need not be the case in general.

With the conventions just introduced we are now in a position to regard every experimental setup of interest to us as consisting of a preparation and a measurement apparatus. The instruments that enter into their definition, which we assume include a clock, can be constructed from instructions expressed in plain language, and their configurations, as well as the reference time τ, can *a fortiori* be described in such a language. The *configuration* of an apparatus will be taken as given by a finite set of numbers (which we can think of as the setting of the instrumental adjustments or the reading of the dials). The theoretical description of a given experiment will depend on the configuration of the apparatus as well as on the reference time chosen. Theoretical descriptions corresponding to different choices of the reference time must, of course, yield identical experimental predictions.

In the body of this book we shall make extensive use of the framework just explained to analyze idealized experiments chosen in such a way as to explain the physical concepts clearly. At this stage however, at the risk of being somewhat less clear, it is preferable to use a realistic example to illustrate how it can be cast into the above framework, for the theory is meant to explain actual experiments.

The example that we shall consider is the "molecular diffraction" experiment schematized in Figure 1.1. The experimental instruments consist of a clock, a supply of helium, a chamber that is equipped with a sensitive pressure gauge, a collimating tube and a shutter, two parallel disks each with a slit cut near its circumference and mounted on a common shaft with their slits offset with respect to each other, a motor that can rotate the shaft with adjustable angular velocity, a crystal of lithium fluoride, another chamber with a hole which is also equipped with a sensitive pressure gauge, and a circular track, on which the second chamber slides, graduated in degrees and centred at the crystal. The first chamber when

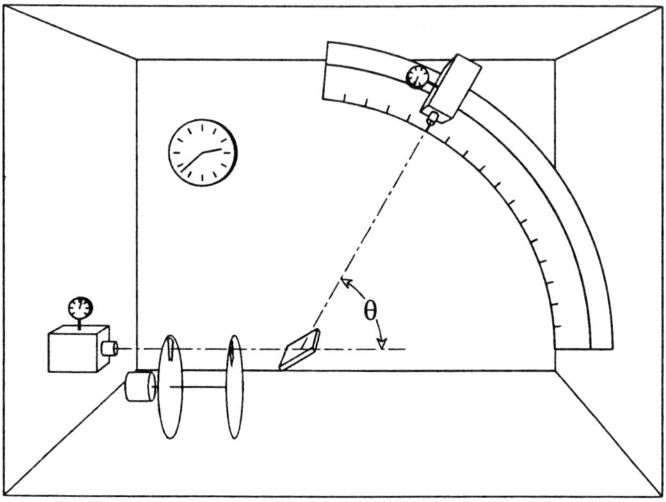

Figure 1.1 Molecular beam diffraction experiment.

filled with helium will be called "the source", the second chamber, "the detector". The disposition of the instruments is as shown in the figure.

To start the experiment, the experimenter sets up the source with helium gas at the desired pressure, adjusts the angle θ that determines the position of the detector, starts the motor, adjusts the angular velocity of the disks and opens the shutter of the source. A fixed interval of time later the pressure in the detector is noted and the run completed.

One possible choice of reference time would be at a moment just before the pressure in the detector starts to rise. With this choice, the source, rotating disks, crystal and the second chamber are sufficient to define the preparation apparatus. Its configuration is given by the gas pressure in the source, the angular velocity of the disks and the angle θ. The only instrument needed to define the measurement apparatus is then the pressure gauge in the detector.

Another choice would be to select the reference time τ by doing some preliminary experiments in which a detector is placed behind the slitted rotating disks and the interval δ between the opening of the shutter and the activation of the detector is noted. The time τ can then be taken to be a time δ after the opening of the shutter. With this choice the crystal and the second chamber would be required to define the measurement apparatus, but note that the angle θ that determines the position of the chamber will still form part of the configuration of the preparation apparatus.

In the preceding discussion we have implicitly made a distinction between the experimental setup and the particular settings of the instruments during the preparation phase. Strictly speaking such distinction does not exist. There is no conceptual difference between, say, adjusting the temperature of the oven of our molecular diffraction experiment and replacing it by a different helium filled oven, or by any other device that produces a beam of helium atoms. It is in the extended sense illustrated by this example that we should understand the term "configuration" in reference to the preparation apparatus.

The preceding remarks cannot be extended to the measurement apparatus. When we talk about different configurations of the latter we mean different readings of the measurement dials in a given experimental setup. Experience shows that these readings may be different in experimental runs performed with identical configurations of the preparation apparatus.

Quantum mechanics explains the results of an experiment in the sense that it correctly predicts the correlations between the configurations of a preparation and of a measurement apparatus defined for arbitrary values of the reference time τ. In Chapters II and III the mathematical theory necessary for the description of the correlations between the configurations of both apparatus for a *given* τ will be presented, in a sense these theory corresponds to the theory of classical kinematics. In Chapter IV the theory that gives the connection between the descriptions of an experiment for *different* values of τ will be introduced. It corresponds to the theory of classical dynamics.

Physics does not only deal with experiments of the type described above. Very often, notably in such fields as astrophysics or meteorology, a preparation apparatus is not constructed. Rather, measurements are performed at different times on, say, the stars or the atmosphere. One can cast this type of observations into the structure that we have described by imagining nature itself to provide a preparation apparatus whose configuration is given by the readings resulting from a group of early observations while the rest of the observations constitute the readings of the measurement apparatus.

There are also experiments in which each run consists of a preparation and several successive sets of measurements. They can be cast into our framework in several ways. Often the most convenient one is to consider the experiment as a series of partial experiments in which each intervening measurement is regarded as a measurement for the previous partial experiment and as a preparation for the next one.

II STATES AND OBSERVABLES

In this chapter we regard an experiment as consisting of a preparation and a measurement phase and take the reference time that separates them as fixed. We introduce the appropriate mathematical structure to describe the correlation between preparation and measurement and use it to define the concepts of state and observable.

1. States and amplitudes

Following the point of view discussed in Chapter I, we regard an experiment as consisting of two phases, a preparation phase to which is associated a preparation apparatus and a measurement phase performed with the help of a measurement apparatus. The set b of numbers formed by the readings of the measurement dials is the configuration of the measurement apparatus. Similarly, the configurations of the preparation apparatus are labelled by a set a of numbers which we can think of as specifying the setting of its adjustments (clearly we can take a and b to be single numbers). If in a particular experiment the preparation apparatus has configuration a and the subsequent configuration of the measurement apparatus is b, we *say* that a system has been prepared in a state labelled by a and has been measured in a state labelled by b. In any given experiment it is the object of quantum mechanics to establish the correlations between any pair of configurations a and b. As explained before, the description of these correlations will in general depend on the reference time τ chosen to separate the preparation from the measurement phases but the correlations themselves are independent of τ. In this chapter τ is regarded as fixed and will not be explicitly written.

As already remarked in Section I.3 the "system" should be thought of as the physical agent responsible for these correlations. The expression "state of the system", to which we have referred above, is sanctioned by general usage. We should be careful, however, not to attribute to it more meaning that it has. Strictly speaking the expression is superfluous, for it could be substituted by "configuration of the apparatus". To attribute a state to the system independently of any preparation or measurement may lead to unwarranted conclusions.

Clearly there can be degrees of refinement in both the preparation and the measurement. For example if the preparing instrument produces a beam of atoms the beam can be more or less well collimated and its energy more or less well selected and similarly the resolution of the measuring instrument may be more or less fine. For the time being we shall assume that the instruments are as good as they can be in the sense, to be made precise later, that the configuration of both of them gives maximal information about the system. For reasons that will appear clear in due time[1] a maximal preparation apparatus is said to prepare the system in a *pure state*. The state measured by a maximal measurement apparatus is also said to be pure. We shall refer to an experiment involving maximal preparation and measurement of a system as a maximal experiment. The appropriate mathematical structure for the description of the states of a system in a maximal experiment can be realized by the introduction of a Hilbert space $\mathbb{H}$, referred to as the Hilbert space of the system, and the following rules.

Rule A. – Given a reference time τ*, to each configuration* a_i *of a maximal preparation apparatus there corresponds a distinct one-dimensional subspace, i.e. a ray, in a Hilbert space* $\mathbb{H}$.

This subspace is called the state of the system prepared by the apparatus in configuration a_i.

A ray is determined by one of its unit vectors. If the configuration of the preparation apparatus is a_i, we denote the state by an arbitrary unit vector $|a_i\rangle$ in the corresponding ray. For brevity $|a_i\rangle$ is also called a state. Two unit vectors that differ by a complex factor of magnitude one – a *phase factor* – represent the same state.

Maximal measurement apparatus obey a postulate similar to Rule A:

Rule B. – Given a reference time τ*, to each configuration* b_j *of a maximal measurement apparatus there corresponds a distinct one-dimensional subspace, i.e. a ray, in a Hilbert space* $\mathbb{H}$.

This subspace is called the state of the system measured by the apparatus when a measurement results in the configuration b_j.

[1] We are here ignoring the possible existence of *superselection rules* which will be defined later in this chapter.

As before we follow general practice and label the ray by one of its unit vectors $|b_j\rangle$ which we also call the measured state.

A complete specification of the correlation between the configurations of the preparation and measurement apparatus in any given experiment is given by the following

Rule C. – Let a_i be the configuration of the preparation apparatus in a given experiment. The probability that a measurement results in the configuration b_j is

$$\mathcal{P}(b_j|a_i) = |\langle b_j|a_i\rangle|^2, \tag{1.1}$$

where $|a_i\rangle$ and $|b_j\rangle$ are the prepared and measured states.

Note that this probability is independent of the arbitrariness of the phase factors of the unit vectors $|a_i\rangle$ and $|b_j\rangle$, while the scalar product $\langle b_j|a_i\rangle$, called the *amplitude* between the states $|a_i\rangle$ an $|b_j\rangle$, is not.

It is important to stress again that the expression "the system has been prepared (measured) in state $|c\rangle$" means no more and no less than "the configuration of the preparation (measurement) apparatus is c".

Although the *state* of a system is defined with reference to a particular experiment, the *system* itself can, of course, be defined independently of any particular experimental setup. We are not yet in a position to give a precise definition of system but we can already state that if two different experimental setups share identical (or equivalent, in the sense to be defined shortly) maximal measuring instruments, then the systems produced in both belong to the same type.

We shall assume that if $|b\rangle$ is a unit vector associated with the configuration b of a measurement apparatus then there exists a preparation apparatus that has the same unit vector $|b\rangle$ associated with one of its configurations,[2] and conversely: if $|a\rangle$ is a unit vector associated with the configuration a of a preparation apparatus then there exists a measurement apparatus with a configuration to which the vector $|a\rangle$ is associated. In other words, the set of states that can be prepared is identical to the set of states that can be measured. Quantum mechanics does not provide a general prescription on how to construct an apparatus with given associated vectors. In practice it may be very difficult, or even unknown how, to make an apparatus that prepares a system in a given state. The validity of the above assumption, and of

[2] We are using here and throughout this book an imprecise way of speaking that is generally accepted. As mentioned earlier the unit vectors should be thought of as labelling rays. Thus the unit vector of the measurement apparatus could be taken to be any vector of the form $\alpha|b\rangle$, $|\alpha| = 1$, while the corresponding vector of the preparation apparatus is any unit vector of the form $\beta|b\rangle$, $|\beta| = 1$.

others that we shall make regarding the existence of apparatus, rests not on their direct empirical verification but on the verification of their consequences.

Suppose that a maximal measurement apparatus B has a set of configurations b_j with associated states $|b_j\rangle$. It is found that if a system is prepared in one of those states $|b_i\rangle$ then a measurement by the apparatus B will invariable yield the configuration b_i. By Rule C this means that $\mathcal{P}(b_j|b_i) = |\langle b_j|b_i\rangle|^2 = \delta_{ij}$, or $\langle b_j|b_i\rangle = \delta_{ij}$, that is, the vectors associated with B are orthonormal. This observation is incorporated in the following rule.

Rule D. – The set of unit vectors $|b_j\rangle$ associated with the configurations b_j of a maximal measurement apparatus constitute an orthonormal basis on $\mathbb{H}$.

This rule states not only that the set $\{|b_j\rangle\}$ is orthonormal but that it forms a basis on the Hilbert space of the system. Although the latter property has not been motivated, it is characteristic of *maximal* measurement apparatus. The rule may then be taken as a mathematical definition of such an apparatus. In Section 3 we shall see that not all measurement apparatus are maximal and we shall discuss there how is it determined in practice whether a given one is maximal or not.

It follows from the above rules that a maximal apparatus is entirely determined by a set of unit vectors in $\mathbb{H}$. Maximal apparatus which are described by the same set of vectors[3] will be regarded as equivalent. A trivial example of two equivalent apparatus is a pair of otherwise identical ones that have different numbers in the graduations of the adjustments or dials. But two equivalent apparatus need not be identical in this sense. In the following, by distinct maximal apparatus we mean apparatus described by different sets of rays.

Without further analysis the introduction of the Hilbert space appears quite unnecessary, for no use of the linearity of $\mathbb{H}$ has yet been made. Its importance will become apparent when we discuss the example in the next section but one can already see one important aspect of Rule C when it asserts that the relation between the configuration and the measurement apparatus is statistical, in contrast with the situation in classical mechanics where maximal experiments yield deterministic results. As a result the predictions of quantum mechanics can only be tested in detail by performing sufficient number of identically prepared experimental runs. The statistical nature of quantum mechanics is connected with the fact that distinct maximal measurements, corresponding to different bases on the Hilbert space, can be performed on a system. Indeed, suppose that all measurements were equivalent, which means that there is only one basis corresponding to a maximal measurement.

[3] Or to be more precise, the same orthogonal set of rays, since two vectors in the same ray represent the same state.

Then, from Rule C and the assumption that the states that can be prepared and those can be measured belong to the same set $\{|a_i\rangle\}$, follows that the matrix of the probabilities would be diagonal, $\mathcal{P}(a_j|a_i) = \delta_{ij}$, yielding deterministic predictions. This is the situation in classical mechanics where all maximal measurements are equivalent since they yield the full set of canonically conjugate coordinates and momenta.

It is important to insist in the fundamental difference between preparation and measurement apparatus. The possible configurations of a preparation apparatus are entirely under the control of the experimenter. We cannot expect, therefore, that there be a quantum rule that constrains them. Once the system has been prepared in a given state, the results of the measurement are beyond the control of the experimenter and it is not surprising that they should be constrained by a physical law such as Rule D.

The linear properties of $\mathbb{H}$ become important when we wish to compare the results of experiments on the same system involving distinct apparatus. Together with B, let us consider a distinct measurement apparatus C with associated basis vectors $|c_i\rangle$. By expressing the vectors $|a_i\rangle$ in terms of them we have

$$\langle b_j|a_i\rangle = \sum_k \langle b_j|c_k\rangle\langle c_k|a_i\rangle \tag{1.2}$$

and by taking the square of the magnitude of this expression we find by (1.1)

$$\mathcal{P}(b_j|a_i) = \sum_k \mathcal{P}(b_j|c_k)\mathcal{P}(a_i|c_k) + \sum_{k\neq l}\langle b_j|c_k\rangle\langle c_k|a_i\rangle\langle c_l|b_j\rangle\langle a_k|c_l\rangle$$

which shows that the probabilities $\mathcal{P}(b_j|c_i)$ and $\mathcal{P}(a_j|c_i)$ are not sufficient to determine $\mathcal{P}(b_j|a_i)$. By writing the complex numbers $\langle b_j|c_i\rangle$ and $\langle a_j|c_i\rangle$ as $r_{ji}e^{i\xi_{ji}}$ and $s_{ji}e^{i\eta_{ji}}$ we see that the *relative phases* $\xi_{jk} - \xi_{jl}$ and $\eta_{jk} - \eta_{jl}$ of the corresponding amplitudes are also needed. The relative phases are therefore determinable in principle by appropriate experiments called *interference* experiments.

2. An example: spin 1/2

To illustrate the preceding considerations it is instructive to think of an idealized Stern-Gerlach experiment. It is assumed that the principle of this experiment is known. We simply remind the reader of the experimental setup which consists of a source that produces a collimated beam of atoms, a photographic plate used as a detector and, between source and detector, a magnet that provides an inhomogeneous magnetic field. Upon development, a number n of more or less well defined

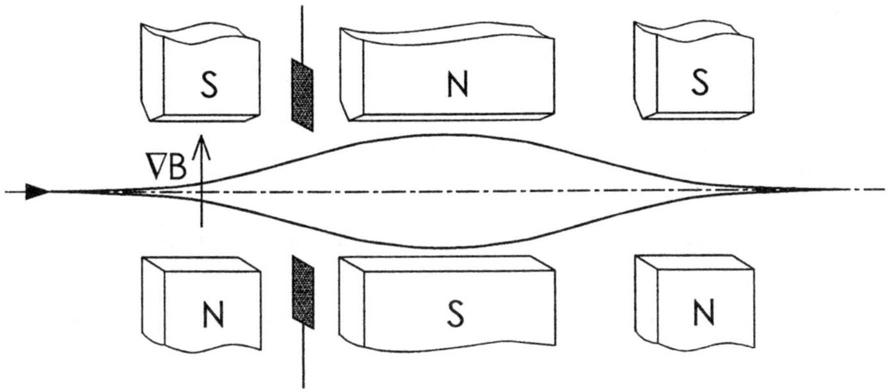

Fig. 2.1 Idealized Stern-Gerlach device.

dots is observed on the plate. In generally accepted shorthand language, the experiment is described by saying that an inhomogeneous magnetic field splits a beam of atoms into n components. We shall use this shorthand language with care, for, as we shall see, taking it too literally may lead to false conclusions. The particles of a beam that splits into n components are said to have spin $(n-1)/2$.

We consider here a beam which is split into two components, that is, a beam of particles of spin 1/2. If the intensity of the beam is reduced sufficiently, it is possible with suitable detectors to detect individual particles. The experiment on the beam is regarded as a succession of experimental runs each performed on one particle. Particles that are detected as having been deflected towards (away from) the direction of the field gradient are said to have a component of the spin equal to $+1/2$ $(-1/2)$ along that direction. It is convenient for our purpose to imagine that the magnet producing the field gradient is followed, as sketched in Figure 2.1, by an appropriate configuration of magnets that reunite the beam. Together with an appropriate oven and collimator, to produce an initially monoenergetic beam, the resulting device can be used both as part of a preparing or of a measuring instrument. For example, to prepare the particles of the emerging beam in a state with the component of the spin along the field gradient it is only necessary to block the opposite half of the magnets gap, while to measure the direction of the spin component of a particle it suffices to put particle detectors that occupy the whole width of the gap.

The full definition of the preparation and measurement apparatus requires

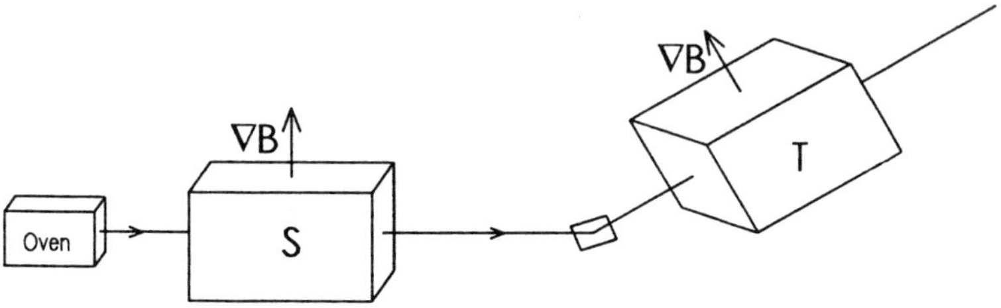

Fig. 2.2 Two Stern-Gerlach devices in different orientations.

the specification of the reference time τ. In each run, this time can be any time at which a particle has already left the preparation magnets but has not yet entered the measurement magnets. The last sentence uses shorthand language, the reader may wish to phrase it in a way that describes what the experimenter will actually do to determine the bounds of the time τ.

In the course of our considerations it will be necessary to imagine that the particles of the beam are made to pass through successive Stern-Gerlach devices in different orientations as depicted in Figure 2.2. To visualize this, one can imagine some kind of ideal mirror-like device that changes the direction of the beam through non-magnetic interactions.

As we shall see in Chapter IX, maximal measurement apparatus for the particles in the beam should measure besides the spin some other quantity such as the momentum. Experience shows however that, in the experiment under consideration, the other quantity can be ignored, that is, we can think of the preparation apparatus as preparing the systems "spin of the particles" rather than the particles themselves. If we adopt this point of view, a Stern-Gerlach measuring device as described above constitutes a maximal measuring instrument while the corresponding preparation device is also maximal and prepares a pure state.

To each given device S we associate a cartesian, orthogonal, right-handed coordinate system with the z-axis along the field gradient and the y-axis along the direction of propagation of the beam. Whether the device is used as a preparing or as a measuring instrument, it has two possible configurations. In the former case they correspond to particles being allowed to pass only through the upper $(+z)$ or through the lower $(-z)$ path. In the latter to either the upper or the lower detector been activated. There follows from Rule D that the Hilbert space $\mathbb{H}$ associated with our system is two-dimensional and we shall label the basis vectors associated with the upper or lower paths of a measuring instrument T by $|T_i\rangle$, $i = 1$ or 2. Given the similarity in the construction of the measuring and preparing Stern-Gerlach

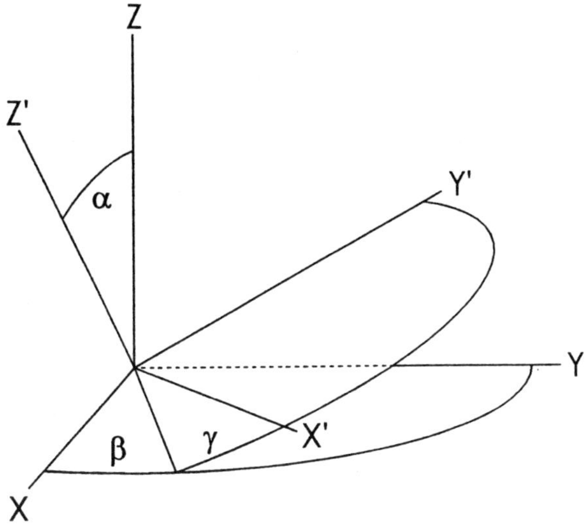

Fig. 2.3 Definition of the Euler angles.

devices we shall take the vectors $|S_i\rangle$ associated with the two configurations of a Stern-Gerlach preparation apparatus to also form an orthonormal basis on $\mathbb{H}$.

The next task is to determine the two-by-two matrix

$$M(T,S) = (\langle T_i|S_j\rangle)$$

of the amplitudes, for an experiment in which the spin is prepared with an apparatus S and measured with T. It is easy to verify that M is a unitary matrix and that therefore $|\det(M)| = 1$. We shall use some general concepts to determine M keeping in mind that a complete determination is not possible because only the relative phases of the amplitudes are of physical significance.

Since the disposition of the magnets is identical whether a device is used to prepare a state or to measure it, we take two apparatus with the same orientation to have the same associated basis on $\mathbb{H}$, which, with appropriate selection of phase, gives for this case $M = I$. Consideration of the isotropy of space leads us to assume that a general matrix $M(T,S)$ depends only and continuously on the Euler angles α, β, γ, of Figure 2.3 defining the orientation of T with respect to S. We take $\det(M) = 1$ by multiplying all $|T_i\rangle$ by a common phase factor. It is left as an exercise for the reader to show that since M is a two-dimensional unitary matrix with positive determinant, its general form is

$$M = \begin{pmatrix} e^{i\mu}\cos\xi & e^{i\nu}\sin\xi \\ -e^{-i\nu}\sin\xi & e^{-i\mu}\cos\xi \end{pmatrix}, \tag{2.1}$$

where μ, ν, ξ are arbitrary real constants with $0 \leq \xi \leq \pi/2$. To determine them we use the following argument.

Let U be a third apparatus with associated basis vectors $|U_i\rangle$, by particularizing (1.2) and writing it in matrix form we have

$$M(T,S) = M(T,U)M(U,S). \tag{2.2}$$

The matrices M are seen to obey under multiplication the same composition rules as three-dimensional rotations. We can obviously write $M(T,S) = M(\alpha,\beta,\gamma)$. When the matrix represents a rotation about a coordinate axis we write it in simplified notation, e.g., $M_x(\zeta) = M(\zeta,0,0)$ for a rotation by the angle ζ about the x-axis. One can decompose a matrix for a general rotation as follows

$$M(\alpha,\beta,\gamma) = M_z(\gamma)M_y(-\pi/2)M_z(\alpha)M_y(\pi/2)M_z(\beta) \tag{2.3}$$

and we need only to be concerned with general rotations about the z-axis and rotations of $\pi/2$ about the y-axis.

Consider first a measurement apparatus rotated an angle φ about the z-axis with respect to a preparation device. Since the direction of the field gradient is the same in both apparatus, we expect that if a particle has been prepared in one path by S it will be measured in the same path by T. Thus according to Rule C, $|\langle T_i|S_j\rangle| = \delta_{ij}$ and the matrix (2.1) reduces in this case to

$$\begin{pmatrix} e^{i\mu(\varphi)} & 0 \\ 0 & e^{-i\mu(\varphi)} \end{pmatrix}.$$

The product rule (2.2) shows that $M_z(\varphi_1+\varphi_2) = M_z(\varphi_1)M_z(\varphi_2)$ so that $\mu(\varphi)$ must be a linear function of φ and since we have taken $M = I$ for two devices in the same orientation, must vanish with φ. Thus $\mu(\varphi) = c\varphi$. To maintain the physical identity of a rotation by φ and a rotation by $\varphi + 2n\pi$ for any integer n the relative phase factor $e^{2ic\varphi}$ of the two non-zero amplitudes must be periodic in φ with period 2π so that $c = 1/2$. Thus we obtain

$$M_z(\varphi) = \begin{pmatrix} e^{\frac{1}{2}i\varphi} & 0 \\ 0 & e^{-\frac{1}{2}i\varphi} \end{pmatrix}. \tag{2.4}$$

Note that $M_z(2\pi) = -I$, which differs from the identity by a phase factor that does not affect the probabilities. Its presence, however, should not be surprising because the two experimental arrangements described by $M_z(0)$ and $M_z(2\pi)$ are different in that the former is achieved by simply placing T at the output of S, while the latter requires a mirror-like device between S and T to make the beam rotate a complete loop.

To determine $M_y(\pi/2)$ it is easier to start with $M_y(\pi)$ corresponding to two apparatus along the same beam direction but with "up" and "down" inverted. It is then clear that one should have $\mathcal{P}(T_i|S_j) = 0$ for $i = j$ so that in the general expression (2.1) $\xi = \pi/2$ and, by arbitrarily taking the phase factor with $\nu = 0$, one has

$$M_y(\pi) = \begin{pmatrix} 0 & 1 \\ -1 & 0 \end{pmatrix}.$$

Since $M_y(\pi) = M_y(\pi/2)M_y(\pi/2)$, taking arbitrarily one of the square roots of $M_y(\pi)$ with positive determinant yields for $M_y(\pi/2)$ and its inverse $M_y(-\pi/2)$

$$M_y(\pi/2) = \begin{pmatrix} 1/\sqrt{2} & 1/\sqrt{2} \\ -1/\sqrt{2} & 1/\sqrt{2} \end{pmatrix}, \qquad M_y(-\pi/2) = \begin{pmatrix} 1/\sqrt{2} & -1/\sqrt{2} \\ 1/\sqrt{2} & 1/\sqrt{2} \end{pmatrix}.$$

These results together with (2.4) and (2.3) yield the general amplitude matrix

$$M(\alpha, \beta, \gamma) = \begin{pmatrix} e^{\frac{1}{2}i(\beta+\gamma)}\cos(\frac{1}{2}\alpha) & ie^{-\frac{1}{2}i(\beta-\gamma)}\sin(\frac{1}{2}\alpha) \\ ie^{\frac{1}{2}i(\beta-\gamma)}\sin(\frac{1}{2}\alpha) & e^{-\frac{1}{2}i(\beta+\gamma)}\cos(\frac{1}{2}\alpha) \end{pmatrix}. \tag{2.5}$$

This matrix constitutes a complete description of the ideal Stern-Gerlach experiment. Needless to say that since it has been obtained from theoretical assumptions its correctness rests ultimately on experimental verification.

It is instructive to examine the experiment in some detail because for the sake of clarity we included some unnecessary features in its description. First of all the statistical nature of the predictions of quantum mechanics requires that they be compared with the results of a set of experimental runs performed with identical configurations of the apparatus. In the example the identical runs are performed on different particles of the beam under the underlying assumptions that the particles are independent and that the experimental conditions do not change with time.

Because, as mentioned before, the experiment is regarded as an experiment on the spin of the particles, rather than on the particles themselves, the direction of the beam before and after passing through one of the devices should be irrelevant. The results of the experiment should depend only on the relative orientation of the magnetic field gradient of the devices S and T which can be expressed by the two Euler angles α and β alone. This is in fact what we have found, for the third angle γ appears in the matrix (2.5) as a phase factor that can be eliminated by multiplying $|T_1\rangle$ and $|T_2\rangle$ by $e^{i\gamma/2}$ and $e^{-i\gamma/2}$ respectively.

The angle β cannot be eliminated in the same way for it would be necessary to multiply the states $|S_i\rangle$ corresponding to the apparatus S by phase factors depending on β, that is on the orientation of the *other* apparatus T. Since β affects only the relative phase of the amplitudes $\langle T_i|S_j\rangle$ the β-dependence cannot be detected by determining the probabilities $|\langle T_i|S_j\rangle|^2$ in an experiment involving only a pair of devices S and T. To detect it one needs to perform an experiment that would involve the use of more than two devices. This conclusion can be easily understood: as we have seen, only the direction of the field gradient (the $+z$-direction) in an apparatus is of importance in our experiment. Given two apparatus S and T with their field gradients in different directions one can choose the $X-Z$ plane of the coordinate system to be any plane through the z-axis of S thus yielding an arbitrary value for β. Once a β for T has been selected, however, the β corresponding to the direction of the field gradient of a third apparatus is, in general, determined.

3. Observables

Let us consider a maximal measurement apparatus B with associated basis $\{|b_i\rangle\}$ in one-to-one correspondence with the apparatus readings $\{b_i\}$. The unitary amplitude matrix

$$M_b = (\langle b_i|c_j\rangle),$$

where $\{|c_j\rangle\}$ is an orthogonal basis in $\mathbb{H}$, together with the correspondence $b_i \leftrightarrow |b_i\rangle$ provides a complete quantum mechanical description of B. Indeed, if M_b is known, one can, in accordance with Rule C, determine the probability $\mathcal{P}(b_i|f) = |\langle b_i|f\rangle|^2$ of a measurement yielding the configuration b_i of B given that the system was prepared in an arbitrary preparation state $|f\rangle = \sum_j f_j|c_j\rangle$. The average value, in the statistical sense, of the reading of B in a set of measurements performed on identical systems prepared in the state $|c_k\rangle$ is

$$\langle B\rangle = \sum_i b_i\mathcal{P}(b_i|c_j) = \sum_i b_i|\langle b_i|c_j\rangle|^2 = \langle c_j|\Big(\sum_i |b_i\rangle b_i\langle b_i|\Big)|c_j\rangle.$$

We define the hermitean operator

$$B = \sum_i |b_i\rangle b_i\langle b_i| \tag{3.1}$$

and call it the *observable* measured by the apparatus B. With this definition we can rewrite the average value of the readings of B (or simply the average of B) resulting from measurements performed on a system prepared in the pure state $|c_j\rangle$, as $\langle B\rangle = \langle c_j|B|c_j\rangle$. Since this result holds for all the elements $|c_j\rangle$ of a basis on $\mathbb{H}$ it also holds if the system is prepared in an arbitrary pure state $|\psi\rangle$. It is clear that the observable B determines all the statistics of a measurement by the corresponding apparatus on a system in the state $|\psi\rangle$, for one can easily verify that the n-th moment of the readings is given by $\langle B^n\rangle = \langle\psi|B^n|\psi\rangle$. Thus the operator B provides a complete description of the measurement apparatus which is independent of the preparation state.

Up to now we have been considering only maximal measurements. As a prototype of an apparatus whose measurements on a given system are not maximal we imagine the apparatus B of the previous paragraph but with the instrument readings modified so that to each reading b_i there correspond $m_i \geq 1$ basis vectors (and at least one of the m_i is greater than 1). It can readily be verified that the correct statistics is then provided by the hermitean operator B with spectral resolution (see A.3.4.2)

$$B = b_i E_i, \qquad E_i = \sum_{r=1}^{m_i} |b_i^r\rangle\langle b_i^r|$$

and probabilities

$$\mathcal{P}(b_i|\psi) = \langle\psi|E_i|\psi\rangle.$$

We cannot, of course, assume that all measurement apparatus are such modifications of maximal ones but experience allows us to generalize the previous results in the form of the following

Rule E. – Given a reference time τ, to every measurement apparatus there corresponds a hermitean operator B, called an observable, *such that (i) the only values that can be obtained in a single measurement are the eigenvalues b_i of B and (ii) the probability of obtaining the value b_i if the system is in the pure state $|\psi\rangle$ is*

$$\mathcal{P}(b_i|\psi) = \langle\psi|E_i|\psi\rangle = \parallel E_i|\psi\rangle \parallel^2 . \tag{3.2}$$

where E_i is the eigenprojection of B belonging to the eigenvalue b_i.

It follows that the average value of the observable is

$$\langle B\rangle = \langle\psi|B|\psi\rangle \tag{3.3}$$

as can be verified by using the probabilities (3.2) and writing the spectral resolution of B

$$B = \sum_i b_i E_i.$$

It is apparent that the observables corresponding to non-maximal measurement apparatus are operators with degenerate spectra while the spectra of observables corresponding to maximal measurement apparatus are non-degenerate. We shall call the latter *maximal observables*. In an obvious extension of the concept of equivalent apparatus, we shall call *equivalent* two non-maximal measurement apparatus if their corresponding observables have the same set of eigenprojections. Such observables will also be called *equivalent*. It follows from the definition A.3.5 of function of a hermitean operator that two equivalent observables are related by an invertible functional relation.

The question arises whether the converse of Rule E is true: is every hermitean operator on the Hilbert space of a system an observable? Or, in other words, is it in principle possible to construct an experimental setup with a measurement apparatus corresponding to any given hermitean operator? It is known that the answer to this question is negative. There are some systems, said to be subject to a *superselection rule*, where some linear combinations of states cannot be realized, for example, a baryon is not found in a state which is a linear combination of a proton state and a neutron state. Suppose that $|a\rangle$ is an allowed and $|b\rangle$ a forbidden state of such a system, it is clear that operators that transform $|a\rangle$ into a non-zero vector in the ray determined by $|b\rangle$ are not observables. Since there is no evidence to the contrary we will feel free, in the absence of a superselection rule, to assume that every hermitean operator is an observable even if it is not known how to construct the necessary instruments. We shall return to discuss superselection rules in Chapter V but, for the time being, we shall ignore their existence in order not to complicate the exposition.

How does one know that a measurement apparatus is maximal? As with many other questions in physics this one is resolved by making a hypothesis and verifying whether its consequences are in agreement with observation. For example, using our classical intuition we may start by assuming that for a given particle the position is a maximal observable. The consequences of this assumption may contradict the observations and show perhaps that, for the given particle, a maximal observable consists of both the position and the spin. One may interpret in this light the discovery of the electron spin by Pauli, Uhlenbeck and Goudsmit.

A question closely related to the one just discussed is: how does one determine the Hilbert space of a system? It will be addressed in detail in later chapters. For the time being the following remark is sufficient. As we know, the eigenvectors $|b_j\rangle$ of a maximal observable B belonging to its different eigenvalues form a basis in the Hilbert space. The space is completely determined if it is known, at least in principle, how to obtain the matrix elements $\langle b_i|C|b_j\rangle$ of an arbitrary observable C.

The concepts just introduced can be illustrated using our spin 1/2 example. We refer all directions to the coordinate system associated with the Stern-Gerlach preparation apparatus S and call s_{e} the component of the spin observable measured by the apparatus T with field gradient along the direction of the unit vector **e** given by the polar angles $\theta = \alpha$ and $\varphi = \beta - \pi/2$. Equation (3.1) applied to the present case gives

$$s_{\mathrm{e}} = \tfrac{1}{2}|T_1\rangle\langle T_1| - \tfrac{1}{2}|T_2\rangle\langle T_2|.$$

In the representation $\{|\mathrm{S}_i\rangle\}$ corresponding to the basis associated with the preparation apparatus S, the matrix elements of s_{e} are

$$(s_{\mathrm{e}})_{ij} = \tfrac{1}{2}\langle S_i|T_1\rangle\langle T_1|S_j\rangle - \tfrac{1}{2}\langle S_i|T_2\rangle\langle T_2|S_j\rangle$$

and can be computed immediately from the elements of the matrix (2.5). After writing α and β in terms of θ and φ and noting that the dependence on γ disappears, as it should, we obtain

$$s_{\mathrm{e}} = \frac{1}{2}\begin{pmatrix} \cos\theta & e^{-i\varphi}\sin\theta \\ e^{i\varphi}\sin\theta & -\cos\theta \end{pmatrix}. \tag{3.4}$$

In this representation the observables of the spin components along the three coordinate axes can be obtained directly from (3.4) by particularizing it for $\theta = \pi/2$, $\varphi = 0$; $\theta = \varphi = \pi/2$ and $\theta = \varphi = 0$, respectively,

$$s_x = \tfrac{1}{2}\sigma_1, \qquad s_y = \tfrac{1}{2}\sigma_2, \qquad s_z = \tfrac{1}{2}\sigma_3, \tag{3.5}$$

where the matrices

$$\sigma_1 = \begin{pmatrix} 0 & 1 \\ 1 & 0 \end{pmatrix}, \qquad \sigma_2 = \begin{pmatrix} 0 & -i \\ i & 0 \end{pmatrix}, \qquad \sigma_3 = \begin{pmatrix} 1 & 0 \\ 0 & -1 \end{pmatrix}$$

are called the Pauli spin matrices.[4] Note that the spin components obey the commutation relations

$$[s_x, s_y] = is_z, \qquad [s_y, s_z] = is_x, \qquad [s_z, s_x] = is_y \tag{3.6}$$

which, as we shall see in Chapter VII, are characteristic of angular momentum. It is for this reason that the, apparently arbitrary, choice of $\pm 1/2$ for the eigenvalues was made.

From (3.5) and (3.4) one can verify that the spin components transform like an euclidean vector

$$s_{\mathrm{e}} = s_x \sin\theta\cos\varphi + s_y \sin\theta\sin\varphi + s_z\cos\theta.$$

---

[4] The set of 2×2 matrices $A, B, \ldots$ form a unitary vector space with addition and multiplication by a complex number defined as for matrices, and with scalar product defined by $(A, B) = \frac{1}{2}\mathrm{Tr}\,(AB)$. In this space the matrices I, σ_1, σ_2, σ_3, form an orthonormal basis. It follows that any 2×2 matrix can be written as a linear combination of the Pauli matrices and the identity.

A relation that suggests the following definition of the *spin vector observable* s

$$\mathbf{s} = \mathbf{i}s_x + \mathbf{j}s_y + \mathbf{k}s_z,$$

an expression whose meaning is obvious.

The eigenvectors, or eigenstates, of $s_{\mathbf{e}}$ can be calculated from (3.4) by solving the eigenvalue equation. More directly, they can be obtained from the amplitude matrix (2.5) by noting that

$$|T_i\rangle = |S_1\rangle\langle S_1|T_i\rangle + |S_2\rangle\langle S_2|T_i\rangle.$$

After putting $\alpha = \theta$, $\beta = \varphi + \pi/2$ and choosing conventional phase factors we get in the representation of the prepared states

$$|T_1\rangle = \begin{pmatrix} e^{-\frac{1}{2}i\varphi}\cos\frac{1}{2}\theta \\ e^{\frac{1}{2}i\varphi}\sin\frac{1}{2}\theta \end{pmatrix}, \qquad |T_2\rangle = \begin{pmatrix} -e^{-\frac{1}{2}i\varphi}\sin\frac{1}{2}\theta \\ e^{\frac{1}{2}i\varphi}\cos\frac{1}{2}\theta \end{pmatrix}. \tag{3.7}$$

As is customary we shall occasionally represent the s_z basis states with the notation $|+\rangle = |S_1\rangle$, $|-\rangle = |S_2\rangle$.

Returning now to the general case, we observe that the spectral theorem for hermitean operators A.3.4.2 tells us that projection operators are the building blocks of observables. Since a projection has two eigenvalues, 0 and 1, they can be thought as observables corresponding to yes-no measurements. Given the observable $B = \sum_i b_i E_i$, the projection E_i corresponds to determining whether a measurement of B yields the value b_i or not. The study of observables can, therefore, be reduced without any loss of generality to the study of projection operators. We shall often bear in mind this remark in the course of our considerations.

4. The density operator

Up to now we have been denoting a pure state, i.e. a ray in the Hilbert space $\mathbb{H}$ of the system, by one of its unit vectors $|\psi\rangle$. It is, of course, possible to denote the ray by the one-dimensional projection operator

$$\rho_\psi = |\psi\rangle\langle\psi| \tag{4.1}$$

that projects on it. This operator notation is conceptually more appropriate than the previously used vector notation because there is a one-to-one correspondence between the one-dimensional projections and the rays on $\mathbb{H}$. Although the operator notation has the practical advantage that the indeterminacy of the phase factors is not present, the vector notation is usually preferred because a pure state represented in it by the linear combination $a|\psi\rangle + b|\phi\rangle$ would be represented in operator notation by a projection operator with the rather awkward expression $aa^*|\psi\rangle\langle\psi|+ab^*|\psi\rangle\langle\phi|+a^*b|\phi\rangle\langle\psi|+bb^*|\phi\rangle\langle\phi|$. In this book we shall use whatever notation is more convenient in each context. In this section, which deals with the generalization of the concept of state, the operator notation is preferable.

By choosing the vector $|\psi\rangle$ in equation (4.1) as a member of an orthonormal basis one can immediately verify that the expressions (3.2) and (3.3) for the probability of the outcome of a measurement of the observable B and its average value become

$$\mathcal{P}(b_i|\psi) = \mathrm{Tr}\,(\rho_\psi E_i), \qquad \langle B\rangle = \mathrm{Tr}\,(\rho_\psi B), \tag{4.2}$$

where E_i is the eigenprojection of B belonging to the eigenvalue b_i. For future reference note that since ρ_ψ is a one-dimensional projection operator it is hermitean, positive, idempotent (i.e., $\rho_\psi^2 = \rho_\psi$) and has unit trace.

We have been considering experiments in which the preparation apparatus was maximal. In the same way as we generalized the theory for the case of a non-maximal measurement, we would like to generalize it to include the possibility of a non-maximal preparation. As before we can imagine a non-maximal preparation apparatus A by modifying the instrumental adjustments of a maximal one in such a way that the same setting a corresponds to any of the m states $|a_r\rangle$, $r = 1,\ldots,m$, which form a subset of set of unit vectors associated to the apparatus. Upon adjusting the instruments the experimenter will not know in which of those states the system would have been prepared and all he can do on the basis of his knowledge of them is to assign some probabilities p_r to the different possible preparation states. If the observable $B = \sum_i b_i E_i$ is then measured, one can calculate the probability $\mathcal{P}(b_i|a)$ of obtaining the value b_i given that the reading of the preparation apparatus was a by the usual probability rule: $\mathcal{P}(b_i|a) = \sum_r p_r \mathcal{P}(b_i|a_r)$. Substituting for the last probability the value given by (4.2) and writing $\rho_r = |a_r\rangle\langle a_r|$ one obtains

$$\mathcal{P}(b_i|a) = \sum_r p_r \mathrm{Tr}\,(\rho_r E_i) = \mathrm{Tr}\left[\Bigl(\sum_r p_r\rho_r\Bigr)E_i\right]. \tag{4.3}$$

If one now defines the *density operator* ρ of the system prepared by the preparation apparatus A in configuration a as

$$\rho = \sum_r p_r \rho_r, \tag{4.4}$$

one can write for the probability (4.3)

$$\mathcal{P}(b_i|a) = \mathrm{Tr}\,(\rho E_i),$$

an expression formally identical to the first of (4.2).

Because they are probabilities, the numbers p_r are non-negative and add to one. It follows that the density operator ρ defined in (4.4) shares with the density operator (4.1), which constitutes the operator representation of a pure state, the properties of being hermitean, positive, and having unit trace. On the other hand ρ is not idempotent [unless only one p_r is different from zero which is the case of the pure state (4.1)]. Note that the expression (4.4) is *not*, in general, the spectral resolution of ρ because, although the ρ_r are one-dimensional projection operators, they need not be orthogonal.

We have no experimental evidence that all the non-maximal preparations are of the simple type described. In fact, as far as experience tells us, it might very well be that for some systems it is not possible to design a maximal preparation apparatus. Guided by the preceding considerations we are then led to the following generalization of Rule A.

Rule 1. – Given a reference time τ, to each configuration of a preparation apparatus there corresponds a positive hermitean operator ρ with unit trace on a Hilbert space $\mathbb{H}$. This operator is called the state of the system prepared by the apparatus.

If the apparatus is maximal then ρ is a projection and the state is said to be pure.

Part (ii) of Rule E must also be modified. We write the generalization as

Rule 2. – Given a reference time τ, to every measurement apparatus there corresponds a hermitean operator B, called an observable, *such that (i) the only values that can be obtained in a single measurement are the eigenvalues b_i of B and (ii) the probability of obtaining the value b_i, if the system has been prepared in the state ρ, is*

$$\mathcal{P}(b_i|\rho) = \mathrm{Tr}\,(\rho E_i), \tag{4.5}$$

where E_i is the eigenprojection of B belonging to the eigenvalue b_i.

If the apparatus is maximal then the spectrum of B is simple and the observable is also said to be maximal.

It follows by a simple calculation that the average value of B is

$$\langle B\rangle = \mathrm{Tr}\,(\rho B). \tag{4.6}$$

If A and B are two observables, their averages obey the following properties: (1) $\langle A\rangle$ is real, (2) if A is positive then $\langle A\rangle \geq 0$, (3) $\langle cA\rangle = c\langle A\rangle$ for all complex numbers c, (4) $\langle A+B\rangle = \langle A\rangle + \langle B\rangle$, (5) $\langle I\rangle = 1$. Property (1) follows because the hermitean operator $\frac{1}{2}(\rho A + A\rho)$ has the same trace as ρA. If A_{ij} are the matrix elements of A in a representation in which ρ is diagonal then $\mathrm{Tr}\,(\rho A) = \sum_i w_i A_{ii}$, if A is positive then $A_{ii} \geq 0$ and, since $w_i \geq 0$, (2) follows. (3) and (4) result from the linearity of the trace and (5) from the definition of ρ.

The first four properties appear eminently reasonable on physical grounds and so does the fifth when one realizes that the identity is the yes-no observable corresponding to the question: is the system in any state? which always has an affirmative answer.

For systems where every hermitean operator A is an observable, the above five properties determine the existence of a unique density operator ρ such that, for every A, $\langle A \rangle = \mathrm{Tr}\,(\rho A)$. To see this, observe that the operators on a finite-dimensional space $\mathbb{H}$ form a unitary space $\mathbb{B}$ with scalar product $(B, A) = \mathrm{Tr}\,(B^\dagger A)$.[5] An orthogonal basis $\{|\psi_i\rangle\}$ on $\mathbb{H}$ induces an orthogonal basis on $\mathbb{B}$ formed by the hermitean operators $|\psi_i\rangle\langle\psi_j| + |\psi_j\rangle\langle\psi_i|$ and $i(|\psi_i\rangle\langle\psi_j| - |\psi_j\rangle\langle\psi_i|)$. Properties (3) and (4) state that $\langle A \rangle$ is a linear functional of A and therefore a scalar product $\mathrm{Tr}\,(\rho^\dagger A)$ on $\mathbb{B}$. (1) and (2) imply that ρ is hermitean and positive while (5) shows that it has unit trace. The uniqueness of ρ follows because, if all elements of the basis on $\mathbb{B}$ are observables, substituting them for A in the expression $\langle A \rangle = \mathrm{Tr}\,(\rho A)$ determines all the components of ρ with respect to that basis or, if one prefers, all the matrix elements of ρ.

In a way the previous result can be used as a justification to consider the density operator ρ to be the most general representation of the state.

If the state ρ is a projection then, as we know, it is a pure state otherwise it is a *mixture*. A mixture is called such because its density operator can be written as a sum of two or more pure states (i.e., one-dimensional projections) with real coefficients smaller than unity. One such sum is the spectral resolution of the operator ρ but there are many other sums involving non-orthogonal pure states as shown by (4.4). A pure state, being itself a one-dimensional projection, admits no such decomposition.[6] It is not difficult to show that ρ is pure if and only if

[5] cf. footnote[3].

[6] One should not confuse the decomposition of the vector representation of a pure state as a linear combination of basis vectors with the decomposition of mixture just referred to. To stress this point consider a pure state with vector representation $|a\rangle$ and operator representation $|a\rangle\langle a|$. Given a basis $\{|\phi_i\rangle\}$ one can write

$$|a\rangle = \sum_i |\phi_i\rangle\langle\phi_i|a\rangle, \qquad |a\rangle\langle a| = \sum_{ij} \langle\phi_i|a\rangle\langle a|\phi_j\rangle|\phi_i\rangle\langle\phi_j|$$

The latter expression is not a sum of pure states but simply the $\{|\phi_i\rangle\}$ representation of the one-dimensional projection $|a\rangle\langle a|$.

$\mathrm{Tr}\,(\rho^2) = 1$.

To illustrate the generalization of the notion of state just introduced let us return to our example of a beam of particles of spin 1/2. Suppose that all that is known is that repeated measurements of s_z on the particles of the beam yield the value 1/2 a fraction p of the times (and the value $-1/2$ a fraction $1-p$ of the times), what can we say about the preparation state ρ of the particles? Let us use the s_z-representation with basis states $|+\rangle$ and $|-\rangle$. The most general form of the state ρ is

$$\rho = \begin{pmatrix} a & c \\ c^* & 1-a \end{pmatrix}$$

with a real. Use of (4.5) with the known probabilities for the two basis states gives

$$p = \mathrm{Tr}\,(\rho|+\rangle\langle+|) = a, \qquad 1-p = \mathrm{Tr}\,(\rho|-\rangle\langle-|) = 1-a.$$

Thus, as a result of the measurements we conclude that the state must be of the form

$$\rho = \begin{pmatrix} p & c \\ c^* & 1-p \end{pmatrix} \tag{4.7}$$

with c undetermined but restricted by the condition that ρ be positive or, equivalently, that the solutions of the eigenvalue equation $\lambda^2 - \lambda + p(1-p) - |c|^2 = 0$ be non-negative, which gives

$$|c|^2 \leq p(1-p).$$

We see that ρ is not in general determined by the measurement of the maximal observable s_z. Only if p is 0 or 1, is c determined to equal zero. In that case ρ is the pure state $|-\rangle$ or $|+\rangle$, respectively, i.e an eigenstate of s_z.

Suppose that after having found that the original beam has the state (4.7) we measure on it the component s_x of the spin. What can we predict about the average $\langle s_x\rangle$ that will result from the measurement? The expressions (3.5) and (4.6) give

$$\langle s_x\rangle = \mathrm{Tr}\,(\rho s_x) = \frac{1}{2}\mathrm{Tr}\left\{\begin{pmatrix} p & c \\ c^* & 1-p \end{pmatrix}\begin{pmatrix} 0 & 1 \\ 1 & 0 \end{pmatrix}\right\} = \frac{1}{2}\mathrm{Tr}\begin{pmatrix} c & p \\ 1-p & c^* \end{pmatrix} = \mathrm{Re}\,(c),$$

so that, except in the case when ρ is a pure eigenstate of s_z, $\langle s_x\rangle$ is undetermined but restricted by the inequality

$$|\langle s_x\rangle| \leq \sqrt{p(1-p)}.$$

Assume now that we do perform a measurement of s_x on the original beam and we find the value 1/2 a fraction q of the times (and naturally a value $-1/2$ the rest of the time). Because in this particular case there are only two possible outcomes of the measurement, measuring the value of s_x is equivalent to measuring $\langle s_x \rangle$. Clearly $\langle s_x \rangle = \frac{1}{2}[q - (1-q)] = q - \frac{1}{2} = \mathrm{Re}(c)$. What can we say now about the average $\langle s_y \rangle$ resulting from repeated measurements of s_y on the original beam? By using the same procedure as in the case of s_x we obtain readily

$$\langle s_y \rangle = -\mathrm{Im}(c)$$

so, in spite of having measured two maximal observables, we still have not determined the state ρ of the initial beam sufficiently to predict the outcome of the present measurement. Except in the pure case, $\langle s_y \rangle$ is still undetermined though subject to the restriction

$$|\langle s_y \rangle| \leq \sqrt{p(1-p) - (q - \tfrac{1}{2})^2}.$$

A measurement of s_y will determine $\mathrm{Im}(c)$ and ρ will then be completely known.

In the general case of a system with an n-dimensional space $\mathbb{H}$ the density operator ρ has $n-1$ independent real elements in the diagonal and $\frac{1}{2}(n^2 - n)$ complex off-diagonal independent elements for a total of $n^2 - 1$ real parameters. To determine it completely $n^2 - 1$ independent measurements (three, as we have seen, for $n = 2$) are therefore needed.

Let us stress that the measurements described in the example are not successive measurements. They are performed repeatedly on the beam emerging from the preparing instrument. What would happen if the measurements were performed in succession will be discussed in the next chapter.

III COMMENSURABILITY

The existence of inequivalent observables poses the question of what happens if one attempts to setup an experiment with instruments that measure two such observables. The answer to this question is the main subject of this chapter.

1. The spin example continued

Let us return to the idealized Stern-Gerlach devices of the examples in Chapter II. As pointed out there, the statement that a beam of spin 1/2 particles is split into two components by the inhomogeneous magnetic field of a device must be looked at carefully. We say that the beam is split because when we perform a measurement with a Stern-Gerlach device, configured as a measuring instrument by the insertion of appropriate particle detectors, the detectors respond as if some of the particles were deflected towards the field gradient direction and others away from it (unless the beam particles are prepared in an eigenstate of the spin component along the field gradient). For example if the detectors consist of a photographic plate it will be exposed in two spots. The response of the detectors together with our everyday direct experience of the surrounding world lead us to the conclusion that the inhomogeneous magnetic field of the Stern-Gerlach magnets does split the original beam into two beams consisting of particles with their spin component respectively parallel and anti-parallel to the field gradient. This conclusion is erroneous.

As can be verified by means of the following experiment, the spin state measured by a Stern-Gerlach measuring device is not altered by the insertion of a a set of magnets. A beam completely polarized in the positive z-direction (i.e., in the pure state $|+\rangle$) is directed to an instrument consisting of Stern-Gerlach magnets with no associated particle detectors (we shall call them bare magnets) and with the field gradient in the $+x$-direction. On the beam emerging from the bare magnets a measurement of s_z is performed. Clearly, if these magnets do not alter the measured state, the measurement will show the emerging beam still to be one hundred percent polarized in the $+z$-direction. Assume on the contrary that the bare magnets were to split the beam and denote by $|\Rightarrow\rangle$ and $|\Leftarrow\rangle$ the eigenstates of s_x belonging to the eigenvalues 1/2 and $-1/2$ respectively. Particularizing (II.3.7) for $\theta = \pi/2, \varphi = 0$, we have in the s_z-representation

$$|\Rightarrow\rangle = \frac{1}{\sqrt{2}}\begin{pmatrix}1\\1\end{pmatrix}, \qquad |\Leftarrow\rangle = \frac{1}{\sqrt{2}}\begin{pmatrix}-1\\1\end{pmatrix}.$$

On emerging from the bare magnets a particle would be in either the state $|\Rightarrow\rangle$ or in the state $|\Leftarrow\rangle$ with equal probability $|\langle+|\Rightarrow\rangle|^2 = |\langle+|\Leftarrow\rangle|^2 = 1/2$. Its state would be represented by the mixture

$$\rho = \frac{1}{2}(|\Rightarrow\rangle\langle\Rightarrow| + |\Leftarrow\rangle\langle\Leftarrow|). \tag{1.1}$$

The subsequent measurement of s_z would yield the value 1/2 with probability

$$\mathrm{Tr}\,(\rho|+\rangle\langle+|) = \frac{1}{2}\left(|\langle+|\Rightarrow\rangle|^2 + |\langle+|\Leftarrow\rangle|^2\right) = \frac{1}{2}$$

and the value $-1/2$ with equal probability $\mathrm{Tr}\,(\rho|-\rangle\langle-|)$. Thus the emerging beam would be unpolarized.

Although the experiment just described would be very difficult to perform in practice, the conclusion that the beam emerging from the bare magnets is measured in the state $|+\rangle$ is born by all the verified predictions of quantum mechanics. Thus *in the absence of a measurement* it is erroneous to assert that the beam is split by the inhomogeneous magnetic field of the magnets into two components, one consisting of particles with their spin along the field gradient, the other of particles with spin in the opposite direction. In this sense the sketches of Chapter II are misleading because they suggest the existence of two beam components in the gap of the magnets regardless of whether a measurement is done or not.

It is important to notice the difference between the mixture (1.1) and the actual pure state $|+\rangle = (1/\sqrt{2})\,(|\Rightarrow\rangle - |\Leftarrow\rangle)$ of the particles in the emerging beam which can be written in operator representation as

$$\frac{1}{2}(|\Rightarrow\rangle\langle\Rightarrow| + |\Leftarrow\rangle\langle\Leftarrow| - |\Rightarrow\rangle\langle\Leftarrow| - |\Leftarrow\rangle\langle\Rightarrow|).$$

Note the last two terms in this expression which show the experiment to be an interference experiment in which the relative phase of the amplitudes $\langle+|\Rightarrow\rangle$ and $\langle+|\Leftarrow\rangle$ play a measurable role as remarked at the end of Section II.1.

In order to confirm the somewhat anti-intuitive conclusion drawn from our experiment we can explore, at least in principle, the region between the bare magnet poles with a microscope of sufficient resolving power to determine whether the particles are seen through the microscope to be deflected towards one pole or the other. In agreement with Rule 2 we will find out that this is indeed the case but, then, if we examine the emerging beam by means of a Stern-Gerlach detector with a photographic plate, we will find two spots of equal intensity in accordance with the prediction given by the mixture (1.1). The measurement of the spin of the particles as they go through the magnets alters somehow the state in which they are measured after they emerge, presumably through their interaction with the illuminating light. We can go further and, by decreasing the beam intensity, perform the experiment with one particle at a time. We will then find that every time a particle is observed with the microscope to be deflected in the $+x$-direction ($s_x = 1/2$) it will be found by a subsequent measurement of s_x after emerging from the magnets to be in the state $|\Rightarrow\rangle$ belonging to the eigenvalue $1/2$ of s_x.

It is worth remarking that, in describing the above experiment, we have been using again a somewhat loose language, for we have not bothered to distinguish clearly the preparation and the measurement phases. As in the previous chapter each experimental run is to be regarded as performed on the individual particles of the beam. If we take the preparation phase to end (and the measurement phase to begin) at a time after the particles have left the bare magnets, the latter are to be regarded as part of the preparation apparatus and we can describe the experimental result by saying that the addition of the bare magnets to the preparation apparatus does not alter the preparation state. Clearly, instead of regarding the bare magnets as part of the preparation apparatus we could regard them as part of the measurement apparatus. In general, as pointed one in Section I.2 one can select the reference time in such a way that the presence of the bare magnets affects the definition of both the preparation and the measurement apparatus.

2. Different kinds of measurement

The above discussion shows that a Stern-Gerlach device provided with a microscope can be thought of as a combination of a measuring and a preparing instrument. It serves to measure s_x on the spin system prepared in a first experiment and to prepare the system in the eigenstate belonging to the measured eigenvalue in a second experiment that follows after the first. Such a device is called a *filter*. The device of the example is a maximal filter because it both measures a maximal observable and prepares the system in a pure state.

Clearly one can imagine filters that are not maximal. As an example consider a Stern-Gerlach device with the field gradient in the "up" ($+z$)-direction used on particles of spin 3/2. In a language that we already know to be somewhat loose, a beam of such particles will be "split" into four beams. Two of them (spin components $-3/2$, and $-1/2$) will be deflected down, the other two (spin components 1/2, and 3/2) up. Imagine the detector to consist of a microscope connected to a recorder which will just record whether the particles are deflected up or down. Suppose that an experiment is done on a single particle and the detector reads "up". What can we say about the state of the particle prepared by the filter? The rules introduced up to now do not permit us to find the answer. One would suspect however that a measurement on the emerging particle by an identical filter will also give the result "up" with certainty. This expectation is confirmed by experience. Now, the most general state which will satisfy the condition of yielding under measurement the value "up" with probability one, is any state in the linear subspace spanned by the vectors $|1/2\rangle$ and $|3/2\rangle$. Such a state is the mixture

$$\rho' = a|\tfrac{3}{2}\rangle\langle\tfrac{3}{2}| + c|\tfrac{3}{2}\rangle\langle\tfrac{1}{2}| + c^*|\tfrac{1}{2}\rangle\langle\tfrac{3}{2}| + (1-a)|\tfrac{1}{2}\rangle\langle\tfrac{1}{2}|$$

or, in the s_x-representation

$$\rho' = \begin{pmatrix} a & c & 0 & 0 \\ c^* & 1-a & 0 & 0 \\ 0 & 0 & 0 & 0 \\ 0 & 0 & 0 & 0 \end{pmatrix},$$

where $0 \le a \le 1$ and c is arbitrary.

The preceding discussion points the way to the following generalization of the notion of filter: a *filter* is a combined measurement and preparation apparatus; it measures an observable A and prepares the system in a state on the eigensubspace belonging to the measured eigenvalue of A. Thus, if

$$A = \sum_j a_j E_j, \qquad E_j = \sum_{k=1}^{s_j} |j,k\rangle\langle j,k|,$$

and the measurement yields the degenerate eigenvalue a_i (of multiplicity s_i), the system is prepared in a state of the form

$$\rho_i' = \sum_{j=1}^{s_j} \sum_{k=1}^{s_i} w_{ijk} |i,j\rangle\langle i,k|,$$

where the coefficients satisfy $w_{ijk} = w_{ikj}^*$ and $\sum_j w_{ijj} = 1$ but are otherwise arbitrary. It follows that a measurement of A as the system emerges from the magnets will again yield the value a_i. In particular, if the non-degenerate eigenvalue a_m

is measured by the filter, then the system is prepared in the pure state $|m\rangle$, the eigenvector of A belonging to a_m.

Not all measurement apparatus are filters, in fact in many cases the measurement is *destructive*: the system is absorbed by the apparatus and loses its identity. Such would be the case if a Stern-Gerlach device were provided with a photographic plate as a detector. But even if the system is not destroyed by the measurement, it is often the case that the measurement apparatus is not a filter. A typical example is the measurement of the momentum of a particle by a device that involves the recoil of the particle with part of the detector. Although after the measurement the particle may still be identified as a system and its state could be determined through the laws of quantum dynamics, it is clearly not in the eigenstate belonging to the measured momentum. Measurements made with a filter are often referred to as measurements of the *first kind* while all others, destructive or not, are often called of the *second kind.*

The question arises whether it is possible to analyze a measurement by studying in detail the interaction of the system under consideration with the measuring instruments in accordance with the laws of quantum dynamics. We shall not delve into this difficult and still unresolved question which has become to be known as *the problem of measurement.* We shall be content to describe a measurement apparatus by its corresponding observable. Such a procedure has proven to be perfectly adequate in the incredibly successful development of the applications of Quantum Mechanics.

3. Commensurable observables

As stated in the previous chapter, one of the fundamental differences between classical and quantum mechanics is that, according to the latter, two distinct maximal measuring instruments applied to the same system measure maximal observables whose eigenvectors form different bases. Except in the special case where the measured state is a common eigenvector, the value obtained from the measurement of one observable does not allow us to infer the outcome of a measurement of the other.[1] Consider instead two equivalent maximal measurement apparatus M and N that measure the maximal observables M and N with a common basis $\{|\phi_i\rangle\}$. If $\{m_i\}$ and $\{n_i\}$ are their (non-degenerate) eigenvalues, we have

$$M = \sum_i m_i|\phi_i\rangle\langle\phi_i|, \qquad N = \sum_i n_i|\phi_i\rangle\langle\phi_i|. \tag{3.1}$$

[1] In classical mechanics a maximal measurement gives the values of all coordinates and momenta from which the value of any observable, i.e, any dynamical variable, can be inferred.

A measurement of M that gives the value m_k is a measurement of the state of the system yielding the state $|\phi_k\rangle$ and is in this sense equivalent to a measurement of N that yields the value n_k. One could be tempted to conclude that if instead of M we had measured N we would have found the value n_k. This conclusion is however unwarranted, for we must not forget that the predictions of quantum mechanics are statistical. What we can assert is that regardless of the state ρ in which the system has been prepared, the probabilities of obtaining the values m_k and n_k upon measurements of M and N, respectively, are equal because both are precisely the probability $\mathrm{Tr}\,(\rho|\phi_k\rangle\langle\phi_k|) = \langle\phi_k|\rho|\phi_k\rangle$ of finding the system in the state $|\phi_k\rangle$. More explicitly, if we perform many measurements of M and of N on identically prepared systems the respective relative frequencies of the outcomes $\{m_i\}$ and $\{n_i\}$ of the measurements of M and N are the same. Therefore in any number of measurements of an experimental run designed to measure, say M, we can substitute the apparatus M by the apparatus N if we replace the outcome n_k by m_k. The observables M and N are said to be *commensurable* because they are both measured by the same operation, e.g., the application of the apparatus M to the system. Note that N is an invertible function of M, the functional relation been given by the one-to-one correspondence $m_i \leftrightarrow n_i$.

As can easily be verified from the spectral resolutions (3.1), the two maximal observables M and N commute. One can also show the converse: two maximal observables that commute are equivalent. We shall soon prove instead a more general result.

The notion of commensurability can be extended to general observables. For this purpose it is convenient to examine the relation between the measurement of an observable and the measurement of the state of the system. In the case of a maximal observable M we have identified the measurement of the observable with the measurement of the state of the system: the statement "the observable has the value m_i" is synonymous with the statement "the system is measured to be in the state $|\phi_i\rangle$". We can make a similar identification for a general observable $A = \sum_i a_i E_i$. The statement "the observable has the value a_i" will be regarded as synonymous with the statement "the state of the system is measured to be a mixture of eigenstates of A belonging to the eigenvalue a_i". Such an identification is consistent with the discussion on measurements of the first kind in the previous section.

Besides the observable A consider another general observable $B = \sum_j b_j F_j$ with eigenvalues b_j and eigenprojections F_j.

The observables A and B are commensurable *if there exists an observable $C = \sum_{ij} c_{ij} R_{ij}$ such that there is a one-to-one correspondence between the eigenvalues* *of C and the ordered pairs of eigenvalues of A and B : $c_{ij} \leftrightarrow \{a_i, b_j\}$, and for any preparation state ρ of the system, the probabilities of the outcome of a measurement obey*

$$\mathcal{P}(a_i|\rho) = \sum_j \mathcal{P}(c_{ij}|\rho), \qquad \mathcal{P}(b_j|\rho) = \sum_i \mathcal{P}(c_{ij}|\rho). \tag{3.2}$$

A set of observables is commensurable if every pair in the set is commensurable.

Roughly speaking this definition rests on the notion that if we can apply two measuring instruments A and B to a system, then A and B together might perhaps be regarded as a measuring instrument C whose configuration is given by their combined readings. A measurement of the observable C will determine this configuration and therefore the readings of both A and B. However the measuring instruments form part of the experimental setup and an experiment with A alone is different from an experiment with both A and B present, the presence of B may very well alter the relative frequencies of the readings of A in an experimental run. Because of this possibility the proviso regarding probabilities must be included in the definition. The following important theorem shows that the possibility just envisaged is often realized.

Two observables A and B are commensurable if and only if they commute.

In particular any function $G(A)$ of the observable A is commensurable with A.

Since, as already mentioned, projections are the building blocks of general observables we start by proving the theorem for a pair of projections E and F. We write $\{\lambda, \mu\}$, $\lambda, \mu = 0, 1$ for a pair of eigenvalues of E and F, in this order, and denote by $\mathbb{E}$ and $\mathbb{F}$ the eigensubspaces of E and F belonging to $\lambda = 1$ and $\mu = 1$, respectively. The eigensubspaces belonging to the zero eigenvalues are then $\mathbb{E}^\perp$ and $\mathbb{F}^\perp$.

Assume first that E and F are commensurable. According to the definition, there exists a hermitean operator

$$R = \sum_{i=1}^{4} r_i R_i, \qquad \sum_{i=1}^{4} R_i = I, \qquad R_i R_j = \delta_{ij} R_i, \tag{3.3}$$

where the r_i are four different real numbers in one-to-one correspondence with the pairs $\{\lambda, \mu\}$,

$$r_1 \leftrightarrow \{1,1\}, \qquad r_2 \leftrightarrow \{1,0\}, \qquad r_3 \leftrightarrow \{0,1\}, \qquad r_4 \leftrightarrow \{0,0\}. \tag{3.4}$$

The condition (3.2) on the probabilities requires that for any state ρ,

$$\mathrm{Tr}\,(\rho E) = \mathrm{Tr}\,(\rho(R_1 + R_2))\,, \qquad \mathrm{Tr}\,(\rho F) = \mathrm{Tr}\,(\rho(R_1 + R_3))\,.$$

We shall show at the end of this proof that for any hermitean operator A, if $\mathrm{Tr}\,(\rho A) = 0$ for any state ρ, then $A = 0$. From this result and the linearity of the trace it follows that

$$E = R_1 + R_2, \qquad F = R_1 + R_3 \tag{3.5}$$

Remembering the orthogonality of the projections R_i we obtain by direct calculation $EF = FE = R_1$, which proves the first part of the theorem. Note that from (3.5) we also obtain $R_2 = E(I - F)$, $R_3 = (I - E)F$ and $R_4 = (I - E)(I - F)$.

Assume next that $[E, F] = 0$, then by the properties of projection operators (A.2.8.1), EF, $E(I - F)$, $(I - E)F$, $(I - E)(I - F)$ are orthogonal projections that add to the identity. They project respectively on the orthogonal subspaces $\mathbb{E} \cap \mathbb{F}$, $\mathbb{E} \cap \mathbb{F}^\perp$, $\mathbb{E}^\perp \cap \mathbb{F}$, $\mathbb{E}^\perp \cap \mathbb{F}^\perp$. Choosing four appropriate real numbers r_i one can construct the observable

$$r_1 EF + r_2 E(I - F) + r_3 (I - E)F + r_4 (I - E)(I - F)$$

which is identical to (3.3). Therefore, if we make the correspondence (3.4) we obtain the probabilities (3.2). To complete the proof we show as promised that, *if A is a hermitean operator and* $\mathrm{Tr}\,(\rho A) = 0$ *for any state ρ, then* $A = 0$. It is sufficient to consider the pure states $|\psi_k\rangle\langle\psi_k|$ corresponding to a basis of eigenvectors of A. Then $\mathrm{Tr}\,(|\psi_k\rangle\langle\psi_k|A) = \langle\psi_k|A|\psi_k\rangle = a_k$, where a_k is the corresponding eigenvalue of A. Thus all the eigenvalues of A are zero and so is therefore A.

In terms of the relation between the measurement of observables and states, the theorem can be understood as follows: A measurement of R that yields, say, the value r, is synonymous with finding the state of the system in $\mathbb{E} \cap \mathbb{F}$ and is equivalent to a joint measurement of E and F yielding the pair $\{1, 1\}$, for such an outcome will tell us that the states lies in both $\mathbb{E}$ and $\mathbb{F}$.

To conclude the proof of the theorem we need only show that two observables are commensurable if and only if their eigenprojections are: suppose first that A and B are commensurable, then in accordance with the definition we can construct the observable $C = \sum_{ij} c_{ij} R_{ij}$ such that $\mathrm{Tr}\,(\rho E_i) = \sum_j \mathrm{Tr}\,(\rho R_{ij})$, for any ρ, which implies that $E_i = \sum_j R_{ij}$; similarly $F_j = \sum_i R_{ij}$. We recall (A.3.4.5) that two observables commute if and only if their eigenprojections commute. It follows that, since the R_{ij} commute, so do E_i and F_j which are therefore commensurable. Conversely, if $[E_i, F_j] = 0$, then the observable

$$C = \sum_{ij} c_{ij} E_i F_j \tag{3.6}$$

satisfies the conditions of the definition for A and B to be commensurable. This concludes the proof of the theorem.

The expression (3.6) shows that two commensurable observables have a common basis of eigenvectors. Conversely, if two observables have a common basis of eigenvectors then they are commensurable, for it is then possible to construct an observable of the form (3.6). These conclusions can be easily seen to apply to any number of observables.

In constructing the observable C that jointly measures A and B it is not even necessary to find real numbers c_{ij} that are in one-to-one correspondence with the pairs $\{a_i, b_j\}$, one can simply write instead of (3.6)

$$C = \sum_{ij} \{a_i, b_j\} E_i F_j$$

and regard C as an operator with composite eigenvalues $\{a_i, b_j\}$, in the sense of A.3.6.

Commensurable observables are often called "simultaneously measurable"; we shall not use this term because in our definition the only place where the time appears is in the selection of the instant that separates the preparation from the measurement phase. As everywhere in this chapter, this instant is taken to be the same in all the experiments compared.

4. The uncertainty relation

Consider again the observables A and B of the previous section. If they are commensurable they are measured jointly by the observable C, equation (3.6). If the system is prepared in a mixture of eigenvectors of C belonging to the eigenvalue c_{ij}, measurements of A and B will yield with certainty the values a_i and b_j respectively. If the observables A and B are not commensurable they may still have some eigensubspaces in common and, if the system is prepared in a state ρ in one of them, the results of the measurement of A and B in the same experiment will still be certain. Another way of seeing this is that the restrictions of A and B to a common eigensubspace commute.

If the system is prepared in an arbitrary state ρ, it is possible to find a lower bound for the product of the standard deviations $(\Delta A)_\rho = \left(\langle A^2\rangle_\rho - \langle A\rangle_\rho^2\right)^{1/2}$ and $(\Delta B)_\rho = \left(\langle B^2\rangle_\rho - \langle B\rangle_\rho^2\right)^{1/2}$ of two arbitrary observables in terms of the average $\langle [A, B]\rangle_\rho$ of their commutator. Here we have written $\langle A\rangle_\rho = \mathrm{Tr}\,(\rho A)$, including the state in the notation for the averages to stress that they depend on it. In the derivation to follow we will assume that the observables have zero averages, which can be done without loss of generality because A and $A - \langle A\rangle_\rho$ have the same standard deviation and are equivalent observables (their eigenvalues are simply shifted by $\langle A\rangle_\rho$). With this assumption we have

$$(\Delta A)^2_\rho(\Delta B)^2_\rho = \mathrm{Tr}\,(\rho A^2)\mathrm{Tr}\,(\rho B^2). \tag{4.1}$$

Remember now that $\mathrm{Tr}\,(U^\dagger V)$ can be thought of as the scalar product of U and V in the vector space of linear operators. It therefore obeys Schwarz' inequality: $\mathrm{Tr}\,(U^\dagger U)\mathrm{Tr}\,(V^\dagger V) \geq |\mathrm{Tr}\,(U^\dagger V)|^2$. Since ρ is a positive operator it has a positive square root $\rho^{1/2}$ which like ρ is also hermitean. Putting $U = \rho^{1/2}A$ and $V = \rho^{1/2}B$, the above inequality becomes

$$\mathrm{Tr}\,(\rho A^2)\mathrm{Tr}\,(\rho B^2) \geq |\mathrm{Tr}\,(\rho BA)|^2.$$

Now, BA can be written as the sum of its hermitean, $H = \frac{1}{2}(AB + BA)$, and antihermitean, $K = \frac{1}{2}(BA - AB)$, parts. Since both ρ and H are hermitean so is $\rho H + H\rho$; it follows that $\mathrm{Tr}\,(\rho H) = \frac{1}{2}\mathrm{Tr}\,(\rho H + H\rho)$ is real. Similarly one can see that $\mathrm{Tr}\,(\rho K)$ is pure imaginary. Therefore

$$|\mathrm{Tr}\,(\rho BA)|^2 = |\mathrm{Tr}\,(\rho H) + \mathrm{Tr}\,(\rho K)|^2 = (\mathrm{Tr}\,(\rho H))^2 + |\mathrm{Tr}\,(\rho K)|^2 \geq |\mathrm{Tr}\,(\rho K)|^2.$$

From (4.1) and the above sequence of inequalities we conclude that $(\Delta A)^2_\rho(\Delta B)^2_\rho \geq |\mathrm{Tr}\,(\rho K)|^2$. Noting that $K = -\frac{1}{2}[A, B]$ and extracting the square root we find the estimate

$$(\Delta A)_\rho(\Delta B)_\rho \geq \tfrac{1}{2}|\langle [A, B]\rangle_\rho|$$

which is known as the uncertainty relation.

The fact that the product $(\Delta A)_\rho(\Delta B)_\rho$ be zero for a particular state ρ does not mean that the observables commute, all it means is that when the particular state lies in an eigensubspace of one of them then the average of the commutator *for that particular state* is zero. For example, if a particle of spin 1/2 is prepared in the eigenstate $|+\rangle$ of s_z then $\Delta s_z = 0$ and $|\langle [s_y, s_z]\rangle|$ is just $|\langle s_x\rangle|$ which is zero because a measurement of s_x on a system prepared in the state $|+\rangle$ yields either 1/2 or $-1/2$ with equal probability.

IV DYNAMICS

The description of an experiment that has been presented in the preceding two chapters is essentially complete but, of course, the correct description of experiments is only a small part of a physical theory, a part that in analogy with classical mechanics one could call "kinematics" (although we shall not insist in using the term because in some branches of quantum mechanics it is given a different meaning). The theory should also be able to, for example, analyze composite systems in terms of their constituent subsystems and to predict the correlations between the outcomes of different experiments performed on the same or on related systems. These other aspects form part of quantum dynamics.

The dynamical laws involve time. In the approach taken in this book the time that appears in the laws is the reference time chosen to separate the preparation from the measurement phases of an experiment. In the previous two chapters this time was assumed to be given, in this chapter we shall introduce the dynamical laws by studying how the description of an experiment depends on it.

1. The dynamical postulate

As explained in Chapter I, the theoretical description of the preparation and the measurement in an experiment depends in general on the reference time chosen to separate the two phases. The description corresponding to a value τ of this time will be called the *τ-depiction*. Suppose that in a given experiment one chooses two different values τ and τ' for the reference time, how are the τ- and the τ'-depictions related? The postulates embodied in Rules A-D (Section II.1) are quite strong. Except for the addition of a natural assumption, they already contain the answer to the question.

Imagine an experiment in which maximal preparation and measurement apparatus are used. As before we call $\mathbb{H}$ the Hilbert space of the system. In the τ-depiction, we denote the apparatus by $\{A, \tau\}$ and $\{B, \tau\}$, respectively. The letters A and B are intended to represent the instruments that enter into the definition of the apparatus, which may or may not be distinct in their construction as remarked in Chapter 1. Let $|a^{(\tau)}\rangle, |b^{(\tau)}\rangle \in \mathbb{H}$ be unit vectors determining the rays corresponding to the configurations a, and b of $\{A, \tau\}$ and $\{B, \tau\}$ respectively.[1] The probability $\mathcal{P}(b|a)$ of obtaining the reading b of B upon measurement if A is set in the configuration a is $|\langle b^{(\tau)}|a^{(\tau)}\rangle|^2$.

By using the operator representation of states, the state in which the system is prepared can be written $\rho_a^{(\tau)} = |a^{(\tau)}\rangle\langle a^{(\tau)}|$, while the measured state becomes $\sigma_b^{(\tau)} = |b^{(\tau)}\rangle\langle b^{(\tau)}|$. The latter expression is, of course, identical with the eigenprojection belonging to the eigenvalue b_i of the observable that $\{B, \tau\}$ measures. In terms of these density operators the probability can be written in a way manifestly independent of the particular choice of the unit vectors.

$$\mathcal{P}(b|a) = \mathrm{Tr}\left(\rho_a^{(\tau)}\sigma_b^{(\tau)}\right).$$

In the τ'-depiction, we call the corresponding unit vectors $|a^{(\tau')}\rangle, |b^{(\tau')}\rangle \in \mathbb{H}$, and the density operators $\rho_a^{(\tau')} = |a^{(\tau')}\rangle\langle a^{(\tau')}|$ and $\sigma_b^{(\tau')} = |b^{(\tau')}\rangle\langle b^{(\tau')}|$. This new depiction must give the same experimental predictions as the previous one. We must therefore have

$$\mathrm{Tr}\left(\rho_a^{(\tau)}\sigma_b^{(\tau)}\right) = \mathrm{Tr}\left(\rho_a^{(\tau')}\sigma_b^{(\tau')}\right).$$

In Chapter II we assumed that, in the absence of superselection rules, every hermitean operator corresponds to an observable and that every state that can be measured can be prepared. This assumption implies that the last equation holds if ρ and σ are replaced by an arbitrary pair R_k, R_l of one-dimensional projections. The correspondence between the τ- and the τ'- depictions consists therefore of a one to one mapping of the set of one-dimensional projections (or of rays) of $\mathbb{H}$ onto itself such that if $R_k^{(\tau)} \mapsto R_k^{(\tau')}$ and $R_l^{(\tau)} \mapsto R_l^{(\tau')}$, then

$$\mathrm{Tr}\left(R_k^{(\tau)}R_l^{(\tau)}\right) = \mathrm{Tr}\left(R_k^{(\tau')}R_l^{(\tau')}\right).$$

According to Wigner's theorem A.6.2. this correspondence induces a one-to-one transformation of $\mathbb{H}$ onto itself such that if $|f_k^{(\tau)}\rangle$ is in the range of $R_k^{(\tau)}$, then there is a vector

$$|f_k^{(\tau')}\rangle = U(\tau', \tau)|f_k^{(\tau)}\rangle$$

[1] Remember that these configurations, referring as they do to instrumental settings, do not depend on τ.

in the range of $R_k^{(\tau')}$, where $U(\tau', \tau)$ is either a unitary or an antiunitary operator on $\mathbb{H}$ determined, up to a phase factor, by the mapping of the rays.

It follows that the relation between the above density operators in the two depictions can be written

$$\rho_a^{(\tau')} = U(\tau', \tau)\rho_a^{(\tau)}U^\dagger(\tau', \tau),$$
$$\sigma_b^{(\tau')} = U(\tau, \tau')\sigma_b^{(\tau)}U^\dagger(\tau, \tau')$$

and that the corresponding unit vectors *may* be selected in such a way that[2]

$$|a^{(\tau')}\rangle = U(\tau', \tau)|a^{(\tau)}\rangle, \qquad |b^{(\tau')}\rangle = U(\tau', \tau)|b^{(\tau)}\rangle.$$

Because these equations hold for arbitrary vectors and reference times, one can easily show that

$$U(\tau', \tau) = U(\tau', \tau'')U(\tau'', \tau), \qquad U(\tau, \tau') = U^\dagger(\tau', \tau).$$

When $\tau' = \tau$ the two depictions must coincide, which means that $U(\tau, \tau)$ equals the identity multiplied by an arbitrary phase factor; thus, $U(\tau, \tau)$ is unitary. As is customary we take $U(\tau, \tau) = I$. On physical grounds it is reasonable to assume that $U(\tau', \tau)$ is continuous in τ and τ'. This assumption implies that $U(\tau', \tau)$, being continuously connected with the unitary operator $U(\tau, \tau)$, is also unitary.

The *dynamical postulate* of quantum mechanics generalizes the conclusions of the preceding discussion. Its formulation is embodied in the following rule.

Rule 3. – Given an experimental setup, there exists a continuous and differentiable unitary evolution operator $U(\tau, \tau')$ *on* $\mathbb{H}$ *such that,*

$$U(\tau, \tau) = I, \qquad U(\tau, \tau') = U(\tau, \tau'')U(\tau'', \tau'). \tag{1.1}$$

For any two reference times τ *and* τ', *the* τ*-depiction* $\rho^{(\tau)}$ *and the* τ'*-depiction* $\rho^{(\tau')}$ *of a state of the system are related by*

$$\rho^{(\tau')} = U(\tau', \tau)\rho^{(\tau)}U^\dagger(\tau', \tau). \tag{1.2}$$

Note that the rule applies both to a prepared and to a measured state. If the state is pure, represented by a unit vector $|\psi^{(\tau)}\rangle$, we have

$$|\psi^{(\tau')}\rangle = U(\tau', \tau)|\psi^{(\tau)}\rangle.$$

---

[2] The unit vectors can, of course, be multiplied by arbitrary phase factors without affecting the rays determined by them.

Why we require the differentiability of $U(\tau, \tau')$ will become apparent in the next section. It follows from (1.1) and the unitarity of $U(\tau, \tau')$ that

$$U(\tau', \tau) = U^{\dagger}(\tau, \tau').$$

Also, if we call $B^{(\tau)}$ and $B^{(\tau')}$ the τ- and τ'- depictions of the observable corresponding to the measurement apparatus, we have the relation

$$B^{(\tau')} = U(\tau', \tau) B^{(\tau)} U^{\dagger}(\tau', \tau). \tag{1.3}$$

The theoretical description of the apparatus in terms of an arbitrary reference time is not very convenient. It is more satisfactory to represent the experimental setup in some standard way which does not involve this arbitrariness. This is the subject of the next section.

2. Schrödinger, Heisenberg and mixed pictures

As already mentioned in Section I.3, the preparation time t_0 and the measurement time t there introduced can be defined in terms of the experimental instruments alone. Knowing the time constants of the instruments, t_0 can be defined as a given time after the closing of the starting switch. Similarly t can be defined as a given time before the dial pointers respond. It appears, therefore, that the t_0-depiction is well suited for the preparation, while the t-depiction is well suited for the measurement apparatus. From now on we shall reserve the term "preparing instrument" to denote those instruments that enter into the definition of the preparation apparatus in the t_0-depiction, while the term "measuring instrument" will denote the instruments that enter into the definition of the t-depiction of the measurement apparatus. The remaining instruments, if any, will be referred to as "field sources" in analogy with the classical terminology. Thus, the bare magnets of the example in Section III.1 constitute an example of instruments that are field sources. As noted in Chapter I there may be experimental setups in which a neat classification of the instruments into the above three categories may not be possible but this in no way invalidates the theoretical description.

Schrödinger picture. – The suitability mentioned above is exploited in one standard representation called the *Schrödinger picture* in which the preparation apparatus is given in the t_0-depiction and the measurement apparatus in the t-depiction.

In this picture, the states $\rho^{(t_0)}$, $|\psi^{(t_0)}\rangle$ will be usually called initial states and written respectively as ρ_0, $|\psi_0\rangle$ to stress the fact that in the t_0-depiction the prepared state does only depend on the configuration of the preparing instrument at the time the experiment is started.

Similarly, in the t-depiction, the observable corresponding to a measurement apparatus $\{\mathrm{B}, t\}$ will be usually denoted simply by B instead of $B^{(t)}$, again omitting the time argument to stress the fact that B is determined by the measuring instrument independently of the measurement time.[3] The operator B is called the *Schrödinger representation* of the observable, often simply the *observable*, corresponding to the appropriate measuring instrument. The argument t will also be omitted in the notation for its eigenprojections and eigenvectors. Thus, we write the spectral resolution as

$$B = \sum_j b_j E_j$$

and an eigenvector that in the notation of the previous section would have been denoted by $|b_j^{(t)}\rangle$ will simply be written as $|b_j\rangle$.

The *Schrödinger state at time* t, usually abbreviated as the state at time t, is defined as the t-depiction of the prepared state. In general we write it as $\rho(t)$ instead of $\rho^{(t)}$ to stress that we are thinking in the Schrödinger picture. In analogous notation, if the state is pure, we denote it in vector form as $|\psi(t)\rangle$, instead of $|\psi^{(t)}\rangle$. The relation between the state at time t and the initial state is given, then, by equation (1.2) in Rule 3 as

$$\rho(t) = U(t, t_0)\rho_0 U^\dagger(t, t_0). \tag{2.1}$$

Once $\rho(\mathrm{t})$ has been determined, the probabilities can be found from Rule 2. The probability $\mathcal{P}(b_i, t|\rho_0, t_0)$ of obtaining the eigenvalue b_i if the observable $B = \sum_i b_i E_i$ is measured at time t, given that the initial state was ρ_0, is, according to that rule and the definition of $\rho(t)$,

$$\mathcal{P}(b_i, t|\rho_0, t_0) = \mathrm{Tr}\,(\rho(t) E_i) \tag{2.2}$$

which by (II.4.6) leads to the following expression for the average value of B obtained from measurements at time t

$$\langle B\rangle(t) = \mathrm{Tr}\,(\rho(t) B)\,. \tag{2.3}$$

If the initial state is the pure state $|\psi_0\rangle$, the corresponding expressions simplify to

$$|\psi(t)\rangle = U(t, t_0)|\psi_0\rangle \tag{2.4}$$

---

[3] As explained at the end of this section, B may depend on t if the instrument is constructed with movable parts.

and

$$\langle B\rangle(t) = \langle\psi(t)|B|\psi(t)\rangle. \tag{2.5}$$

Although the times t_0 and t were introduced as being the preparation and the measurement times, from a formal point of view they can be chosen arbitrarily. When we do so, we shall often refer to the operator B as "the Schrödinger observable at time t"; this term means exactly the same as "the t-depiction of the observable". Similarly, the terms "the Schrödinger state at time t" and "the t-depiction of the state" are interchangeable.

Heisenberg picture. – In the picture just discussed, the probabilities were calculated by transforming the initial state into the t-depiction in which the observable was given. Instead we can equally well transform the observable B backwards in time into the t_0-depiction in which the initial state is given. This is what is done in the *Heisenberg picture*. The t_0-depiction of the observable is called its *Heisenberg representation*. We write it as $B(t)$, as is customary. In the notation used in Section 1 this observable is $B^{(t_0)}$ but the customary notation is preferable because it stresses the fact that the Heisenberg observable depends both on the measuring instrument and on the measurement time (it also depends on the preparation time t_0 but this dependence is usually not included in the notation). By equation (1.3), the Heisenberg observable $B(t)$ is related to the Schrödinger observable B by

$$B(t) = U^\dagger(t, t_0)BU(t, t_0). \tag{2.6}$$

Its spectral resolution is

$$B(t) = \sum_i b_i E_i(t)$$

with eigenprojections

$$E_i(t) = U^\dagger(t, t_0)E_i U(t, t_0). \tag{2.7}$$

It follows from Rule 2 that the probability $\mathcal{P}(b_i, t|\rho_0, t_0)$ becomes in the Heisenberg picture

$$\mathcal{P}(b_i, t|\rho_0, t_0) = \mathrm{Tr}\left(\rho_0 E_i(t)\right).$$

Which is, of course, identical to (2.2) since, as can be seen from equations (2.1) and (2.7) and the cyclic property of the trace, both expressions for the probability can be written

$$\mathcal{P}(b_i, t|\rho_0, t_0) = \mathrm{Tr}\left(U(t, t_0)\rho_0 U^\dagger(t, t_0)E_i\right). \tag{2.8}$$

In this picture the average value of the results of the observation of B becomes, therefore,

$$\langle B\rangle(t) = \mathrm{Tr}\left(\rho_0 B(t)\right). \tag{2.9}$$

Or, in case the state is pure

$$\langle B\rangle(t) = \langle\psi_0|B(t)|\psi_0\rangle. \tag{2.10}$$

In the Heisenberg picture the states remain fixed and the observables change with time according to the dynamical evolution given by (2.6).

As in the case of the Schrödinger picture, it is sometimes convenient to select t_0 and t arbitrarily. The corresponding observable will often be called the Heisenberg observable at time t with respect to t_0.

Interaction picture. – Any unitary transformation can be decomposed in an infinity of ways into a product of unitary transformations. Consider one such decomposition of the unitary evolution operator $U(t, t_0)$,

$$U(t, t_0) = U_0(t, t_0)U'(t, t_0). \tag{2.11}$$

Substituting into (2.8) we have after using the cyclic property of the trace

$$\mathcal{P}(b_i, t|\rho_0, t_0) = \mathrm{Tr}\left(U'(t, t_0)\rho_0 U'^{\dagger}(t_0, t)U_0^{\dagger}(t, t_0)E_i U_0(t, t_0)\right).$$

We define a new state

$$\rho_I(t) = U'(t, t_0)\rho_0 U'^{\dagger}(t_0, t), \tag{2.12}$$

new projection operators

$$E_{Ii}(t) = U_0^{\dagger}(t, t_0)E_i U_0(t, t_0)$$

and a corresponding observable $B_I(t) = \sum_i b_i E_{Ii}(t)$, i.e.,

$$B_I(t) = U_0^{\dagger}(t,_0 t)BU_0(t, t_0). \tag{2.13}$$

The average value of B obtained from measurements at time t can then be written

$$\langle B\rangle(t) = \mathrm{Tr}\left(\rho_I B_I(t)\right).$$

The resulting picture of the dynamical evolution is called a *mixed picture*, or, for reasons that will appear clear in Section 5, an *interaction picture*, while the representations (2.12) and (2.13) are *mixed* (or *interaction*) *representations* of the state and the observable, respectively.

In the case of a pure preparation state (2.12) and (2.13) simplify to

$$|\psi_I(t)\rangle = U'(t, t_0)|\psi_0\rangle$$

and

$$\langle B\rangle(t) = \langle\psi_I(t)|B_I(t)|\psi_I(t)\rangle.$$

The mixed pictures are useful when dealing with a system that consists of weakly interacting subsystems.

Although we have not explicitly mentioned it before, nothing of what we have said precludes that the Schrödinger observable B might depend on time. Such will be the case if the measuring instrument B of our experimental setup were constructed with movable parts whose motion would affect its associated basis or the calibration of its dials. The construction of the instruments may intentionally include a time dependence of the observable but, as every experimenter knows, a time dependence is often unintentionally present. For example, B could contain electrical circuits that would be affected by variations in the power supply voltage or diffraction gratings that could be sensitive to variations in temperature. When we want to stress the time dependence of the Schrödinger representation of an observable A, as will be the case in the Section 4, we shall write it as A_t to distinguish it from the dynamical time dependence of its Heisenberg or its mixed representation.

3. An aside on functions of several observables

According to the dynamical postulate, a knowledge of the evolution operator $U(t, t_0)$ as a function of its arguments permits us to predict the probabilities of the outcome of a measurement performed at any time t, given that the system was prepared in a known state at time t_0. When we use the Schrödinger picture we shall say more concisely: $U(t, t_0)$ transforms the state at t_0 into the state at time t; but when using the concise language one must not forget that the notion "state at time t" is defined in terms of measurements or preparations done with experimental apparatus.

The one-to-one mapping of the Hilbert space into itself carried out by the evolution operator is analogous to the canonical transformation of classical mechanics generated by Hamilton's principal function which transforms the initial canonical coordinates and momenta into their values at time t. For many classical systems this transformation is entirely determined by the hamiltonian function which depends on the canonical variables and the time. We shall show in the next section that an analogous result holds in quantum mechanics but, to understand the development to be presented there, it is first necessary to learn some properties of linear operators that are functions of several hermitean operators.

In Appendix A we have defined functions of a hermitean operator $A = \sum_i a_i E_i$ through the relation $G(A) = \sum_i G(a_i) E_i$. We require to extend the definition to functions of a finite set of hermitean operators. It is sufficient to consider a set of two operators A and $B = \sum_i b_i F_i$. If A and B commute, the definition can be reduced to the above one because, as discussed in Section III.3, both A and B are functions of the operator $C = \sum_{ij} c_{ij} E_i F_j$.

Consider the general case in which A and B may not commute, then such a reduction is not possible and another definition is needed. We know what is meant by the simple functions

$$\mu A + \nu B, \qquad A^n, \qquad AB,$$

where μ and ν are any complex numbers and n any non-negative integer (or any integer if A has an inverse). We can then construct any polynomial of A and B. We define the set of functions of A and B as the set of all such polynomials and limits of sequences of polynomials. Thus any function that has a power series expansion is included, for example, for any vector $|f\rangle$ in the Hilbert space $\mathbb{H}$,

$$\exp(A+B)|f\rangle = \sum_{m=0}^{\infty} \frac{(A+B)^m}{m!}|f\rangle$$

(we can include functions with a Laurent series expansion if the appropriate inverses exist). It is not difficult to see that if A and B commute this definition coincides with the above one.

The question arises: given a finite set of hermitean operators commuting or not, and a linear operator R, under what conditions can R be written as a function of the set? The answer is provided by the following important theorem

Let $\mathcal{S} = \{A_m\}$ be a finite set of hermitean operators and $\mathcal{S}' = \{C_n\}$ the set of all linear operators that commute with every A_m. The set of all linear operators that are functions of $\mathcal{S}$ is the set $\mathcal{S}''$ of all operators that commute with every C_n.

The set $\mathcal{S}''$ is called the *operator algebra generated by* $\mathcal{S}$.

We prove the theorem for a set $\mathcal{S}$ consisting of two hermitean operators A, B. The generalization to any finite number of operators is obvious.

For any vector $|f\rangle$ let $\mathbb{M}$ be the subspace spanned by the sequence

$$|f\rangle,\ A|f\rangle,\ B|f\rangle,\ A^2|f\rangle,\ AB|f\rangle,\ B^2|f\rangle,\ A^3|f\rangle,\ A^2B|f\rangle,\ ABA|f\rangle, \ldots$$

i.e., the sequence of the vectors resulting from applying to $|f\rangle$ all monomials formed with A and B (including $A^0 = B^0 = I$ and, if the appropriate inverses exist, monomials with negative exponents). Let M be the projection operator that projects on $\mathbb{M}$. The application of either A or B to any vector of the sequence yields a vector of the sequence. Therefore $MAM = AM$ and taking the adjoint we find that $[A, M] = 0$. Similarly, B and M commute.

To prove the theorem we show that (a) if R is a linear operator that commutes with every operator that commutes with A and B, then R is a function of A and

B, and (b) if R is a function of A and B, it commutes with every operator that commutes with A and B.

Proof of (a): by hypothesis R commutes with M, therefore $R|f\rangle = RM|f\rangle = MR|f\rangle$ so that $R|f\rangle \in \mathbb{M}$ and R is consequently a polynomial, or a limit of a sequence of polynomials,[4] of A and B and, by definition, a function of A and B. The proof of (b) follows because by hypothesis R is a polynomial, or a limit of a sequence thereof, of A and B and, since every operator that commutes with A and B commutes with such a polynomial or limit, it commutes with R.

If a set $\mathcal{S}$ of hermitean operators have no common eigenvectors, the only operators that commute with every member of the set are multiples of the identity. Since every linear operator commutes with the identity it follows that every linear operator is a function of the set.

According to the definition in A.2.9.3, the set $\mathcal{S}$ is an *irreducible* set of hermitean operators, that is, a set that is reduced only by the null subspace and $\mathbb{H}$ itself. This characterization has the advantage that it survives in infinite dimensional Hilbert spaces where the concept of eigenvalue is much less useful. For that reason we prefer to restate the above result in the form of the following corollary:

Every linear operator is a function of an irreducible set of hermitean operators.[5]

Such a strong result does not apply in the infinite-dimensional case. A similar result applies, however, to the observables of interest in quantum mechanics. Although the proof will not be given, we shall feel free to use the generalization in later chapters.

Evidently, if an irreducible set of operators $\mathcal{S}$ is enlarged by adding to it any operator, the resulting set is also irreducible. Such is the case with the algebra $\mathcal{S}''$ generated by $\mathcal{S}$, which, for that reason, is called an *irreducible operator algebra.* The following almost obvious property is characteristic of an irreducible algebra

[4] In an n-dimensional vector space there are only n^2 linearly independent monomials because the space of operators has dimension n^2. Thus every polynomial or limit of a sequence of polynomials can be written as a linear combination of such n^2 monomials.

[5] Perhaps the simplest irreducible set $\mathcal{S}$ one can think of is the set formed by a complete family of one-dimensional orthogonal projections $|e_i\rangle\langle e_i|$ together with another one-dimensional projection $|h\rangle\langle h|$ which is not orthogonal to any projection of the family, that is, $\langle h|e_i\rangle \neq 0$. We know that every operator A can be written in the form $A = \sum_{kl} A_{kl}|e_k\rangle\langle e_l|$. We evidently have

$$|e_k\rangle\langle e_l| = \frac{|e_k\rangle\langle e_k||h\rangle\langle h||e_l\rangle\langle e_l|}{\langle e_k|h\rangle\langle h|e_l\rangle}$$

which when substituted in the expansion of A gives this operator as a function of the operators of $\mathcal{S}$.

$\mathcal{S}''$: given any pair of vectors $|f\rangle, |g\rangle \in \mathbb{H}$, there exists an operator $B \in \mathcal{S}''$ such that $|g\rangle = B|f\rangle$.

We shall say that an irreducible set of operators is *minimal* if none of its elements can be regarded as a function of the others.

Quantum mechanical observables play a role analogous to dynamical variables in classical mechanics. The latter are functions of a set of canonical variables. We expect then that the quantum mechanical analogue of a set of classical canonical variables would be some minimal irreducible set of observables, for, as we have just seen, every observable is a function of such a set. We shall return to this question in Section 5.

To illustrate the concepts introduced in this section, consider the case of spin 1/2. The spin component observables s_z and s_x have no eigenvectors in common, thus every linear operator is a function of them. In this case the functional dependence of a given operator is easy to determine because (see footnote II.4) every linear operator can be written as a linear combination $\alpha s_x + \beta s_y + \gamma s_z + \delta I$, and $s_y = -i[s_z, s_x]$ (in fact, since $s_x s_z = -s_z s_x$, every operator can be written as a linear combination of the four monomials $s_z, s_x, s_z s_x$, and I; cf. footnote 3). The set of spin components $\{s_x, s_y, s_z\}$ is irreducible but not minimal, while the set formed by two components is irreducible and minimal. In the case of spin one, it can be verified that two different spin components also constitute a minimal irreducible set, but the functional dependence of any given operator on, say, s_z and s_x is more difficult to determine because the space of operators has dimension nine. We shall see that identical results hold for any spin. In fact, in spin space, two components of the spin along two different directions always constitute a minimal irreducible set.

4. The hamiltonian

The dynamical postulate, Rule 3, identifies a system as a dynamical system in the sense that, as equation (1.2) shows, the state $\rho(t)$ at time t is determined by the state at any other time. Quantum dynamics is simpler than classical dynamics because the dynamical mapping of the former is a unitary mapping of the Hilbert space into itself which is simpler than the canonical mapping of the classical phase space into itself generated by Hamilton's principal function. As a consequence, the evolution operator $U(t, t')$ obeys a first order linear differential equation which we now derive.

Since the dynamical postulate includes the assumption that $U(t, t')$ is differentiable, let us calculate its derivative with respect to t. By definition of derivative we have with η real

$$\frac{\partial U(t,t_0)}{\partial t} = \lim_{\eta\to 0} \frac{U(t+\eta,t_0)-U(t,t_0)}{\eta} = \lim_{\eta\to 0} \frac{[U(t+\eta,t)-I]U(t,t_0)}{\eta},$$

where to get the last expression we have used the composition rule in (1.1) (generally the time derivative will be written $\partial/\partial t$; the notation d/dt will be reserved for the Heisenberg derivative to be introduced in Section 5). The differentiability of U permits us to define the operator

$$H_t = i \lim_{\eta\to 0} \frac{U(t+\eta,t)-I}{\eta}, \tag{4.1}$$

where we have written the time dependence as a subscript for reasons that will appear clear shortly. In terms of this operator the derivative of $U(t,t_0)$ can be written

$$\frac{\partial U(t,t_0)}{\partial t} = -iH_tU(t,t_0). \tag{4.2}$$

The operator H_t is hermitean, for taking the adjoint of (4.1) and using (1.1) we have

$$H_t^\dagger = i \lim_{\eta\to 0} \frac{I-U^\dagger(t+\eta,t)}{\eta} = i \lim_{\eta\to 0} \frac{U(t+\eta,t)U(t,t+\eta)-U(t,t+\eta)}{\eta}.$$

Simplifying the last numerator we obtain

$$H_t^\dagger = i \lim_{\eta\to 0} \frac{U(t+\eta,t)-I}{\eta} \lim_{\eta\to 0} U(t,t+\eta) = H.$$

Equation (4.1) shows that H_t is an observable depending on the measurement time t only. If $\mathcal{Q}$ is an irreducible set of observables, then, from the theorem of the previous section, H_t can be written as a function of $\mathcal{Q}$ and the variable t. When we want to stress this fact we shall write $H_t(\mathcal{Q})$, using as is customary the same letter for a function and its value.

When the experimental arrangements that we want to consider are not affected by a change in the preparation time t_0, as would be the case if all the experimental apparatus are time independent, the evolution operator can only depend on the difference $t-t_0$, thus $U(t,t_0) = U(t-t_0)$. Then the definition (4.1) says that H is time independent which suggests that it may be related to the classical hamiltonian function. In fact, we will encounter further theoretical arguments to support this suggestion. More importantly, as is well known, a wealth of experimental evidence shows that the eigenvalues of the operator $\hbar H$, where $2\pi\hbar \approx 6.626 \times 10^{-34}$Js is Planck's constant, are energies. For this reason the observable $\hbar H_t(\mathcal{Q})$ is called the *hamiltonian.* In this book, unless explicitly stated, we will use a system of units such that $\hbar = 1$ and write the hamiltonian as H_t or $H_t(\mathcal{Q})$.

The knowledge of the hamiltonian is all that is needed to determine completely the dynamical behaviour of a system, for by (4.1) and (1.1) the evolution operator can be obtained as the solution of the initial value problem

$$\frac{\partial U}{\partial t} = -iH_t U, \qquad U(t_0, t_0) = I. \tag{4.3}$$

In general the solution $U(t, t_0)$ of this problem cannot be written in closed form, but *if the hamiltonian is such that* $[H_t, H_{t'}] = 0$ *for all* $t, t' \geq t_0$, then the solution of (4.3) is

$$U(t, t_0) = \exp\Big[-i \int_{t_0}^{t} dt'\ H_{t'}\Big]. \tag{4.4}$$

That this expression satisfies the initial condition in (4.3) is obvious. To verify that it also satisfies the differential equation, note that in this case there is a family of time independent orthogonal projections E_n such that

$$H_t = \sum_n \epsilon_{t,n} E_n,$$

where the eigenvalues $\epsilon_{t,n}$ are functions of time. By definition of function of an operator, (4.4) can be written

$$U(t, t_0) = \sum_n \exp\Big[-i \int_{t_0}^{t} dt'\ t' \epsilon_{t',n}\Big] E_n. \tag{4.5}$$

Differentiating we obtain

$$\frac{\partial U(t, t_0)}{\partial t} = -i \sum_n \epsilon_{t,n} \exp\Big[-i \int_{t_0}^{t} dt'\ \epsilon_{t',n}\Big] E_n.$$

Since $E_m E_n = E_n \delta_{mn}$, we can rewrite the right member as

$$-i \sum_m \epsilon_{t,m} E_m \sum_n \exp\Big[-i \int_{t_0}^{t} dt'\ \epsilon_{t',n}\Big] E_n = -iH_t \exp\Big[-i \int_{t_0}^{t} dt'\ H_{t'}\Big]$$

which is just $-iH_t U(t, t_0)$. Equation (4.5) shows that in the case been considered the evolution operator and the hamiltonian commute.

If the hamiltonian H is time independent, we talk of a *conservative* system. The evolution operator (4.4) becomes then simply

$$U(t, t_0) = \exp[-i(t - t_0)H].$$

Equation (4.5) simplifies to

$$U(t, t_0) = \sum_n e^{-i\epsilon_n(t-t_0)} E_n \tag{4.6}$$

and a knowledge of the (time independent) spectral resolution of the hamiltonian leads through the last equation to a complete solution of the dynamical problem.

When the hamiltonians at different times do not commute, it might still be possible in specific cases to find a solution of (4.3) in closed form, as in the example that we will work out in Section 6. In general, however, we must content ourselves with solutions that can only be obtained approximately. In such cases it is often convenient to replace (4.3) by the integral equation

$$U(t, t_0) = I - i \int_{t_0}^{t} dt' \; H_{t'} U(t', t_0)$$

which by differentiation is seen to be equivalent to the initial value problem. This integral equation is particularly useful for obtaining series solutions of (4.3).

5. Equations of motion in the different pictures

It is often convenient to transform the differential equation in (4.3) into a differential equation for the states or observables, as the case may be, in the different pictures introduced in Section 4.2.

We start with the *Schrödinger picture*. Differentiating the expression (2.1) for the Schrödinger representation of the state $\rho(t)$ at time t and using the equation of motion (4.3) for U, as well as its adjoint

$$\frac{\partial U^\dagger}{\partial t} = iU^\dagger H_t$$

to eliminate the derivatives of U, we obtain

$$\frac{\partial \rho}{\partial t} = -iH_t U \rho_0 U^\dagger + iU \rho_0 U^\dagger H_t$$

and substituting (2.1) into the right member we obtain the *von Neumann equation*

$$\frac{\partial \rho}{\partial t} = -i[H_t, \rho] \tag{5.1}$$

which describes the evolution of a state in the Schrödinger picture. In case the state is a pure state $|\psi\rangle$, then from (2.4) we obtain by the same procedure

$$\frac{\partial|\psi\rangle}{\partial t} = -iH_t|\psi\rangle$$

which is known as the (time dependent) *Schrödinger equation.* In the Schrödinger picture no equation is needed for the observables since, although they may have an intrinsic time dependence built into the corresponding experimental instrument, they have no dynamical time dependence.

In the *Heisenberg picture* the state remains constant and equal to ρ_0. For the Heisenberg representation of the observable whose Schrödinger representation is B_t – a possible intrinsic time dependence is included – we use the definition (2.6) and, to avoid confusion, write $B_H(t)$ for the Heisenberg representation of B_t,

$$B_H(t) = U^\dagger(t, t_0)B_tU(t, t_0). \tag{5.2}$$

Differentiating as before we obtain after inserting the identity operator $UU^\dagger$ in appropriate places

$$\frac{dB_H}{dt} = iU^\dagger H_tUU^\dagger B_tU + U^\dagger\frac{\partial B_t}{\partial t}U - iU^\dagger B_tUU^\dagger H_tU.$$

As is customary we have written d/dt for the time derivative of an observable in the Heisenberg picture or, in short, the *Heisenberg derivative.* Using again (5.2), we obtain

$$\frac{dB_H}{dt} = i[H_H, B_H] + \Big(\frac{\partial B_t}{\partial t}\Big)_H,$$

where for brevity the time argument on the Heisenberg representation of the observables has been omitted. Note the close analogy of this equation with the expression for the derivative along the motion of a dynamical variable in classical mechanics (D.1.3.3). The quantum mechanical expression is obtained by formally replacing the classical Poisson bracket with i times the commutator.

In analogy with classical mechanics, an observable whose Heisenberg picture $B_H(t)$ is time independent is called a *conserved* observable or a *constant of the motion.*

To understand the content of the last equation we choose n time independent observables, denoted by X_r, $r = 1, \ldots, n$, in the Schrödinger picture, which constitute a minimal irreducible set $X = \{X_r\}$ and write B_t and H_t as functions of them: $B_t = B(X, t)$, $H_t = H(X, t)$. As we know from Section 3. these functions are defined as polynomials or limits of polynomials of the X_r. Consider a typical monomial term $M = \alpha(t)X_k^a X_l^b X_m^c$ in the definition; we have, $U^\dagger MU = \alpha(t)U^\dagger X_k^a X_l^b X_m^c U$, which can also be written

$$\alpha(t)U^\dagger X_k^a UU^\dagger X_l^b UU^\dagger X_m^c U = \alpha(t)X_{Hk}^a(t)X_{Hl}^b(t)X_{Hm}^c(t) = M_H(t).$$

It follows that $B_H(t) = B(X_H(t),t)$, $H_H(t) = H(X_H(t),t)$. With this explicit functional dependence of the hamiltonian, the expression for the time derivative of B_H reads

$$\frac{dB_H}{dt} = i\left[H\left(X_H(t),t\right),B_H\right] + \left(\frac{\partial B_t}{\partial t}\right)_H. \tag{5.3}$$

In conservative systems the hamiltonian commutes with the evolution operator, therefore $H_H = H_t$, that is, the functions $H(X)$ and $H(X(t))$ have the same value. If furthermore B_t has no time dependence then (5.3) simplifies to

$$\frac{dB_H}{dt} = i\left[H\left(X_H(t)\right),B_H\right], \tag{5.4}$$

an equation that should not be confused with the somewhat similar von Neumann equation (5.1).

A comparison of equations (5.3) and (5.4) with the classical mechanical equations (see Section D.1) that express the motion derivative of a dynamical variable in terms of its Poisson bracket with the hamiltonian, shows that the latter reduce to the former if the Poisson bracket is replaced by i times the commutator.

The expression (5.3) [or (5.4)] is not, in general, an equation of motion for B_H for it contains the unknown time dependence of the observables $X_{Hr}(t)$. However if we replace B_t by each of the operators X_r we obtain

$$\frac{dX_{Hr}}{dt} = i[H\left(X_H(t),t\right),X_{Hr}], \qquad r = 1,...,n \tag{5.5}$$

which does constitute a system of n differential equations for the X_{rH} [6] known as *Heisenberg's equations of motion.*

Guided by the close analogy between the Heisenberg equations of motion and the Hamilton equations of classical mechanics we are now in a position to make more precise the concept of quantum mechanical system. As was recalled in Chapter I, the theoretical characterization of the type to which a classical mechanical system belongs consists in the specification of its coordinates. A specific system within

[6] The same situation arises in classical mechanics. If $q = \{q_r\}$ and $p = \{p_r\}$ are canonical coordinates and momenta, and $H(p,q)$ is the hamiltonian, the expression $df/dt = (H,f)$, where $(\cdot,\cdot)$ is the Poisson bracket, is not in general an equation for f. However if f is taken to be a canonical variable, we obtain Hamilton's equations of motion $dq_r/dt = \partial H/\partial p_r$, $dp_r/dt = -\partial H/\partial q_r$.

this type is then defined by its intrinsic parameters (such as mass, charge, etc.). In the hamiltonian formalism one must also specify the momenta conjugate to the coordinates. The dynamical variables are functions of the canonical variables (coordinates and momenta). In any given experimental situation, the dynamical behaviour of the system is embodied in the knowledge of the hamiltonian as a function of the canonical variables.

We have just seen that the behaviour of a quantum mechanical system is determined if one knows its hamiltonian as a function of an irreducible set of observables. Inspired by classical mechanics, we characterize the type of system by specifying a maximal observable (which may be composite in the sense of A.3.6) that formally plays the role of the set of classical coordinates.[7] A specific system within a given type is then determined by a set of intrinsic parameters (some of these parameters like mass and charge may have a classical analogue, others like spin may not). Systems that belong to the same type and have the same intrinsic parameters are called *identical systems*. In the same way as the set of classical coordinates is a subset of the set of canonical variables, the specified maximal observable is a subset of a minimal irreducible set of observables, which, in this context, will be called a *fundamental set of observables*. The other members of this set (that play the role of the momenta of classical mechanics) must also be specified. All observables can then be written as functions of the observables of a fundamental set (they also depend, in general, on the intrinsic system parameters). In a particular experimental setup, the knowledge of the hamiltonian as a function of the observables of a fundamental set completely determines the dynamical evolution of the system.

It is clear from the definition that there is not a unique fundamental set of observables belonging to a system. We shall see however that considerations of symmetry uncover particular sets of non-commuting maximal observables that can be taken to constitute a "natural" fundamental set.

As an example, the system of the type "spin systems" that we have been discussing are characterized by a maximal observable consisting of the spin component along a given direction. It is part of a fundamental set of observables consisting of the spin components along two different directions. These spin components were obtained in Chapter II from considerations of rotational symmetry. A particular system within the type is defined by giving as intrinsic parameters its spin, say 1/2, and, as we shall see in the example of Section 6, its gyromagnetic ratio. Another

[7] There is no implication that the maximal observable has any physical analogy to the classical coordinates. The analogy may exist sometimes, as is the case with the position observable of a spinless particle, but not in others, such as a spin component in the example that we have been using in earlier chapters.

example, which shall be discussed in Chapter VI, are systems of the type "particle without internal degrees of freedom" for which the position is a maximal observable which together with the momentum forms a fundamental set. We shall see that these observables arise in a natural way from the consideration of galilean invariance. Particular spinless particles are characterized by their mass and their electric charge.

We turn now to the *mixed* or *interaction picture*. As already hinted this picture is most useful when the hamiltonian can be written

$$H_t = H_{0t} + H'_t$$

and the solution $U_0(t, t_0)$ of the problem

$$\frac{\partial U_0}{\partial t} = -iH_{0t}U_0, \qquad U_0(t_0, t_0) = I \tag{5.6}$$

is known. Often H_{0t} is the hamiltonian of a system formed by non-interacting subsystems and H'_t represents the interaction between them, hence the term "interaction" picture. Once $U_0(t, t')$ is found, a particular mixed picture is defined by (2.11),

$$U(t, t_0) = U_0(t, t_0)U'(t, t_0).$$

Differentiating this expression and using the evolution equation (4.2) together with the definition (5.6) of U_0 we get

$$U_0\frac{\partial U'}{\partial t} = -iH'_tU_0U'$$

and multiplying on the left by $U_0^\dagger$ we obtain an equation of motion for U',

$$\frac{\partial U'}{\partial t} = -iH'_IU', \qquad H'_I = H'_I(t) = U_0^\dagger H'_tU_0, \tag{5.7}$$

which together with the initial condition $U'(t_0, t_0) = I$ determines $U'(t, t_0)$. The equation of motion for the interaction representation $\rho_I(t)$ of the state is obtained by differentiating (2.11) and simplifying the result with the aid of (5.6) and (5.7). We get an equation similar to von Neumann's

$$\frac{\partial \rho_I}{\partial t} = -i[H_I, \rho_I]$$

which if the state is pure simplifies to an equation similar to Schrödinger's

$$\frac{\partial|\psi_I\rangle}{\partial t} = -iH'|\psi_I\rangle.$$

The interaction time derivative of an observable is obtained from (2.13) by the use of a procedure entirely similar to the one used to obtain the Heisenberg derivative. One gets

$$\frac{dB_I}{dt} = i[H_{0I}, B_I] + \left(\frac{\partial B_t}{\partial t}\right)_I, \qquad H_{0I} = U_0^\dagger H_{0t} U_0,$$

where, as in the case of the Heisenberg picture, the hamiltonian should be written as a function of the observables X_r of the fundamental set $X : H_{0I} = H\left(X_I(t), t\right)$. More explicitly, the preceding equation becomes

$$\frac{dB_I}{dt} = i\left[H\left(X_I(t), t\right), B_I\right] + \left(\frac{\partial B_t}{\partial t}\right)_I, \qquad X_{Ir} = U_0^\dagger X_{Ir} U_0.$$

Again if the "non-interacting" hamiltonian H_0 is time independent and the observable B_t has no intrinsic time dependence, this expression simplifies to

$$\frac{dB_I}{dt} = i\left[H_0\left(X_I(t)\right), B_I\right],$$

which, as in the Heisenberg case, yields equations of motion for the observables X_r:

$$\frac{dX_{rI}}{dt} = i\left[H_0\left(X_I(t)\right), X_{rI}\right], \qquad r = 1, ..., n.$$

This completes the general presentation of the different pictures. Usually the Schrödinger picture is found to be the most convenient for exactly soluble problems when explicit solutions are needed. The Heisenberg picture is often very useful in the context of general theoretical considerations, while, as already mentioned, the interaction picture is useful for the approximate solutions of problems that are not exactly soluble.

6. An example

To illustrate some aspects of the discussion in this chapter, we return to the spin 1/2 system. Imagine an experimental setup including a Stern-Gerlach preparation apparatus that prepares a beam of spin 1/2 particles at $t = 0$, a source that produces a time dependent magnetic field

$$\mathbf{B} = \mathbf{k}B_0 + (\mathbf{i}\cos\omega t - \mathbf{j}\sin\omega t)B_1$$

and a Stern-Gerlach measuring instrument.

What the hamiltonian for such a system might be is, of course, a matter that can ultimately only be settled by experiment but the isotropy of space already says much about the form of the hamiltonian. We know that in the two-dimensional space $\mathbb{H}$ of spin 1/2, the hamiltonian must be a linear combination of the spin components s_x, s_y, s_z and the identity. It will also depend on the external field $\mathbf{B}$. The assumption on the isotropy of space implies that the hamiltonian must be a scalar with respect to rotations of the coordinate system. The only scalars that can be formed with the spin and $\mathbf{B}$ are scalar functions of $\mathbf{s}\cdot\mathbf{B}$. Moreover the hamiltonian should vanish with the external field. We, thus, conclude that the hamiltonian must have the form $f(\mathbf{s}\cdot\mathbf{B})\mathbf{s}\cdot\mathbf{B}$. Experimentally it is found that the scalar function f is a negative constant, thus

$$H_t = -\beta \mathbf{s}\cdot\mathbf{B}.$$

The positive constant β is called the gyromagnetic ratio. Of course, as was discussed in Section 3, H_t can also be written as a (non-linear) function of any two spin components.

We calculate now the time evolution operator $U(t)$. Since the hamiltonian

$$H_t = -\beta B_0 s_z - \beta B_1 (s_x \cos\omega t - s_y \sin\omega t) \tag{6.1}$$

is such that its values at two arbitrary times do not commute, we cannot use the form (4.4) and must use an *ad hoc* procedure. If we substitute H_t into the differential equation (4.3) for $U(t)$ and put for brevity $\omega_0 = \beta B_0$, $\omega_1 = \beta B_1$, we have

$$\frac{\partial U}{\partial t} = i\left(\omega_0 s_z + \omega_1 s_x \cos\omega t - \omega_1 s_y \sin\omega t\right) U.$$

If we work in the s_z-representation, this equation becomes the matrix equation

$$\begin{pmatrix} \partial U_{11} & \partial U_{12} \\ \partial U_{21} & \partial U_{22} \end{pmatrix} = \frac{i}{2}\begin{pmatrix} \omega_0 & \omega_1 e^{i\omega t} \\ \omega_1 e^{-i\omega t} & -\omega_0 \end{pmatrix}\begin{pmatrix} U_{11} & U_{12} \\ U_{21} & U_{22} \end{pmatrix},$$

where we have used the notation $\partial U = \partial U/\partial t$. Multiplying the matrices and equating corresponding elements we obtain a system of differential equations for the U_{ij}

$$\begin{aligned}
-2i\partial U_{11} &= \omega_0 U_{11} + \omega_1 e^{i\omega t} U_{21}, \\
-2i\partial U_{21} &= \omega_1 e^{-i\omega t} U_{11} - \omega_0 U_{21}, \\
-2i\partial U_{22} &= \omega_1 e^{-i\omega t} U_{12} - \omega_0 U_{22}, \\
-2i\partial U_{12} &= \omega_0 U_{12} + \omega_1 e^{i\omega t} U_{22}.
\end{aligned}$$

We need only solve the first two equations, for the other two are obtained from the first pair by changing the indices of the matrix elements, $1 \leftrightarrow 2$, and the signs of ω_0 and ω. This is not surprising because the elements of the matrix U are not all independent: they are linked by the condition of unitarity. To solve the system formed by the first two equations we try a solution of the form

$$U_{11} = a e^{\frac{1}{2}i(\eta+\omega)t}, \qquad U_{21} = b e^{\frac{1}{2}i(\eta-\omega)t}.$$

After substitution in the system we obtain a homogeneous system of two linear equations for a and b whose compatibility condition is

$$\begin{vmatrix} \eta+\omega-\omega_0 & -\omega_1 \\ -\omega_1 & \eta+\omega+\omega_0 \end{vmatrix} = 0,$$

yielding two values for η

$$\eta = \pm\gamma, \qquad \gamma = \sqrt{\omega^2 + (\omega_0-\omega)^2}.$$

The general solution of the system of two equations is, therefore

$$U_{11} = e^{\frac{1}{2}i\omega t}\left(a_1 e^{\frac{1}{2}i\gamma t} + a_2 e^{-\frac{1}{2}i\gamma t}\right)$$

$$U_{21} = \frac{e^{-\frac{1}{2}i\omega t}}{\omega_1}\left[a_1(\omega-\omega_0+\gamma)e^{\frac{1}{2}i\gamma t} + a_2(\omega-\omega_0-\gamma)e^{-\frac{1}{2}i\gamma t}\right].$$

The initial condition $U_{ij} = \delta_{ij}$ at $t = 0$ determines the constants, and the other two matrix elements can be obtained by the appropriate replacements, leading to the final result

$$U(t) = \begin{pmatrix} e^{\frac{1}{2}i\omega t}\left[\cos\frac{1}{2}\gamma t + \frac{i}{\gamma}(\omega_0-\omega)\sin\frac{1}{2}\gamma t\right] & \frac{i}{\gamma}\omega_1 e^{\frac{1}{2}i\omega t}\sin\frac{1}{2}\gamma t \\ \frac{i}{\gamma}\omega_1 e^{-\frac{1}{2}i\omega t}\sin\frac{1}{2}\gamma t & e^{-\frac{1}{2}i\omega t}\left[\cos\frac{1}{2}\gamma t - \frac{i}{\gamma}(\omega_0-\omega)\sin\frac{1}{2}\gamma t\right] \end{pmatrix}.$$

If the system was prepared at $t = 0$ in the pure state $|+\rangle$, then in the Schrödinger picture the state at time t is

$$|\psi(t)\rangle = U(t)|+\rangle = \begin{pmatrix} e^{\frac{1}{2}i\omega t}\left[\cos\frac{1}{2}\gamma t + \frac{i}{\gamma}(\omega_0-\omega)\sin\frac{1}{2}\gamma t\right] \\ \frac{i}{\gamma}\omega_1 e^{\frac{1}{2}i\omega t}\sin\frac{1}{2}\gamma t \end{pmatrix} \tag{6.2}$$

and the probability of finding the eigenvalue $-1/2$ upon measurement of s_z at time t (the probability of "flip") is

$$|\langle -|\psi(t)\rangle|^2 = \frac{\omega_1^2}{\omega_1^2 + (\omega_0-\omega)^2}\sin^2\left[\tfrac{1}{2}\sqrt{\omega^2+(\omega_0-\omega)^2 t}\right],$$

where we have used the explicit form of γ. The resonance that occurs when $\omega = \omega_0$ is the basis of the experimental technique of magnetic resonance. The average value of s_z obtained from such measurements is by (2.5) and (6.2)

$$\langle s_z\rangle(t) = \langle\psi(t)|s_z|\psi(t)\rangle = 1 - \frac{2\omega_1^2}{\gamma^2}\sin^2\gamma t. \tag{6.3}$$

It is instructive to write the Heisenberg equations of motion for the observables of a fundamental set, say $s_{Hz}(t)$ and $s_{Hx}(t)$. Since the mapping from the Schrödinger to the Heisenberg representations is unitary, the commutation relations (II.3.6) are valid for the Heisenberg operators. With those commutation relations and the hamiltonian (6.1), the Heisenberg equations (5.5) become

$$\frac{ds_{Hz}}{dt} = -\omega_1\left[i(s_{Hx}s_{Hz} - s_{Hz}s_{Hx})\cos\omega t + s_{Hx}\sin\omega t\right],$$

$$\frac{ds_{Hx}}{dt} = -i\omega_0(s_{Hz}s_{Hx} - s_{Hx}s_{Hz}) + \omega_1 s_{Hz}\sin\omega t,$$

a system of two non-linear equations which is not *a priori* obvious how to attack. Of course, by solving the linear equation for $U(t)$ we have already solved this system. For example, for the Z-component of the spin we get from the definition (2.6) of a Heisenberg observable

$$s_{Hz}(t) = \frac{1}{2}\begin{pmatrix} a(t) & b(t) \\ b^*(t) & -a(t)\end{pmatrix},$$

where

$$a(t) = 1 - \frac{2\omega_1^2}{\gamma^2}\sin^2\tfrac{1}{2}\gamma t,$$

$$b(t) = \frac{2\omega_1^2}{\gamma^2}\sin\tfrac{1}{2}\gamma t\Big[(\omega_0 - \omega)\sin\tfrac{1}{2}\gamma t + i\gamma\cos\tfrac{1}{2}\gamma t\Big].$$

If we now calculate the average value of s_z for measurements at time t on a system prepared in the state $|+\rangle$ at $t = 0$, according to (2.10) we obtain

$$\langle s_z\rangle(t) = \langle +|s_{Hz}(t)|+\rangle = a(t) = 1 - \frac{2\omega_1^2}{\gamma^2}\sin^2\tfrac{1}{2}\gamma t$$

in agreement with (6.3).

It should appear clear that the direct solution of the, in general non-linear, Heisenberg equations of motion is usually more difficult than the solution of the linear equation for the evolution operator or the, equally linear, Schrödinger equation.

V GENERAL FORMULATION

We have seen in the first part of this book how Quantum Mechanics is a theory that predicts the correlations between the configuration of a preparation apparatus and the ensuing configuration of a measurement apparatus. Up to now, the theory presented was restricted to the case when the number of possible configurations of the apparatus is finite. As a consequence, the Hilbert spaces on which states and observables are defined were finite-dimensional. For the full development of the theory it is necessary to introduce idealized models of instruments that may have an infinite number of possible configurations. The generalization of the rules of Quantum Mechanics that are necessary to discuss these idealized instruments is presented in the first part of this chapter. It requires the use of infinite-dimensional Hilbert spaces whose properties are reviewed in Appendices B and C. Once the theory has been generalized we shall be in a position to examine what restrictions does galilean invariance impose on the observables of a system.

1. Generalization of the basic postulates

It is, of course, true that any actual measuring instrument is capable of giving only a finite number of distinct readings. Nevertheless it has turned out to be extremely fruitful to regard many physical quantities as having an infinity of possible values which can be measured by idealized instruments, thought of as some kind of limiting form of actual instruments. An obvious example is a coordinate of a particle in classical mechanics. Its possible values are the real numbers, even if an actual instrument, because of its limited range and resolution, can only measure a finite number of distinct values.

In quantum mechanics, as in other physical theories, many observables are regarded as capable of yielding upon measurement an infinite number of possible values; which means that the associated idealized instruments have an infinite number of possible configurations.

It follows from the discussion in Chapter II that the dimension of the Hilbert space of a system equals the number of possible outcomes of the measurement of a maximal observable. In the finite-dimensional case treated there, this number equals the number of eigenvalues. One would, therefore, expect that to discuss observables with an infinity of possible outcomes it would be necessary to introduce infinite-dimensional Hilbert spaces, since only in these can an operator have an infinity of eigenvalues.

The postulates introduced in the previous chapters, embodied in Rules 1 and 2 of Chapter II and Rule 3 of Chapter IV, make no reference to the dimension of the Hilbert space, and one might suppose that they remain valid, as they stand, even in cases where there are observables which have to be described by operators on an infinite-dimensional Hilbert space. In the case of Rules 1 and 3 the supposition is correct. There is, however, a complication that requires Rule 2 to be generalized. Among the observables that one is led to consider, there are some very important ones that correspond to idealized apparatus with a continuum of possible configurations. This is the case, for example, of the observables position and momentum of a particle. The results in B.2.4 show that the set of eigenvalues of an operator in a Hilbert space is always discrete; therefore, the continuous readings of the idealized instruments cannot be the eigenvalues of an operator and Rule 2 cannot hold true for the corresponding observables.

The idealized instruments with a continuum of configurations are conceptual limits of actual instruments. For example, as we shall see in Chapter VI, the position of a particle can be measured with a counter of vanishingly small dimensions whose position in space is determined with a device that measures length with absolute precision. To the actual instruments are associated hermitean operators with discrete spectra. To the idealized instruments we associate appropriate limits of sequences of these operators, which tend to operators with a continuous spectrum. Note in this respect that, as stated in B.2.3.3, any hermitean operator can be regarded as the limit of such a sequence. These considerations point the way to the generalization of Rule 2.

As reviewed in Section B.2, every hermitean operator B on a Hilbert space $\mathbb{H}$ has a unique spectral resolution

$$B = \int_{-\infty}^{\infty} \beta \, dE_B(\beta) = \int_{\Sigma} \beta \, dE_B(\beta), \tag{1.1}$$

where Σ is the region of the real line which constitutes the spectrum of B, and the integral is the Stieltjes integral of B.2.1.4. The projections $E_B(\beta)$ of the spectral family of B, defined in B.2.2, are such that, for any vector $|f\rangle \in \mathbb{H}$, the scalar product $\langle f|E_B(\beta)|f\rangle$ constitutes a probability distribution function. We are now in a position to state the following postulate that generalizes Rule 2.

Rule 2(g). – Given a reference time τ, to every measurement apparatus there corresponds a hermitean operator B, called an observable *such that (i) the only values that can be obtained in a single measurement are the points of the spectrum Σ of B and (ii) the probability of obtaining a value b lying in the region Δ of the real axis, if the system has been prepared in the state ρ, is*

$$\mathcal{P}(\Delta|\rho) = \operatorname{Tr}(\rho E_B[\Delta]), \tag{1.2}$$

where $E_B[\Delta]$ is the projection

$$E_B[\Delta] = \int_\Delta dE_B(\beta).$$

In particular, the probability of obtaining a value in the interval $(b_1, b_2]$ is $\operatorname{Tr}\{\rho[E_B(b_2) - E_B(b_1)]\}$. Thus, according to probability theory, the probability distribution function is $\operatorname{Tr}[\rho E_B(\beta)]$, and the average value of the observable is

$$\langle B\rangle = \int_{-\infty}^{\infty} \beta\, d\big\{\operatorname{Tr}[\rho E_B(\beta)]\big\} = \operatorname{Tr}\Big\{\rho \int_{-\infty}^{\infty} \beta\, dE_B(\beta)\Big\}.$$

Taking into account the spectral resolution (1.1), this equation becomes

$$\langle B\rangle = \operatorname{Tr}(\rho B),$$

an equation identical to (II.4.6).

We shall be concerned only with operators for which the distribution function $\operatorname{Tr}[\rho E_B(\beta)]$ is differentiable (see B.3.2.1). One can then define a probability density $\pi(\beta|\rho)$ for the outcome β of the measurement by the formula

$$\pi(\beta|\rho) = \frac{d}{d\beta}\operatorname{Tr}[\rho E_B(\beta)]. \tag{1.3}$$

If the spectrum of the observable is pure discrete it coincides with the set of eigenvalues β_i. Then, according to B.2.4.2, the eigenprojection belonging to β_i is given by $E_i = E_B(\beta_i) - E_B(\beta_i-)$; the spectral resolution becomes $B = \sum_i \beta_i E_i$, and equation (1.2) yields

$$\mathcal{P}(\beta_i|\rho) = \operatorname{Tr}(\rho E_i),$$

a result identical to (II.4.5)

If the prepared state is pure and represented by the vector $|\psi\rangle$, then $\rho = |\psi\rangle\langle\psi|$ and formula (1.2) yields

$$\mathcal{P}(\Delta|\rho) = \langle\psi|E_B[\Delta])|\psi\rangle, \tag{1.4}$$

while the average value of the observable is given by a formula identical to (II.3.3)

$$\langle B\rangle = \langle\psi|B|\psi\rangle.$$

Again, in the case of an observable with a pure discrete spectrum, equation (1.4) gives

$$\mathcal{P}(\beta_i|\rho) = \langle\psi|E_i|\psi\rangle,$$

as in the finite-dimensional case.

Although Rule 1 in Chapter II regarding the preparation of systems, remains valid in the general case, we must note that it is not possible to prepare a system in a pure state such that a subsequent measurement of an observable B with a continuous spectrum yield with certainty a given value b, for B has no eigenvectors.

Nothing has to be remarked regarding the dynamical postulate, Rule 3 in Chapter IV. It applies in the general case, and so do all the results of that chapter with minor modifications to be mentioned in due course.

Following the classic formulation of von Neumann, many authors work with the postulates embodied in Rules 1 and 3 together with the generalization of Rule 2 just presented. Many others, including the majority of physicists, prefer to use the formulation of Dirac which, for this reason, has been adopted in this book. Its general form is the object of the next section.

2. Dirac's formulation

As discussed in Section 1, actual instruments are finite in the sense of having but a finite number of possible configurations, if the extension of the theory to idealized infinite instruments is performed appropriately, the rules for dealing with the theoretical description of these instruments should be simple generalizations of the rules that we have studied in the preceding chapters. Dirac's formulation of quantum mechanics constitutes such an extension.

The mathematical basis of Dirac's formulation is reviewed in Appendix C. We shall be using the notation in Section C.3. As discussed in the appendix we require to distinguish a set of "restricted vectors" of the Hilbert space $\mathbb{H}$, whose components with respect to a given basis "vanish sufficiently fast at infinity" in the sense of C.2.1. The operators that enter in the formalism are the "good operators", referred to in C.2.2.1. As explained there, they are continuous operators defined on the restricted vectors.

The following physical considerations may provide some insight on why Dirac's formulation is well suited to quantum mechanics.

Suppose that in an experiment a system is prepared in a pure state $|\psi\rangle$. Let $\{|\beta_i\rangle\}$ be a basis of eigenvectors of an observable $B = \sum_i \beta_i |\beta_i\rangle\langle\beta_i|$ with a discrete spectrum that, for simplicity, we shall assume non-degenerate. The probability of obtaining the value β_i upon measurement of B is $|\psi_i|^2$, where $\psi_i = \langle\beta_i|\psi\rangle$ is the component of $|\psi\rangle$ along the basis vector $|\beta_i\rangle$. If B is an idealization of an observable measured with an actual physical instrument, then, the probability will tend to zero as β_i tends to infinity, for an actual instrument is unable to register arbitrarily large values. This observation suggests that the pure states in which a system can be prepared belong to a set of vectors that "vanish sufficiently fast at infinity" with respect to any basis formed by the eigenvectors of a physical observable, that is, an observable that corresponds to actual instruments. Moreover only observables acting on those states need to be considered, which makes it reasonable to regard these observables as belonging to the set of "good operators".

These remarks suggest that Dirac's formulation is very appropriate to describe the states and observables of quantum mechanics. That this is so is confirmed by the considerable success it enjoyed even before its mathematical soundness was established.

Without further ado we shall assume that all the operators we are concerned with are "good operators" and shall use throughout this book the results and the notation of Section 3 of Appendix C. As mentioned in C.3.2.2 a most important consequence of this assumption is that one does not have to worry about the domain of definition of the operators because they are all defined, and continuously so, in the spaces of both restricted and generalized vectors.

For example, if the operator B has a pure continuous spectrum, we have

$$B = \int_{-\infty}^{\infty} d\beta\ \beta|\beta\rangle\langle\beta|.$$

In this notation the corresponding family of projections $E_B(\beta)$ of its spectral family are

$$E_B(\beta) = \int_{-\infty}^{\beta} d\beta'\ |\beta'\rangle\langle\beta'|$$

so that the probability density (1.3) becomes

$$\pi(\beta|\rho) = \mathrm{Tr}\,(\rho|\beta\rangle\langle\beta|). \tag{2.1}$$

If $\{|w_i\rangle\}$ is a basis of eigenvectors of ρ, we can write

$$\rho = \sum_i w_i |w_i\rangle\langle w_i|.$$

Calculating the trace in (2.1) with respect to this basis, we have

$$\pi(\beta|\rho) = \sum_i \langle w_i|\rho|\beta\rangle\langle\beta|w_i\rangle = \sum_i w_i\langle\beta|w_i\rangle\langle w_i|\beta\rangle,$$

that is,

$$\pi(\beta|\rho) = \langle\beta|\rho|\beta\rangle = \rho(\beta,\beta),$$

where we have written $\rho(\beta,\beta') = \langle\beta|\rho|\beta'\rangle$ for the representation of ρ in the function space induced by B (see C.3.4.3), which we shall simply call the B-space.

If the state ρ is pure, $\rho = |\psi\rangle\langle\psi|$, the preceding formula becomes

$$\pi(\beta|\rho) = |\langle\beta|\psi\rangle|^2 = |\psi(\beta)|^2,$$

where $\psi(\beta) = \langle\beta|\psi\rangle$ is the B-space representation of $|\psi\rangle$, which is also called the B-wave function.

As can be seen from the above expressions, the Dirac formalism permits a transcription of the formulae used in the case of an observable with a discrete spectrum to the case of an observable with a non-degenerate continuous one. The transcription essentially consists in replacing eigenvalues by generalized eigenvalues, one-dimensional projections by the generalized operators of the type $|\beta\rangle\langle\beta|$, and probabilities by probability densities.

More complicated expressions result if the operator B has a mixed or a continuous degenerate spectrum. They can be obtained from the results in C.3.5 and C.3.6 and shall not be reproduced here.

It is important to stress that Dirac's formalism, though extremely useful because of its simplicity, is not necessary. All problems that can be treated using it can also be treated using von Neumann's formulation without the introduction of generalized vectors. Indeed, because of the restriction to observables represented by good operators and to prepared states represented (in the pure case) by restricted vectors, Dirac's formulation of quantum mechanics is less general than von Neumann's. The success of the former indicates, however, that its lesser generality has a physically sound basis.

3. Kinematical symmetries

The rules of quantum mechanics, as we have formulated them, involve only operators on a Hilbert space $\mathbb{H}$. Consider a set of experiments in which an observable with a discrete spectrum is measured. For a given choice τ of the reference time, let ρ be the preparation state and A the measured observable. The experimental predictions are embodied in the expression $\langle A \rangle = \mathrm{Tr}\,(\rho A)$ for the average value of the observable. Suppose that there is a one to one mapping $\mathbf{V}$ of the hermitean operators on $\mathbb{H}$ onto themselves that preserves the trace of the product of any state ρ times any observable A; that is, from $\bar{\rho} = \mathbf{V}\rho$, $\bar{A} = \mathbf{V}A$, it follows that $\mathrm{Tr}\,(\rho A) = \mathrm{Tr}\,(\bar{\rho}\bar{A})$. Then, the "unbarred" description of the experiments in terms of the states ρ and the observables A is equivalent to their "barred" description in terms of the states $\bar{\rho}$ and the observables $\bar{A}$.

A mapping with the above properties is called a *kinematical symmetry*.[1] For $\mathbf{V}$ to be a kinematical symmetry it is necessary and sufficient that, if R, R', are any two one-dimensional projections, then

$$\mathrm{Tr}\,(\bar{R}\bar{R}') = \mathrm{Tr}\,(RR'). \tag{3.1}$$

The necessity is obvious. The sufficiency can be proven as follows. By Wigner's theorem A.6.2, a mapping of one-dimensional projections that satisfies (3.1) is induced by a transformation of $\mathbb{H}$ onto itself generated by an operator V which is either unitary or antiunitary. As can be seen from A.2.11.2 and A.5.4.1, if B is a hermitean operator, so is $\bar{B} = VBV^\dagger$, and they both have the same trace. Moreover, for any pair of operators B, B', we have $(\overline{BB'}) = \bar{B}\bar{B}'$. Thus, if ρ is a state and A an observable, we have $\mathrm{Tr}\,(\bar{\rho}\bar{A}) = \mathrm{Tr}\,(\rho A)$.

Not only have we proven the desired result but we have also shown that every kinematical symmetry is induced by a transformation of $\mathbb{H}$ onto itself generated by an operator which is either unitary or antiunitary. Conversely, every unitary or antiunitary operator generates a kinematical symmetry.

We already have met an important example of kinematical symmetry, namely the dynamical mapping: for fixed t and t_0, the unitary evolution operator $U(t, t_0)$ generates a kinematical symmetry transformation that maps the t_0-depiction of states and observables onto their t-depiction through the formulae of Section IV.1.

[1] In classical mechanics, the word kinematics means description of the motion or of the state of a system at any time. The adjective "kinematical" is used here in the related sense of pertaining to the description of experiments. In the latter sense, the notion of motion, contained in the etymology of the word, is lost. The adjective is necessary in order to distinguish the symmetry from the more restrictive concept of "dynamical symmetry" to be discussed in Chapter VIII.

The last remark brings to mind the question: what happens to the evolution operator under the kinematical mapping **V** generated by the (unitary or antiunitary) operator V? Let H_t be the hamiltonian of a system in the "unbarred" description, and $U(t, t_0)$ its time evolution operator, which, as we know, is the solution of the initial value problem

$$\frac{\partial U}{\partial t} = -iH_t U, \qquad U(t_0, t_0) = I. \tag{3.2}$$

In case V is unitary and time independent one can easily verify that the evolution operator and the hamiltonian in the "barred" description of the system are simply $\bar{U}(t, t_0) = VU(t, t_0)V^\dagger$ and $\bar{H}_t = VH_tV^\dagger$, respectively. In fact, multiplication of (3.2) by V on the left and $V^\dagger$ on the right yields

$$\frac{\partial \bar{U}}{\partial t} = -iH_t\bar{U}, \qquad \bar{U}(t_0, t_0) = I.$$

In the general case neither is $\bar{U}$ the evolution nor $\bar{H}_t$ the hamiltonian in the "barred" description. We shall look into this matter in the next section of this chapter.

In Section IV.5, we saw that the knowledge of a fundamental set of observables, together with the hamiltonian as a function of this set, embodies the complete theoretical description of a system in a given experimental situation. In principle all experimental results can be obtained without reference to a Hilbert space if one knows how to perform the necessary algebraic operations with the observables of the fundamental set (this is the reason why their commutation relations are so important). Such an approach has its usefulness in some applications of quantum mechanics to infinite systems where the introduction of a Hilbert space is not possible. Except in such difficult cases, it is in practice much more useful to use a description in which the observables are represented by operators on a Hilbert space rather than to deal directly with their algebra. One should, however, keep in mind that, as we have seen in this section, there are infinitely many possible such descriptions namely all those that are connected by unitary or antiunitary transformations of the space onto itself.

4. The dynamics under a kinematical symmetry

It was pointed out in the preceding section that under a time dependent kinematical symmetry the evolution operator, – and, as a consequence, the hamiltonian – of a system does not in general transform in the same way as states and observables do. In this section this question is examined in more detail.

We describe the system in the Schrödinger picture by a fundamental set X of observables. Let $\mathbf{V}$ be an, in general time dependent, kinematical symmetry, and let V_t be a unitary or antiunitary operator that it induces on $\mathbb{H}$ [V_t is a function of X but, as we do with $U(t,t_0)$, we omit this dependence]. The time dependence of $\mathbf{V}$ is to be understood in the following sense: under $\mathbf{V}_t$ the Schrödinger state $\rho(t)$ is transformed into

$$\bar{\rho}(t) = V_t \rho(t) V_t^\dagger \tag{4.1}$$

and the Schrödinger observable A_t into

$$\bar{A}_t = V_t A_t V_t^\dagger.$$

From equation (4.1) and the relation

$$\rho(t) = U(t,t_0)\rho(t_0)U^\dagger(t,t_0)$$

we obtain

$$\bar{\rho}(t) = V_t U(t,t_0)\rho(t_0)U^\dagger(t,t_0)V_t^\dagger. \tag{4.2}$$

If $\bar{U}'(t,t_0)$ is the evolution operator in the "barred" description of the system, one has

$$\bar{\rho}(t) = \bar{U}'(t,t_0)\bar{\rho}(t_0)\bar{U}'^\dagger(t,t_0).$$

From this equation, together with (4.1) and (4.2), we find

$$\bar{U}'(t,t_0) = V_t U(t,t_0) V_{t_o}^\dagger. \tag{4.3}$$

The hamiltonian $\bar{H}_t'$ in the "barred" description can be found by differentiating (4.3) with respect to t. Putting $\partial = \partial/\partial t$, we find

$$\partial \bar{U}' = [\partial V_t U \mp i V_t H_t U]\, V_{t_o}^\dagger,$$

where the upper (lower) sign applies if V is unitary (antiunitary). Inverting (4.3) and substituting the result into the last equation, one gets

$$\partial \bar{U}' = \mp i \bar{H}_t' \bar{U}' \tag{4.4}$$

with

$$\bar{H}_t' = \bar{H}_t \pm i \partial V_t V_t^\dagger, \qquad \bar{H}_t = V_t H_t V_t^\dagger. \tag{4.5}$$

In the above equations the prime has been used to stress that, in general, neither the evolution operator nor the hamiltonian in the barred description are just the transforms by V_t of the corresponding operators.

As mentioned in Section 3, if $V_t = V$ is unitary and time independent, equation (4.4) gives the dynamics of the system with the same hamiltonian as (3.2) does in the sense that the hamiltonian $\bar{H}'_t = \bar{H}_t = VH_tV^\dagger$ represents in the "barred" description the same observable that H_t represents in the "unbarred" one.

If V_t is unitary but time dependent, the hamiltonian $\bar{H}'_t$ in the "barred" description is an observable which is different from the observable H_t. If the latter is the energy, the former is not.

In any case, for unitary V_t, the transformed evolution operator is given, according to (4.4), as the solution of the initial value problem

$$\partial\bar{U}' = -i\bar{H}'_t\bar{U}', \qquad \bar{U}'(t_0, t_0) = I$$

which is of the standard form (3.2).

If on the other hand V_t is antiunitary, then the new evolution operator is the solution of

$$\partial\bar{U}' = i\bar{H}'_t\bar{U}', \qquad \bar{U}'(t_0, t_0) = I$$

with opposite sign in the differential equation. As before $\bar{H}'_t$ differs from $\bar{H}_t$ if the transformation is time dependent. In particular if $\bar{H}'_t = H_t$ then the dynamics is described by the equation

$$\partial\bar{U}' = iH_t\bar{U}' \tag{4.6}$$

which differs from (3.2) in the sign only.

The selection of sign in the definition (IV.4.1) of the hamiltonian was arbitrary, for the dynamical equations (3.2) and (4.6) represent the dynamics of the same system in two different descriptions linked by an antiunitary transformation. To show this it is sufficient to show that one can always select an antiunitary V_t such that $\bar{H}'_t = H_t$.

Consider first the case in which $H_t = H$, time independent. We require, then, an antiunitary V such that $VHV^\dagger = H$. Let $\{|\psi_i\rangle\}$ be a basis of eigenvectors of H. The time independent antiunitary operator defined by $V|\psi_i\rangle = |\psi_i\rangle$, for all i, satisfies our requirement.

If H$_t$ is time dependent, the condition $\bar{H}'_t = H_t$ becomes by (4.5)

$$i\partial V_t = V_tH_t - H_tV_t.$$

One can verify without difficulty that the antiunitarity of V_t is preserved by this equation. Thus, a solution with an antiunitary initial condition V_{t_o} will give us the desired transformation.

The transformation of an observable in the Heisenberg picture under V_t has not yet been discussed. It can be obtained from the requirement

$$\mathrm{Tr}\left(\bar{\rho}(t_0)\bar{A}_H(t)\right) = \mathrm{Tr}\left(\rho(t_0)A_H(t)\right).$$

In fact, by (4.1) the right member becomes

$$\mathrm{Tr}\left(V_{t_o}^{\dagger}\bar{\rho}(t_0)V_{t_o}\bar{A}_H(t)\right) = \mathrm{Tr}\left(\bar{\rho}(t_0)V_{t_o}A_H(t)V_{t_o}^{\dagger}\right)$$

from which we conclude that

$$\bar{A}_H(t) = V_{t_o}A_H(t)V_{t_o}^{\dagger}.$$

This somewhat unpleasing result can be understood if one remembers that, as discussed in Section IV.2, the Heisenberg picture of an observable is identical with its t_0-depiction.[2]

In fact under a time dependent kinematical symmetry the character of the Schrödinger and Heisenberg pictures is not preserved. A clear example is the dynamical mapping which transforms one picture into the other.

5. The galilean transformations as kinematical symmetries

In future chapters we shall consider experiments that are galilean invariant and see how this invariance restricts the form of the hamiltonian of the systems prepared in them. In this and the following section we explore the structure that the galilean invariance induces on the Hilbert space $\mathbb{H}$ of any such system.

Consider two experimental setups $\mathcal{E}$ and $\mathcal{E}'$ that have identically arranged instruments but differ from each other in the following respects: (a) all the instruments of $\mathcal{E}'$are rotated by a constant amount with respect to those of $\mathcal{E}$, (b) at the starting time of $\mathcal{E}'$ its instruments are displaced by $\mathbf{d}$ from those of $\mathcal{E}$, (c) the setup $\mathcal{E}'$ has a constant velocity $\mathbf{v}$ relative to $\mathcal{E}$, and (d) $\mathcal{E}'$ is started a time s after $\mathcal{E}$. To describe the disposition of the instruments of $\mathcal{E}$ we choose a reference frame – the $\mathcal{E}$-frame – defined by some features of the experimental setup (e.g., the walls of a room). A frame defined by corresponding features of $\mathcal{E}'$ – the $\mathcal{E}'$ frame – will be

[2] One could have defined instead of V a transformation V′ by requiring that the transform of a Heisenberg operator be given by $\bar{A}'_H(t) = V_t A_H(t) V_t^{\dagger}$. The definition (4.1) is however more consistent with the point of view taken in this book because the argument t in V_t can be interpreted as the reference time. The operators $\bar{A}_H(t)$ and $\bar{A}'_H(t)$ represent, of course, the same observable. They are linked by a kinematical symmetry defined by the unitary operator $V_{t_0}V_t^{\dagger}$. Needless to say that when V is time independent the two definitions are identical.

used to describe the disposition of its instruments. The time in each experiment is read from its own clock which is set to zero at the start of the experiment to which it belongs. The frame and the clock of an experiment constitute its space-time frame.

The *principle of galilean invariance* as applied to the situation at hand can be formulated as follows: *if the experimental setups $\mathcal{E}$ and $\mathcal{E}'$ can be regarded as being free from external influences, the results in both experiments, when referred to their respective space-time frames, must be identical.* Since quantum mechanics makes only statistical predictions, the term identical results must be interpreted in a statistical sense, that is, the statistics of the outcomes of sufficiently many experimental runs in $\mathcal{E}$ and $\mathcal{E}'$ must coincide.

We now assume that the experiments $\mathcal{E}$ and $\mathcal{E}'$ obey the principle of galilean invariance and investigate its consequences. To this end we shall examine the spatio-temporal relation between the instruments of both experiments and use the results to infer how their quantum mechanical descriptions are related.

The arrangement of the instruments set up by the experimenter is given at any time by a set of coordinates and velocities (the velocities must be included because the arrangement may be changing with time).[3] Let $\mathbf{x}$, and $\dot{\mathbf{x}}$ be the $\mathcal{E}$-frame position and velocity of a particular feature F of $\mathcal{E}$ at the $\mathcal{E}$-clock time τ which will be selected as reference time for the quantum description of $\mathcal{E}$. Because the experimental setups are identical, the corresponding feature F' of $\mathcal{E}'$ must have the same $\mathcal{E}'$-frame position and velocity at the $\mathcal{E}'$-clock time τ which we take as the reference time for $\mathcal{E}'$. According to the $\mathcal{E}$ clock the reference time of $\mathcal{E}'$ is $\tau + s$ and, at this time, the $\mathcal{E}$-frame position and velocity of F' are

$$\mathbf{x}' = \mathbf{R}\mathbf{x} + \mathbf{v}s + \mathbf{d}, \qquad \dot{\mathbf{x}}' = \mathbf{R}\dot{\mathbf{x}} + \mathbf{v},$$

where $\mathbf{R}$ is the tensor that represents the constant amount by which $\mathcal{E}'$ is rotated with respect to $\mathcal{E}$. We shall denote the constant vector $\mathbf{v}s + \mathbf{d}$ by $\mathbf{a}$ and call it the displacement of $\mathcal{E}'$ with respect to $\mathcal{E}'$. The vector $\mathbf{a}$ is the position of F' at the $\mathcal{E}'$-reference time, relative to the position of F at the $\mathcal{E}$-reference time. In terms of $\mathbf{a}$, the preceding equations become

$$\mathbf{x}' = \mathbf{R}\mathbf{x} + \mathbf{a}, \qquad \dot{\mathbf{x}}' = \mathbf{R}\dot{\mathbf{x}} + \mathbf{v}. \tag{5.1}$$

These equations give the relation between the $\mathcal{E}$-frame descriptions of both experimental setups at corresponding stages in the progress of the respective experiments.

---

[3] If among the instruments there are field sources, we shall assume the fields to be determined by the configuration of the sources.

Our choice of parameters is such that the transformations do not depend on the difference between the reference times of both experiments. Therefore we can imagine without loss of generality that both times coincide.

For different values of the parameters $\mathbf{R}$, $\mathbf{a}$, and $\mathbf{v}$, the equations (5.1) define a family of transformations $\mathbf{t} : \mathbf{x} \mapsto \mathbf{x}'$, $\dot{\mathbf{x}} \mapsto \dot{\mathbf{x}}'$, which we will call *kinematical galilean transformations.*

A particular $\mathbf{t}$ is determined by the values of nine parameters (three for each of the vectors $\mathbf{v}$ and $\mathbf{s}$, and three for the rotation $\mathbf{R}$). We write

$$\mathbf{t} = \{\mathbf{R}, \mathbf{a}, \mathbf{v}\}.$$

These transformations form a group $\mathfrak{G}_K$, which we call the *kinematical Galilei group*, with product operation

$$\{\mathbf{R}_2, \mathbf{a}_2, \mathbf{v}_2\}\{\mathbf{R}_1, \mathbf{a}_1, \mathbf{v}_1\} = \{\mathbf{R}_2\mathbf{R}_1, \mathbf{a}_2 + \mathbf{R}_2\mathbf{a}_1, \mathbf{v}_2 + \mathbf{R}_2\mathbf{v}_1\},$$

unit element $\mathfrak{e} = \{\mathbf{1}, 0, 0\}$, and inverse

$$\mathbf{t}^{-1} = \{\mathbf{R}, \mathbf{a}, \mathbf{v}\}^{-1} = \{\mathbf{R}^{-1}, -\mathbf{R}^{-1}\mathbf{a}, -\mathbf{R}^{-1}\mathbf{v}, \}$$

(we shall meet the full Galilei group in Chapter VIII in connection with the study of dynamical symmetries). One can verify without difficulty that the rotations $\mathfrak{j}(\mathbf{R}) = \{\mathbf{R}, 0, 0\}$; the displacements: $\mathfrak{p}(\mathbf{a}) = \{\mathbf{1}, \mathbf{a}, 0\}$; and the homogeneous galilean transformations (sometimes called boosts): $\mathfrak{g}(\mathbf{v}) = \{\mathbf{1}, 0, \mathbf{v}\}$ are subgroups of $\mathfrak{G}_K$; and that any $\mathbf{t}$ can be written as a product of them

$$\{\mathbf{R}, \mathbf{a}, \mathbf{v}\} = \mathfrak{g}(\mathbf{v})\mathfrak{p}(\mathbf{a})\mathfrak{j}(\mathbf{R}). \tag{5.2}$$

Because $\mathbf{t}$ is an analytic function of its parameters, the group is an example of a nine-parameter *Lie group*. A one-parameter subgroup of $\mathfrak{G}_K$ consists of a subset of its elements $\mathfrak{t}(\lambda)$ that depend on a single parameter and obey the relation $\mathfrak{t}(\lambda)\mathfrak{t}(\lambda') = \mathfrak{t}(\lambda + \lambda')$. It follows that $\mathfrak{t}(0) = \mathfrak{e}$ and $\mathfrak{t}^{-1}(\lambda) = \mathfrak{t}(-\lambda)$. Examples are the homogeneous galilean transformations along a fixed axis and the displacements in a fixed direction. Another example consists of the combined displacements and homogeneous galilean transformations along $\mathbf{e}$: $\mathfrak{t}(\lambda) = \{\mathbf{1}, \lambda\alpha\mathbf{e}, \lambda\varphi\mathbf{e}\}$ where φ, α are constants and $\mathbf{e}$ a fixed unit vector.

We turn now our attention to the relation between the quantum-mechanical descriptions of the two experiments.

To fix ideas let us consider the case in which $\mathbf{R} = \mathbf{1}$ and $\mathbf{v} = 0$. That is, $\mathcal{E}'$ is displaced from $\mathcal{E}$ by $\mathbf{a}$, or, as we shall say, $\mathcal{E}'$ is connected to $\mathcal{E}$ by $\{\mathbf{1}, \mathbf{a}, 0\}$. To describe the experiment $\mathcal{E}$ we choose in the Hilbert space $\mathbb{H}$ a description in which the state corresponding to the configuration of the preparation apparatus is ρ and the observable corresponding to the measurement apparatus is $A = \sum_r \alpha_r E_r$. We call this description the $\mathcal{E}$-description. To describe $\mathcal{E}'$ we could choose a description in which the state and the observable are given by the same operators but we do not want to do that. What we want to do is to find the operators $\rho(\mathbf{a})$ and $A(\mathbf{a}) = \sum_r \alpha_r E_r(\mathbf{a})$ that give the state and the observable of $\mathcal{E}'$ in the $\mathcal{E}$-description.

The principle of galilean invariance requires that the experiments $\mathcal{E}$ and $\mathcal{E}'$ give the same results, that is, that the probabilities of obtaining the reading r in each experiment be equal. We must therefore have

$$\mathrm{Tr}\,(\rho E_r) = \mathrm{Tr}\,[\rho(\mathbf{a})E_r(\mathbf{a})].$$

Since this relation must hold for any prepared state ρ and any measured observable, it follows from the discussion in Section V.3 that the mapping $\rho \mapsto \rho(\mathbf{a}), E_r \mapsto E_r(\mathbf{a})$ is a kinematical symmetry, and is therefore generated by an operator $V(\mathbf{a})$ on $\mathbb{H}$, determined up to a phase factor, which is either unitary or antiunitary. It is reasonable to assume, and we shall do so, that $\rho(\mathbf{a})$ and $E_r(\mathbf{a})$ are continuous in $\mathbf{a}$. The operator $V(\mathbf{a})$ is then (up to a phase factor) continuously connected with the identity and must, therefore, be unitary.

The preceding considerations apply equally well to a pair of identical experiments $\mathcal{E}$ and $\mathcal{E}'$ linked by an arbitrary kinematical galilean transformation $\mathbf{t}$, that is, if A and A$'$ are two corresponding apparatus in $\mathcal{E}$ and $\mathcal{E}'$, then, in a given representation, the operators A and A' corresponding to them obey the relation

$$A' = V(\mathbf{t})AV^\dagger(\mathbf{t}),$$

where $V(\mathbf{t})$ is a unitary operator determined by $\mathbf{t}$ only up to a phase factor, and is said to represent[4] $\mathbf{t}$ [for brevity we shall often use the expression "the transformation $V(\mathbf{t})$" to mean "the transformation represented by $V(\mathbf{t})$"].

As a result of the indeterminacy of $V(\mathbf{t})$ we can not assert that $V(\mathbf{t}\mathbf{t}')$ is the product $V(\mathbf{t})V(\mathbf{t}')$, but only that

$$V(\mathbf{t}\mathbf{t}') = e^{i\theta}V(\mathbf{t})V(\mathbf{t}'),$$

where θ is a real number that may depend on $\mathbf{t}$ and $\mathbf{t}'$. The phase factors are to a large extent arbitrary but not entirely so: they are restricted by the fact that the product operation of the group is associative; as a result they cannot always be taken to equal one.

---

[4] Because of the indeterminacy of the phase factors, the correspondence between group elements and the unitary operators on $\mathbb{H}$ is not one-to-one. The representation that we are studying is an example of what is called a *projective* representation to distinguish it from a *vector* representation in which that correspondence is one-to-one. An example of a vector representation is the representation of the rotations by the matrices $\mathbf{R}$.

In the theory of representations of Lie groups it is shown that the phase factors can be selected in such a way that (a) the group unit element $\mathfrak{e}$ is represented by the identity operator I on $\mathbb{H}$, (b) that $V(\mathfrak{t}\mathfrak{e}) = V(\mathfrak{t})$ and (c) that for all *one-parameter* subgroups the phase factors are unity. We shall assume from now on that the phases have been selected in the above manner.

Consider, then, a one-parameter subgroup $\mathfrak{T}(\lambda)$ of the kinematical Galilei group represented by $V(\lambda)$. According to (c) in the preceding paragraph the group property, $\mathfrak{t}(\lambda + \lambda') = \mathfrak{t}(\lambda)\mathfrak{t}(\lambda')$, translates into the relation $V(\lambda + \lambda') = V(\lambda)V(\lambda')$. An argument similar[5] to the one used to introduce the hamiltonian in Section IV.4 leads us to infer the existence of a hermitean operator L given by

$$L = i \lim_{\epsilon \to 0} \frac{V(\epsilon) - I}{\epsilon}$$

such that

$$V(\lambda) = \exp(-iL\lambda).$$

The operator L is called the *generator* of the transformations represented by $V(\lambda)$.

For future reference we give next the explicit formulas, in terms of the generator L, for the relations satisfied by the states and observables of two experiments $\mathcal{E}$ and $\mathcal{E}'$ connected by kinematical galilean transformations belonging to the one-parameter subgroup $\mathfrak{t}(\lambda)$. The state ρ prepared in experiment $\mathcal{E}$ is related to the state $\rho(\lambda)$ prepared identically in $\mathcal{E}'$ by

$$\rho(\lambda) = e^{-iL\lambda}\rho e^{iL\lambda}.$$

If the states are pure and represented by the vectors $|\psi\rangle$ and $|\psi(\lambda)\rangle$, respectively, this relation becomes

$$|\psi(\lambda)\rangle = e^{-iL\lambda}|\psi\rangle, \qquad |\psi(\lambda)\rangle = e^{-iL\lambda}|\psi\rangle.$$

Suppose that to the measurement apparatus of experiment $\mathcal{E}$ belongs the observable A, then to the identical apparatus in $\mathcal{E}'$ will belong the observable $A(\lambda)$ given by

$$A(\lambda) = e^{-iL\lambda}Ae^{iL\lambda}. \tag{5.3}$$

---

[5] Actually, in Section IV.4 we assumed the differentiability of $U(t,t')$ but this is not necessary in this case, for a theorem due to Stone states: *Let $V(\lambda)$ be a unitary operator on a Hilbert space $\mathbb{H}$ satisfying the relation $V(\lambda+\lambda') = V(\lambda)V(\lambda')$ for all real λ, λ' and such that, for all $|\phi\rangle, |\psi\rangle \in \mathbb{H}$, $\langle\psi|V(\lambda)|\phi\rangle$ is a continuous function of λ. Then there exists a hermitean operator L such that $V(\lambda) = \exp(-iL\lambda)$.* Differentiability could also have been dispensed with in Section IV.4 for a *time independent* hamiltonian since, in that case, U satisfies the relation $U(t+t') = U(t)U(t')$.

The study of the one-parameter subgroups is important because any transformation $\mathfrak{t}(\lambda)$ can be decomposed into a product of transformations belonging to them. We select next convenient one-parameter subgroups to perform the decomposition. To this end we introduce a coordinate system with axes along orthogonal unit vectors $\mathbf{e}_1, \mathbf{e}_2, \mathbf{e}_3$, and write the component of a vector $\mathbf{b}$ as $b_i = \mathbf{b} \cdot \mathbf{e}_i$, $i = 1, 2, 3$. We select nine one-parameter subgroups as follows.

(i) The three subgroups of displacements

$$\mathfrak{p}_i(a) = \{\mathbf{1}, a\mathbf{e}_i, 0\}$$

along the three coordinate axes. We denote the corresponding generators by P_i. Clearly any displacement can be decomposed into the product of three elements, one in each subgroup,

$$\mathfrak{p}_1(a_1)\mathfrak{p}_2(a_2)\mathfrak{p}_3(a_3) = \mathfrak{p}(\mathbf{a}) = \{\mathbf{1}, \mathbf{a}, 0\}.$$

(ii) The three subgroups of rotations $\mathfrak{j}_i(\varphi)$ by an angle φ about the three coordinate axes.

$$\mathfrak{j}_i(\varphi) = \{\mathbf{R}_i(\varphi), 0, 0\},$$

where $\mathbf{R}_i(\varphi)$ are the three rotation matrices about the coordinate axes

$$\mathbf{R}_1(\varphi) = \begin{pmatrix} 1 & 0 & 0 \\ 0 & \cos\varphi & -\sin\varphi \\ 0 & \sin\varphi & \cos\varphi \end{pmatrix},$$

$$\mathbf{R}_2(\varphi) = \begin{pmatrix} \cos\varphi & 0 & \sin\varphi \\ 0 & 1 & 0 \\ -\sin\varphi & 0 & \cos\varphi \end{pmatrix},$$

$$\mathbf{R}_3(\varphi) = \begin{pmatrix} \cos\varphi & -\sin\varphi & 0 \\ \sin\varphi & \cos\varphi & 0 \\ 0 & 0 & 1 \end{pmatrix}.$$

We call the corresponding generators J_i. If, as in Chapter II, we denote an arbitrary rotation matrix by the three Euler angles α, β, γ, and write $\mathbf{R} = \mathbf{R}(\alpha, \beta, \gamma)$, one can verify without difficulty that

$$\mathfrak{j}_3(\gamma)\mathfrak{j}_2(\alpha)\mathfrak{j}_3(\beta) = \mathfrak{j}(\alpha, \beta, \gamma) = \{\mathbf{R}(\alpha, \beta, \gamma), 0, 0\}$$

so that an arbitrary rotation can be decomposed into a product of rotations belonging to our subgroups (in fact, there are infinitely many such decompositions, another example was given in Section II.2).

It is shown in most textbooks in classical mechanics that an arbitrary rotation can be achieved by rotating a certain angle φ about a fixed axis $\mathbf{e}$ so that it can properly be denoted $\mathfrak{j}_{\mathbf{e}}(\varphi)$. We shall use this notation occasionally.

(iii) The three subgroups of homogeneous galilean transformations

$$\mathfrak{g}_i(v) = \{\mathbf{1}, 0, v\mathbf{e}_i\}$$

with velocities along the three coordinate axes, and generators $-G_i$ (note the minus sign introduced for convenience). Again we have the decomposition of an arbitrary homogeneous transformation

$$\mathfrak{g}_1(v_1)\,\mathfrak{g}_2(v_2)\,\mathfrak{g}_3(v_3) = \{\mathbf{1}, 0, \mathbf{v}\}$$

into a product of elements of the one-parameter subgroups.

From the decomposition (5.2) and those under (i)–(iii), it follows that any arbitrary transformation $\mathfrak{t}$ of $\mathfrak{G}_K$ can be decomposed into a product of transformations belonging to the one-parameter subgroups just introduced.

6. The generators of kinematical galilean transformations

We shall now be concerned with the infinitesimal transformations of $\mathfrak{G}_K$. Suppose that an element $\mathfrak{t}$ is given by the nine parameters θ_i, a_i, v_i just introduced. For small values of the parameters, $\mathfrak{t}$ differs little from the identity. To first order in the parameters we write $\mathfrak{t} = \mathfrak{e} + \delta\mathfrak{t}$ and call $\delta\mathfrak{t}$ the infinitesimal transformation corresponding to $\mathfrak{t}$. In the theory of Lie groups it is shown that the structure of the infinitesimal transformations completely determines the structure of the group in a neighbourhood of $\mathfrak{e}$. Note that the object $\delta\mathfrak{t}$ is not itself a transformation. This is not the place to give a precise definition of $\delta\mathfrak{t}$; it is sufficient to our purposes to see how to determine an infinitesimal transformation in a concrete example which is of interest for what follows.

Consider the subgroup of rotations. As we know from (ii) any rotation $\mathfrak{j}(\mathbf{R})$ can be written as a product of rotations $\mathfrak{j}_i(\varphi) : \mathbf{x} \mapsto \mathbf{R}_i(\varphi)\mathbf{x}$, about the coordinate axes. Two rotations about different axes i and j do not commute, that is,

$$\mathfrak{j}_{ij} = \mathfrak{j}_i(\varphi)\,\mathfrak{j}_j(\varphi)\,\mathfrak{j}_i^{-1}(\varphi)\,\mathfrak{j}_j^{-1}(\varphi) \tag{6.1}$$

differs from $\mathfrak{e}$, and we are interested in finding out by how much. We do it for infinitesimal rotations $\mathfrak{e} + \delta\mathfrak{j}_i(\varphi) : \mathbf{x} \mapsto [\mathbf{1} + \delta\mathbf{R}_i(\varphi)]\mathbf{x}$, where $\mathbf{1} + \delta\mathbf{R}_i(\varphi)$ is the approximation to $\mathbf{R}_i(\varphi)$ to lowest significant order in the small parameter φ. As it turns out the lowest significant order is φ^2. By direct calculation one gets

$$\delta\mathbf{R}_i(\varphi) = \mathbf{M}_i\varphi + \tfrac{1}{2}\mathbf{M}_i^2\varphi^2, \tag{6.2}$$

where

$$\mathbf{M}_1 = \begin{pmatrix} 0 & 0 & 0 \\ 0 & 0 & -1 \\ 0 & 1 & 0 \end{pmatrix}, \qquad \mathbf{M}_2 = \begin{pmatrix} 0 & 0 & 1 \\ 0 & 0 & 0 \\ -1 & 0 & 0 \end{pmatrix}, \qquad \mathbf{M}_3 = \begin{pmatrix} 0 & -1 & 0 \\ 1 & 0 & 0 \\ 0 & 0 & 0 \end{pmatrix}. \tag{6.3}$$

These matrices satisfy the commutation relation

$$\mathbf{M}_i\mathbf{M}_j - \mathbf{M}_j\mathbf{M}_i = \epsilon_{ijk}\mathbf{M}_k, \tag{6.4}$$

where ϵ_{ijk} is the antisymmetric symbol ($\epsilon_{ijk} = 0$ if two indices are equal, $\epsilon_{ijk} = 1$ (-1) if i, j, k is an even (odd) permutation of 1,2,3). The matrix product

$$\mathbf{R}_{ij} = \mathbf{R}_i(\varphi)\mathbf{R}_j(\varphi)\mathbf{R}_i^{-1}(\varphi)\mathbf{R}_j^{-1}(\varphi)$$

corresponding to (6.1) can now be calculated to give

$$\mathbf{R}_{ij} = \mathbf{1} + \delta\mathbf{R}_{ij} + o(\varphi^3), \qquad \delta\mathbf{R}_{ij} = \epsilon_{ijk}\mathbf{M}_k\varphi^2.$$

Or, by (6.2) and (6.4), to lowest order in the parameter,

$$\delta\mathbf{R}_{ij} = \delta\mathbf{R}_i(\varphi)\,\delta\mathbf{R}_j(\varphi) - \delta\mathbf{R}_j(\varphi)\,\delta\mathbf{R}_i(\varphi) = \epsilon_{ijk}\delta\mathbf{R}_k,$$

an equation that can be written in terms of the infinitesimal transformations

$$\delta\mathfrak{j}_{ij} = \delta\mathfrak{j}_i(\varphi)\,\delta\mathfrak{j}_j(\varphi) - \delta\mathfrak{j}_j(\varphi)\,\delta\mathfrak{j}_i(\varphi) = \epsilon_{ijk}\delta\mathfrak{j}_k(\varphi). \tag{6.5}$$

By putting $\mathfrak{j}_{ij} = (\mathfrak{e} + \delta\mathfrak{j}_{ij})$ we have an approximation for $\mathfrak{j}_{ij}$ valid for small values of φ.

We shall now study the representations of infinitesimal transformations as operators on $\mathbb{H}$. To save space let us, for now, denote by C_m, $m = 1, \ldots, 9$, the generators $P_i, J_i, -G_i$ (no particular order implied) of the one-parameter groups introduced above, and by $\mathfrak{t}_m(\lambda)$ the transformation represented by $e^{-iC_m\lambda}$. Consider the product

$$\mathfrak{t}_{mn} = \mathfrak{t}_n(\lambda)\,\mathfrak{t}_m(\lambda)\,\mathfrak{t}_n^{-1}(\lambda)\,\mathfrak{t}_m^{-1}(\lambda). \tag{6.6}$$

It is represented up to a phase factor by

$$e^{-iC_m\lambda}e^{-iC_n\lambda}e^{iC_m\lambda}e^{iC_n\lambda} \tag{6.7}$$

(if we are worried about using the same parameter for physical quantities that should have different dimensions, we can think of the units having been changed appropriately). Expanding the exponentials to lowest significant order we find an approximation to (6.7)

$$I - \lambda^2[C_m, C_n].$$

Now, being an element of $\mathfrak{G}_K$, (6.6) can be decomposed into a product of the one-parameter subgroup transformations which, denoting by μ_m, $m = 1, \ldots, 9$, the parameters, are represented by a unitary operator of the general form $W = \prod_m e^{-iC_m \mu_m}$. Except for a phase factor $e^{i\theta}$, this product must equal the product (6.7). The parameters μ_m and the phase θ must be analytic functions of λ that, at $\lambda = 0$, must vanish (because the transformations equal the identity) and have vanishing λ-derivatives [because so does (6.7)]. It follows, that to order λ^2, the equality of the transformations yields

$$[C_m, C_n] = i \sum_l c^l_{mn} C_l + i\beta_{mn} I. \tag{6.8}$$

The constants β_{mn} (which we call the local phases) arise from the expansion of the phase factor. One can see from the hermiticity of the generators that the β_{mn} must be real. So do the constants c^l_{mn} (called the structure constants).

The latter can be determined from the multiplication law of the group. In particular, if $\mathfrak{t}_m(\lambda)$ and $\mathfrak{t}_n(\lambda)$ commute, then (6.6) equals $\mathfrak{e}$, and (6.7) must be a multiple of the identity. In this case $c^l_{mn} = 0$ for all l. In the general case one needs only to consider the infinitesimal transformations. In fact, as suggested by our previous example, (6.8) is nothing more than the representation of the equation

$$\delta\mathfrak{t}_{mn} = \delta\mathfrak{t}_m(\lambda)\delta\mathfrak{t}_n(\lambda) - \delta\mathfrak{t}_m(\lambda)\delta\mathfrak{t}_n(\lambda) = \sum_l c^l_{mn} \delta\mathfrak{t}_l(\lambda) \tag{6.9}$$

among infinitesimal transformations. The operator $-i\lambda C_m$ can be seen to represent $\delta\mathfrak{t}_m(\lambda)$. Since this object is not an element of $\mathfrak{G}_K$, it is not surprising that its representative is not unitary (it is antihermitean).

We have selected the phase factors to be unity for all the one-parameter subgroups, and represented the unit element by the identity. It follows that $\beta_{mm} = 0$. More generally, from the antisymmetric property of the commutators, $\beta_{nm} = -\beta_{mn}$. Moreover, the phase factors, and therefore the local phases β_{mn}, are restricted by the associative law of the group multiplication. This law translates into the Jacobi identity for the commutators of the generators

$$\Big[[C_l, C_m], C_n\Big] + \Big[[C_m, C_n], C_l\Big] + \Big[[C_n, C_l], C_m\Big] = 0. \tag{6.10}$$

In spite of these restrictions there remains a considerable degree of arbitrariness in the selection of the β_{mn}.

The structure constants and, in so far as possible, the local phases of a representation of $\mathfrak{G}_K$ will next be determined. To do so it is convenient to return to our original notation.

The displacements $\mathfrak{p}_i(a)$, along different coordinate axes commute with each other. So do the corresponding homogeneous galilean transformations $\mathfrak{g}_i(v)$. Therefore the corresponding structure constants equal zero, and, using an *ad hoc* notation for the local phases, we have

$$[P_i, P_j] = ia_{ij}I, \tag{6.11}$$

$$[G_i, G_j] = ib_{ij}I. \tag{6.12}$$

For the rotations we have already determined the structure constants. They are given in equation (6.5). We have, then,

$$[J_i, J_j] = \epsilon_{ijk}J_k + ic_{ij}I. \tag{6.13}$$

A rotation about a coordinate axis and a displacement along it commute. The corresponding structure constant is therefore zero. Take now the case of a displacement along $\mathbf{e}_j$ and a rotation about $\mathbf{e}_i, i \neq j$. To find the structure constants we need to compute $\mathfrak{t}_{ij} = \mathfrak{j}_i(\lambda)\mathfrak{p}_j(\lambda)\mathfrak{j}_i^{-1}(\lambda)\mathfrak{p}_j^{-1}(\lambda)$ to order λ^2. From the transformation equations (5.1) and the definitions of $\mathfrak{j}_i$ and $\mathfrak{p}_i$ we find $\mathfrak{t}_{ij}$ to be the transformation

$$\mathbf{x} \mapsto \mathbf{x} - \lambda[1 - \mathbf{R}_i(\lambda)]\mathbf{e}_j.$$

By (6.2) we have to order λ^2

$$\mathbf{x} \mapsto \mathbf{x} + \lambda^2\mathbf{M}_i\mathbf{e}_j.$$

From the definition (6.3) of $\mathbf{M}_1$ we have in matrix notation

$$\mathbf{M}_1\mathbf{e}_2 = \begin{pmatrix} 0 & 0 & 0 \\ 0 & 0 & -1 \\ 0 & 1 & 0 \end{pmatrix} \begin{pmatrix} 0 \\ 1 \\ 0 \end{pmatrix} = \begin{pmatrix} 0 \\ 0 \\ 1 \end{pmatrix} = \mathbf{e}_3$$

so that to this order, $\mathfrak{t}_{12}$ is a displacement of λ^2 along $\mathbf{e}_3$: $\delta\mathfrak{t}_{12} = \delta\mathfrak{p}_3$. Similarly we have $\delta\mathfrak{t}_{23} = \delta\mathfrak{p}_1$, or, in general, $\delta\mathfrak{t}_{ij} = \epsilon_{ijk}\delta\mathfrak{p}_k$, that is,

$$\delta\mathfrak{j}_i(\lambda)\delta\mathfrak{p}_j(\lambda) - \delta\mathfrak{p}_j(\lambda)\delta\mathfrak{j}_i(\lambda) = \epsilon_{ijk}\delta\mathfrak{p}_k(\lambda).$$

According to (6.9) the structure constant corresponding to $\delta\mathfrak{p}_k$ is ϵ_{ijk} and the others are zero. Hence, the generators J_i and P_i satisfy the relation

$$[J_i, P_j] = i\epsilon_{ijk}P_k + id_{ij}. \tag{6.14}$$

To find the commutator $[J_i, G_j]$ we examine a product of a homogeneous galilean transformation and a rotation, times its inverse: $\mathfrak{t}_{ij} = \mathfrak{j}_i(\lambda)\,\mathfrak{g}_j(\lambda)\,\mathfrak{j}_i^{-1}(\lambda)\,\mathfrak{g}_j^{-1}(\lambda)$.

By direct calculation we find this transformation to be

$$\mathbf{x} \mapsto \mathbf{x} - \lambda[\mathbf{1} - \mathbf{R}_i(\lambda)]t\mathbf{e}_j.$$

There is no need to proceed any further, for this relation is identical in form to the one in the previous paragraph. Replacing P_i by $-G_i$ in (6.14) we have

$$[J_i, G_j] = i\epsilon_{ijk}G_k + ie_{ij}I. \tag{6.15}$$

Since both displacements and homogeneous galilean transformations add a quantity to $\mathbf{x}$, they commute. Therefore,

$$[G_i, P_j] = if_{ij}I. \tag{6.16}$$

The structure constants for all possible pairs of generators are now determined. If the expressions (6.11)–(6.16) of the commutators are replaced in Jacobi's identity one obtains an undetermined system of linear equations for the 36 local phases that one can solve systematically. We shall not do this because in this case, due to the large number of structure constants that are zero, it is easier to follow an *ad-hoc* procedure.

Suppose that we are given a set of local phases that satisfy the conditions $\beta_{mn} = \beta_{nm}$. First of all, we redefine J_i, P_i, G_i by adding to each of them a multiple of the identity to cancel the local phases in equations (6.13)–(6.15). The new sets of generators and local phases is equivalent to the former one. Next we find what do Jacobi's identity conditions (6.10) tell us about the remaining phases.

For the kinematical Galilei group we have seen that the generator commutators fall into two types: (a) those with one structure constant different from zero and (b) those with no structure constant different from zero. Since the identity commutes with every operator, those of Jacobi's identities involving commutators of the second type only, impose no conditions on the local phases. Those involving a commutator of the first type do (although not all conditions are independent). Note that in every expression (6.11)–(6.16) one of the operators of the commutator appears in the right side of another commutator expression. To construct all the conditions on the local phases given by the Jacobi identities we proceed as follows.

Start with a commutator expression, for example,

$$[P_3, P_1] = ia_{31}I,$$

given by (6.11). Look for an expression that contains on the right member one of the operators in the commutator and solve it for the operator. Such is (6.14) which, solved for P_3, yields

$$P_3 = -i[J_1, P_2] - id_{12}I.$$

Substitute the resulting expression into the original commutator and expand the resulting double commutator by use of Jacobi's identity. This gives a condition involving only the local phase in the expression with which we started. In the case at hand,

$$a_{31}I = -\Big[[J_1, P_2], P_3\Big] = \Big[[P_2, P_3], J_1\Big] + \Big[[P_3, J_1], P_2\Big].$$

Since multiples of the identity do not contribute to commutators, no local phases appear on the right member terms: these can be evaluated entirely with the known structure constants. In our case, from (6.11) and (6.14) we find that both terms are zero and, therefore, $a_{31} = 0$. From the commutator just studied we can evaluate two others by circular permutation of the indices. In our case we get $a_{ij} = 0$.

The procedure just explained will next be used to find restrictions on the remaining local phases.

Equations (6.12), (6.15) differ from (6.11), (6.14) in the replacement of G_i for P_i. The previous calculation can be repeated to yield $b_{ij} = 0$.

The last local phases that we have to examine are the f_{ij} of (6.16). Following our procedure we replace G_3 as given by (6.15) into the expression $[G_3, P_1] = if_{31}I$ to obtain

$$f_{31}I = -\Big[[J_1, G_2], P_1\Big] = \Big[[G_2, P_1], J_1\Big] + \Big[[P_1, J_1], G_2\Big] = 0$$

by virtue of (6.16) and (6.14). Circular permutation of the indices yields $f_{ij} = 0$, $i \neq j$. If the two indices are equal we have, for example,

$$f_{11}I = -\Big[[J_2, G_3], P_1\Big] = \Big[[G_3, P_1], J_2\Big] + \Big[[P_1, J_2], G_3\Big].$$

The first term of the last member is again zero but the last equals $-i[G_3, P_3] = f_{33}I$. By circular permutation we conclude that

$$f_{11} = f_{22} = f_{33} = \mu,$$

where μ is a real number. No further restrictions are obtained if we substitute in $[G_i, P_j]$ the expression of P_j obtained from (6.14).

To sum up we have now taken into account all the restrictions on the local phases and we have found the set of them implied in the following commutation relations among the generators of the kinematical Galilei group:

$$[P_i, P_j] = 0, \qquad [G_i, G_j] = 0, \tag{6.17}$$

$$[J_i, J_j] = i\epsilon_{ijk}J_k, \qquad [J_i, P_j] = i\epsilon_{ijk}P_k, \qquad [J_i, G_j] = i\epsilon_{ijk}G_k, \tag{6.18}$$

$$[G_i, P_j] = i\mu\delta_{ij}I. \tag{6.19}$$

As we saw these commutation relations are representations of the corresponding relations of the infinitesimal transformations of the kinematical Galilei group. From them we can find relations among the unitary operators on $\mathbb{H}$ that represent the elements of $\mathfrak{G}_K$. We illustrate this use of the commutation relations with a few important examples.

Let us first find the operator P_i' into which the generator P_i is mapped by the rotation $\mathfrak{j}_3(\varphi)$. Because this rotation is represented by $e^{-iJ_3\varphi}$, we have

$$P_i' = e^{-iJ_3\varphi} P_i e^{iJ_3\varphi}. \tag{6.20}$$

Since by the second equation (6.18) P_3 and J_3 commute, $P_3' = P_3$. Multiplying the same equation on the left by $e^{-iJ_3\varphi}$ and on the right by its inverse we see that $[J_3, P_1'] = iP_2'$ and $[J_3, P_2'] = -iP_1'$. Taking the last two relations into account and differentiating (6.20) with respect to φ we obtain the system of differential equations

$$\frac{dP_1'}{d\varphi} = P_2', \qquad \frac{dP_2'}{d\varphi} = -P_1'$$

whose general solution is

$$P_1' = A\cos\varphi + B\sin\varphi, \qquad P_2' = -A\sin\varphi + B\cos\varphi,$$

where A and B are operators to be determined by the condition $P_1' = P_i$ at $\varphi = 0$. We thus obtain after using again (6.20)

$$\begin{aligned} e^{-iJ_3\varphi} P_1 e^{iJ_3\varphi} &= P_1\cos\varphi + P_2\sin\varphi, \\ e^{-iJ_3\varphi} P_2 e^{iJ_3\varphi} &= -P_1\sin\varphi + P_2\cos\varphi, \\ e^{-iJ_3\varphi} P_3 e^{iJ_3\varphi} &= P_3. \end{aligned}$$

Similar formulas are, of course, obtained if we consider the rotations $\mathfrak{j}_1(\varphi)$ and $\mathfrak{j}_2(\varphi)$ about the other two coordinate axes. We see that the P_i transform like the components of an euclidean vector and we can define the vector operator

$$\mathbf{P} = P_1\mathbf{e}_1 + P_2\mathbf{e}_2 + P_3\mathbf{e}_3 \tag{6.21}$$

which is called the generator of displacements. If $\mathbf{a} = a_1\mathbf{e}_1 + a_2\mathbf{e}_2 + a_3\mathbf{e}_3$ is an arbitrary displacement we have

$$e^{-i\mathbf{P}\cdot\mathbf{a}} = e^{-iP_1a_1} e^{-iP_2a_2} e^{-iP_3a_3}$$

because according to (6.17) P_1, P_2, and P_3 commute with each other. From the last equation it follows that $e^{-i\mathbf{P}\cdot\mathbf{a}}$ does indeed represent the displacement

$$\mathfrak{p}(\mathbf{a}) = \mathfrak{p}_1(a_1)\,\mathfrak{p}_2(a_2)\,\mathfrak{p}_3(a_3).$$

Because of their commutation relations in (6.17) and (6.18), identical considerations apply to the generators $-G_i$. One can define a vector operator

$$\mathbf{G} = G_1\mathbf{e}_1 + G_2\mathbf{e}_2 + G_3\mathbf{e}_3 \tag{6.22}$$

and call $-\mathbf{G}$ the generator of homogeneous galilean transformations, for $e^{i\mathbf{G}\cdot\mathbf{v}}$ represents one such transformation with arbitrary uniform velocity $\mathbf{v}$.

From the first of equations (6.18) we also find that $\mathbf{J} = J_1\mathbf{e}_1 + J_2\mathbf{e}_2 + J_3\mathbf{e}_3$ is a vector operator. Let $\mathbf{e}$ be an arbitrary fixed unit vector. We ask the question: what element of $\mathfrak{G}_K$ does the unitary operator $W(\varphi) = e^{-i\mathbf{J}\cdot\mathbf{e}\varphi}$ represent? Note first that $-i\mathbf{J}\cdot\mathbf{e}\varphi = -i\mathbf{e}\cdot(J_1\mathbf{e}_1 + J_2\mathbf{e}_2 + J_3\mathbf{e}_3)\varphi$ represents the infinitesimal transformation $\delta\mathfrak{j}_1(\mathbf{e}\cdot\mathbf{e}_1\varphi) + \delta\mathfrak{j}_2(\mathbf{e}\cdot\mathbf{e}_2\varphi) + \delta\mathfrak{j}_3(\mathbf{e}\cdot\mathbf{e}_3\varphi)$ which corresponds to the product of three rotations about the coordinate axes by infinitesimal angles $(\mathbf{e}\cdot\mathbf{e}_1)\varphi$. We know (think of angular velocity in classical mechanics) that those three rotations are equivalent to a rotation of φ about $\mathbf{e}$. We conclude by integration of

$$\frac{dW}{d\varphi} = -i\mathbf{J}\cdot\mathbf{e}W$$

that

$$e^{-i\mathbf{J}\cdot\mathbf{e}\varphi} = e^{-i(J_1\mathbf{e}_1+J_2\mathbf{e}_2+J_3\mathbf{e}_3)\cdot\mathbf{e}\varphi}$$

represents $\mathfrak{j}_\mathbf{e}(\varphi)$, a rotation of φ about the axis $\mathbf{e}$. For this reason

$$\mathbf{J} = J_1\mathbf{e}_1 + J_2\mathbf{e}_2 + J_3\mathbf{e}_3 \tag{6.23}$$

is called the generator of rotations.

The image $\mathbf{G}'$ of $\mathbf{G}$ under the mapping induced on $\mathbb{H}$ by a displacement $\mathfrak{p}(\mathbf{a})$ is

$$G_i' = e^{-i\mathbf{P}\cdot\mathbf{a}}G_i e^{i\mathbf{P}\cdot\mathbf{a}}.$$

Differentiating with respect to the component a_j of $\mathbf{a}$, one gets

$$\frac{\partial G_i'}{\partial a} = \mu\delta_{ij}I,$$

where (6.19) and the identity $[G_i', P_j] = [G_i, P_j]$ have been used. Integrating this equation and using the condition $\mathbf{G}' = \mathbf{G}$ for $\mathbf{a} = 0$, one obtains

$$e^{-i\mathbf{P}\cdot\mathbf{a}}\mathbf{G}e^{i\mathbf{P}\cdot\mathbf{a}} = \mathbf{G} - \mu\mathbf{a}I.$$

A similar calculation yields for the transform of $\mathbf{P}$ under a homogeneous galilean transformation $\mathfrak{g}(\mathbf{v})$

$$e^{i\mathbf{G}\cdot\mathbf{v}}\mathbf{P}e^{-i\mathbf{G}\cdot\mathbf{v}} = \mathbf{P} - \mu\mathbf{v}I. \tag{6.24}$$

The relations that we have found among the generators $\mathbf{P}$, $\mathbf{J}$ and $\mathbf{G}$ do not tell us which observables these operators may represent. It is not until the system, and with it the Hilbert space $\mathbb{H}$, is specified that we shall be able to completely define the generators by expressing them as functions of a fundamental set of the system observables.

7. Superselection rules

In many of our arguments we have felt free to use the assumption that to every "good" operator there corresponds an observable (even if we may not know how to realize it experimentally). As was mentioned in Sect. II.3, the assumption is not valid for systems that exhibit superselection rules. In the Hilbert space $\mathbb{H}$ of such a system there are "forbidden" vectors that correspond to pure states that cannot be realized; therefore, all operators that transform an allowed vector into a forbidden one are not observables. We gave the example of a baryon that cannot be found in a state which is a linear combination of a proton state and a neutron state. In this section we examine what happens to our arguments in the case of a system with superselection rules.

Given any set of operators that are observables, we shall still assume that any operator which is a function of them also represents an observable.[6] That is, that the set of observables of the system constitutes an algebra $\mathcal{A}$ of operators. Since superselection rules are present, there are operators that are not members of $\mathcal{A}$ which means (see Sect. IV.3) that the algebra is not irreducible. Let $\mathbb{M} \subset \mathbb{H}$ be a subspace that reduces $\mathcal{A}$, then so does its orthogonal complement $\mathbb{M}^{\perp}$. Let E and $E^{\perp}$ be the respective projections. An operator $A \subset \mathcal{A}$ is of the form

$$A = A_M + A_{M^{\perp}},$$

where, by A.2.9

$$A_M = EA = AE, \qquad A_{M^{\perp}} = E^{\perp}A = AE^{\perp}.$$

It may be that the restrictions $E\mathcal{A}E$ and $E^{\perp}\mathcal{A}E^{\perp}$ of $\mathcal{A}$ to $\mathbb{M}$ and $\mathbb{M}^{\perp}$ are irreducible on their respective subspaces. If this is not the case we can repeat the procedure and eventually arrive at a decomposition of $\mathbb{H}$ into a direct sum of subspaces $\mathbb{M}^{(k)}$ such that the restriction of the algebra of observables to each of them is irreducible. The $\mathbb{M}^{(k)}$ are called *coherent* subspaces. Then, if we call $E^{(k)}$ the corresponding

---

[6] If the operator, say A, is not hermitean we can always write it in the form $A = A_1 + iA_2$ with A_1 and A_2 hermitean and take A_1 and A_2 to be observables.

projection operators, an observable A can be written

$$A = \sum_k A^{(k)}, \qquad A^{(k)} = E^{(k)}A = AE^{(k)}.$$

We can therefore regard A as the direct sum of the observables $A^{(k)}$ on the coherent subspaces $\mathbb{M}^{(k)}$.

The "forbidden" pure states of a system with superselection rules are the linear combinations of states in different coherent subspaces. Mixtures with components in different coherent subspaces are, however, not forbidden. Let ρ be such a mixture, because every observable A commutes with the $E^{(k)}$ and these projections add to the identity, we have

$$\langle A \rangle = \operatorname{Tr}(\rho A) = \operatorname{Tr}\Big(\sum_k \rho E^{(k)} A E^{(k)}\Big) = \operatorname{Tr}\Big(\sum_k E^{(k)} \rho E^{(k)} A\Big).$$

Thus, ρ and $\sum_k E^{(k)}\rho E^{(k)}$ represent the same state in the sense that they give the same expectation value for every observable. Contrary to what happens in the absence of superselection rules, there is not a one-to-one correspondence between density operators and states. This is not surprising because in Section II.4 we used the assumption that every operator represents an observable to prove the uniqueness of the density operator.

If we require that ρ, as the observables do, commute with all the $E^{(k)}$ then the one-to-one correspondence is re-established. To see this observe that, with the added condition, we have

$$\rho = \sum_k E^{(k)} \rho E^{(k)} = \sum_k \rho^{(k)},$$

where

$$\rho^{(k)} = \sum_k E^{(k)} \rho E^{(k)}$$

can be regarded as a density operator of a system with no superselection rules and Hilbert space $\mathbb{M}^{(k)}$ and is, therefore, in one-to-one correspondence with the states of this system.

To recapitulate: we have seen that the Hilbert space $\mathbb{H}$ of a system with superselection rules can be decomposed into a direct sum

$$\mathbb{H} = \bigoplus_k \mathbb{M}^{(k)}$$

and that the observables and states can also be written as direct sums

$$A = \bigoplus_k A^{(k)}, \qquad \rho = \bigoplus_k \rho^{(k)}$$

The $\mathbb{M}^{(k)}$ are the coherent subspaces on which the operators $A^{(k)}$ form an irreducible algebra and the density operators $\rho^{(k)}$ are in one-to-one correspondence to the states.

The decompositions just performed show that all the arguments that depended on the assumption that to every operator there corresponds an observable will hold if rather than applying them to the full space $\mathbb{H}$ we restrict them to each of the coherent subspaces. In fact, it is often convenient to regard the coherent subspaces as been the Hilbert spaces of different systems (for example, the proton and the neutron are frequently regarded as different particles). We shall follow this approach in the remainder of this book although some of the systems that will be studied as different could be studied together as single systems with superselection rules.

VI A PARTICLE IN ONE DIMENSION

One of the simplest systems one can think of is a single particle without any internal structure. Its study constitutes an essential foundation for the understanding of more complex systems. In this chapter we apply the framework presented in the previous ones to the particularly simple idealization in which a single spatial coordinate is of importance. We shall find out what restrictions does galilean invariance impose on the hamiltonian of such a system and solve the dynamical problem for some simple but important cases such as the harmonic oscillator.

Although the principle of galilean invariance is universally valid within the realm of non-relativistic physics, to avoid any misunderstanding we should stress here that there are many occasions when an approximate description that violates it is possible. Often, for example, a subsystem of a system prepared in an experiment can be approximately described in such a manner that its dynamical behaviour is entirely given in terms of a subset of the fundamental set of observables of the overall system. In such a case the approximate hamiltonian of the subsystem does not have to comply with the restrictions imposed by galilean invariance which only apply to the exact hamiltonian of the overall system.

1. A particle without spin

A common measuring instrument is a "particle counter". We think of such a counter as consisting of a detector that occupies a small, say cubic to fix ideas, region Δ_ξ of space which we label by the position vector $\boldsymbol{\xi}$ of its centre. We shall call $\mathbf{e}_1, \mathbf{e}_2, \mathbf{e}_3$, the unit vectors along the axes of an orthogonal coordinate system, and write $\boldsymbol{\xi} = \xi_1\mathbf{e}_1 + \xi_2\mathbf{e}_2 + \xi_3\mathbf{e}_3$. If the counter is activated in an experiment we say that the position of a system has been measured in the region Δ_ξ. A single such counter corresponds to a yes-no observable, which will be denoted by the projection operator $E(\Delta_\xi)$. In a given experiment, imagine that a set of these counters fills the whole space and that in every experimental run exactly one counter is activated. Mathematically this means that the projections belonging to different counters are orthogonal and that they add to the identity. One can then define three hermitean operators

$$Q_{\Delta i} = \sum_{\Delta_\xi} \xi_i E(\Delta_\xi), \qquad i = 1, 2, 3$$

or, equivalently, the vector operator

$$\mathbf{Q}_\Delta = \mathbf{e}_1 Q_{\Delta 1} + \mathbf{e}_2 Q_{\Delta 2} + \mathbf{e}_3 Q_{\Delta 3} = \sum_{\Delta_\xi} \boldsymbol{\xi} E(\Delta_\xi)$$

which can be called the "approximate position" observable. In the above formulas the summations extend over all the counters.

In accordance with an idealization in which the counters become infinitesimally small, we define the observable $\mathbf{Q}$, the *position* by taking the limit $\Delta \to 0$. In an obvious generalization of the notation of C.3.5.4 we write

$$\mathbf{Q} = \int d^3\xi \; \boldsymbol{\xi} \Gamma_Q(\boldsymbol{\xi}),$$

where the integral extends over all space. The $\Gamma_Q(\boldsymbol{\xi})$ are the generalized eigenprojections of Dirac's formalism. They are related to the projections $E(\Delta_\xi)$ by

$$E(\Delta_\xi) = \int_{\Delta_\xi} d^3\xi \; \Gamma_Q(\boldsymbol{\xi}).$$

In the sense of B.2.10, the spectrum of the vector operator $\mathbf{Q}$ is the whole euclidean space. If the spectrum of $\mathbf{Q}$ is simple, that is, if $\mathbf{Q}$ is a maximal observable we can write $\Gamma_Q(\boldsymbol{\xi}) = |\boldsymbol{\xi}\rangle\langle\boldsymbol{\xi}|$, and

$$\mathbf{Q} = \int d^3\xi \; \boldsymbol{\xi} |\boldsymbol{\xi}\rangle\langle\boldsymbol{\xi}|.$$

Equivalently we can say that the components Q_i of Q form a maximal set of observables, and write

$$Q_i = \int d^3\xi \; \xi_i |\xi_1, \xi_2, \xi_3\rangle\langle\xi_1, \xi_2, \xi_3|$$

with

$$|\boldsymbol{\xi}\rangle = |\xi_1, \xi_2, \xi_3\rangle.$$

If in an experimental setup the position observable is maximal, we call the system under experimental investigation a particle without spin, or more generally *a particle without internal degrees of freedom*. Such systems constitute a very good approximation in many experimental situations. For brevity we shall refer to them as spinless particles.

In this chapter we are concerned with a spinless particle in one dimension, that is, with experiments on a spinless particle where the results depend only on one component ξ of the positions of the experimental instruments. We shall, then, be dealing with a single maximal observable Q with spectral resolution

$$Q = \int d\xi \; \xi |\xi\rangle\langle\xi|,$$

where the integral extend over the whole real line. Remember that the statement that Q is a maximal observable is equivalent to the statement: every operator that commutes with Q is a function of Q. As explained in B.3.2 and, in connection with Dirac's notation, in C.3.4.1, the operator Q induces a representation of the Hilbert space of a spinless particle as the $L^2_\xi(-\infty, \infty)$-space of functions $f(\xi) = \langle\xi|f\rangle$, which we shall call position space.

As we have stressed many times, the hamiltonian of a system in a given experimental setup can ultimately be determined only by experiment. However, in previous chapters we have seen how considerations of symmetry have allowed us to restrict the possible forms of the hamiltonian. We shall do the same here by considering, as in Section V.5, two identical experimental setups $\mathcal{E}$ and $\mathcal{E}'$ that obey galilean invariance and shall use the structure of the kinematical Galilei group to find what restrictions does it impose on the hamiltonian of the single particle system prepared in them. To describe the measurements we shall take, as in that section, inertial frames attached to corresponding physical features of $\mathcal{E}$ and $\mathcal{E}'$. They will be called the $\mathcal{E}$-frame and the $\mathcal{E}'$-frame, respectively.

It is important to remember that when we compare the observables of $\mathcal{E}$ and $\mathcal{E}'$, both the observables of $\mathcal{E}$ and those of $\mathcal{E}'$ will be written in what we called in Section V.5 the $\mathcal{E}$-description, that is a structure on the Hilbert space $\mathbb{H}$ determined by a set of irreducible operators representing a fundamental set of observables of experiment $\mathcal{E}$. For brevity we shall refer to this structure as the Hilbert space $\mathbb{H}_\mathcal{E}$.

The structure of the Galilei group depends on the dimensions of space. In this chapter we are not ignoring that space is three-dimensional, rather, we are considering experiments that can be described with one spatial coordinate only. Therefore, the structure of the kinematical Galilei group embodied in the commutation relations (V.6.17)-(V.6.19) can be used.

Rotations are, of course, unimportant in the present case. We only need to take into account displacements and homogeneous galilean transformations along one of the coordinate axes, which, dropping the unnecessary subscript, will be denoted by $\mathfrak{p}(a)$ and $\mathfrak{g}(v)$, respectively. Of the commutation relations that determine the structure of the representations of the Galilei group, only one is relevant to the present discussion. It can be written

$$[G, P] = im, \tag{1.1}$$

where P and $-G$ are the generators of displacements and homogeneous galilean transformations along the axis of interest and m is a real constant (unless needed for clarity, the identity I will not be written explicitly).

2. The momentum operator

In this section we imagine that the experiment $\mathcal{E}'$ is connected to $\mathcal{E}$ by a displacement $\mathfrak{p}(a)$. The observable to be measured is the position of the particle. In experiment $\mathcal{E}$ we denote it by

$$Q = \int d\xi \; \xi|\xi\rangle\langle\xi|. \tag{2.1}$$

This observable is measured by a set of counters whose configuration is labelled by the position ξ of the counter which is activated, referred to the $\mathcal{E}$-frame.

According to the transformation formula (V.5.3), the set of counters of experiment $\mathcal{E}'$ measures an observable which, in the Hilbert space $\mathbb{H}_{\mathcal{E}}$, is given by

$$Q(a) = e^{-iPa} Q e^{iPa}, \tag{2.2}$$

where P is the generator of displacements. The configuration of this set of counters is labelled by the position ξ' of the activated counter referred to the $\mathcal{E}'$-frame. It is clear that this set of counters is identical to the one of experiment $\mathcal{E}$ when the labels of the configuration of the latter are changed from ξ to $\xi - a$. Thus, remembering (2.1), we have

$$Q(a) = \int d\xi \; (\xi - a)|\xi\rangle\langle\xi| = Q - a. \tag{2.3}$$

Substituting this result in the transformation formula (2.2) we obtain

$$Q - a = e^{-iPa} Q e^{iPa}.$$

By evaluating the a-derivative of this expression at $a = 0$ or, equivalently, by expanding the right member into a power series in a to first order, we find the commutation relation

$$[Q, P] = i. \tag{2.4}$$

which, as we shall see next, essentially determines the generator P. Writing (2.4) in position representation we have

$$(\xi - \xi')\langle\xi|P|\xi'\rangle = i\delta(\xi - \xi')$$

whose general solution (see C.1.8.3) is

$$\langle\xi|P|\xi'\rangle = -i\delta'(\xi - \xi') + g(\xi)\delta(\xi - \xi')$$

with $g(\xi)$ an arbitrary function. Or, putting $\langle\xi|f\rangle = f(\xi)$,

$$\langle\xi|P|f\rangle = (Pf)(\xi) = -if'(\xi) + g(\xi)f(\xi).$$

Which shows that P is determined up to an arbitrary function of ξ. Customarily one takes $g(\xi) = 0$, and writes

$$\langle\xi|P|\xi'\rangle = -i\delta'(\xi - \xi')$$

or equivalently

$$\langle\xi|P|f\rangle = (Pf)(\xi) = -if'(\xi). \tag{2.5}$$

For reasons that will appear shortly, the generator P is called the *momentum* of the particle. One can write the preceding equations in ordinary units by making the substitution $P \to P/\hbar$, in which case the commutation relation (2.4) becomes $[Q, P] = i\hbar$.

If we introduce the $L^2_\eta(-\infty, \infty)$-space of Fourier transforms $\tilde{f}(\eta)$ of $f(\xi)$, defined by

$$\tilde{f}(\eta) = (2\pi)^{-1/2} \int d\xi \; e^{-i\eta\xi} f(\xi),$$

and Fourier transform the expression (2.5), we obtain

$$(\widetilde{Pf})(\eta) = \eta \tilde{f}(\eta).$$

Thus, the structure of P in L^2_η is identical to the structure of Q in L^2_ξ. For this reason we call L^2_η the momentum space. We write

$$P = \int d\eta \; \eta|\eta\rangle\langle\eta|$$

and

$$\langle\eta|f\rangle = \tilde{f}(\eta), \qquad \langle\eta|\xi\rangle = (2\pi)^{-1/2}e^{-i\eta\xi}.$$

Since Q is a maximal observable, so is P.

The operators Q and P constitute a fundamental set of observables. To see this we have to show that the only observables that commute with both are multiples of the identity. Take an observable that commutes with Q; because of our definition of spinless particle it must be a function $F(Q)$ of Q and, for any vector $|f\rangle$ in its domain, $(F(Q)f)(\xi) = F(\xi)f(\xi)$. If $F(Q)$ commutes with P, then $([F(Q)P - PF(Q)]f)(\xi) = 0$, and, since $(Pf)(\xi) = -if'(\xi)$, we get $F'(\xi)f(\xi) = 0$ from which we conclude that $F(\xi)$ is a constant c and that, therefore, $(F(Q)f)(\xi) = cf(\xi)$ and $F(Q) = cI$, a multiple of the identity. Since the set $\{Q, P\}$ is fundamental, every operator is a function of its elements.

In concluding this section it is worth remarking that all the results obtained apply no matter which reference time is chosen for the description of the observables.

3. The velocity, the mass and the hamiltonian

We now turn our attention to the one-parameter subgroup of homogeneous galilean transformations $\mathfrak{g}(v)$. Accordingly, we imagine again two identical experimental setups $\mathcal{E}$ and $\mathcal{E}'$, with $\mathcal{E}'$ moving with respect to $\mathcal{E}$ at the constant velocity v. By our definition of homogeneous galilean transformations in Section V.5 we can assume without loss of generality that both setups coincide spatially at their common reference time τ. The results of the previous section tell us then that

$$Q(v) = Q, \tag{3.1}$$

where Q and $Q(v)$ are the Schrödinger position observables of $\mathcal{E}$ and $\mathcal{E}'$, respectively.

Let $Q(t)$ and $Q(v;t)$ be the Heisenberg picture with respect to the time τ of the same observables. According to Section IV.2 we have

$$Q(t) = U^\dagger(t,\tau)QU(t,\tau),$$

where $U(t,t')$ is the (unknown) unitary evolution operator associated with experiment $\mathcal{E}$, whose existence is asserted by the dynamical postulate. If $U(v;t,t')$ is the evolution operator of $\mathcal{E}'$, one gets the similar relation

$$Q(v;t) = U^\dagger(v;t,\tau)QU(v;t,\tau),$$

where the equality (3.1) has been used.

Suppose that instead of τ we use a different reference time t for both experiments. At this time the two experimental setups are a distance $v(t-\tau)$ apart and we can apply the argument that led to equation (2.3) to establish that

$$Q^{(t)}(v) = Q^{(t)} - v(t-\tau),$$

where, as in Section IV.1, $Q^{(t)}$ and $Q^{(t)}(v)$ are the t-depictions of the position observables. The operators $Q(t)$ and $Q(v;t)$ represent the same pair of observables[1] of $\mathcal{E}$ as $Q^{(t)}$ and $Q^{(t)}(v)$ do. We must therefore have

$$Q(v;t) = Q(t) - v(t-\tau).$$

Taking the Heisenberg derivative of this expression with respect to t, we get

$$i[H_t(v;t), Q(v;t)] = i[H_t(t), Q(t)] - v. \tag{3.2}$$

Here, $H_t(t)$ and $H_t(v;t)$ are the hamiltonians of experiments $\mathcal{E}$ and $\mathcal{E}'$, respectively, in the Heisenberg picture.

The velocities of the particle in experiments $\mathcal{E}$ and $\mathcal{E}'$ are defined as the Schrödinger operators $\dot{Q}$ and $\dot{Q}(v)$ on $\mathbb{H}_{\mathcal{E}}$ given, respectively, by

$$\dot{Q} = i[H_t, Q], \qquad \dot{Q}(v) = i[H_t(v), Q(v)], \tag{3.3}$$

where H_t and $H_t(v)$ are the τ-depictions of the respective hamiltonians. In other words, all the operators in these equations are Schrödinger observables at time τ. Letting $t \to \tau$ in equation (3.2) we find that the velocities are related by

$$\dot{Q}(v) = \dot{Q} - v, \tag{3.4}$$

in the same manner as in classical mechanics.

We now apply the transformation formula (V.5.3) to the observables position and velocity. By the position transformation (3.1) we have

$$Q(v) = Q = e^{iGv} Q e^{-iGv},$$

which tells us that the generator $-G$ of homogeneous galilean transformations commutes with Q. Since this observable is maximal it follows that G is a function of Q only

$$G = F(Q). \tag{3.5}$$

The velocity transformation (3.4) yields

$$\dot{Q}(v) = \dot{Q} - v = e^{iGv} \dot{Q} e^{-iGv}$$

[1] It is worth remarking that

$$Q^{(t)} = U(t,\tau)U^{\dagger}(t,\tau)QU(t,\tau)U^{\dagger}(t,\tau) = Q$$

and that

$$Q^{(t)}(v) = U(t,\tau)U^{\dagger}(v;t,\tau)QU(v;t,\tau)U^{\dagger}(t,\tau).$$

from which we obtain in the usual way the commutation relation

$$[G, \dot{Q}] = i. \tag{3.6}$$

We shall use this relation later on to determine restrictions on the hamiltonian H_t but first we determine G as a function of the fundamental set of observables $\{P, Q\}$. It is convenient to repeat here the Galilei group commutation relation (1.1)

$$[G, P] = im.$$

Applying this equation to a vector $|f\rangle$ and remembering (3.5), one gets in position space

$$-iF(\xi)\frac{df(\xi)}{d\xi} + i\frac{dF(\xi)f(\xi)}{d\xi} = imf(\xi)$$

which gives $F'(\xi) = m$ and putting the integration constant equal to zero one obtains $F(\xi) = m\xi$, that is,

$$G = mQ \tag{3.7}$$

which shows that m cannot be put equal to zero for this will contradict the velocity transformation (3.4). Substituting (3.7) into (3.6) we see that the observable $m\dot{Q}$ obeys the same commutation relation (2.4) as Q does, therefore it is of the form $m\dot{Q} = P + g(Q)$.

It is not difficult to show that any given function $g(Q)$ can be removed by a unitary transformation $S(Q)$ that depends only on Q. Customarily one takes the representation in which $g(Q) = 0$ and writes

$$m\dot{Q} = P. \tag{3.8}$$

The transformation $S(Q)$ commutes with Q and does not therefore change the position observable which is experimentally fixed by our set of counters and their labels. The above relation means that there is a position space representation in which the position is represented by multiplication by ξ, the generator of displacements by $-id/d\xi$, and the velocity by $-(i/m)d/d\xi$. One reserves the name "momentum" for the generator of displacements in this representation. Equation (3.8) further confirms the correctness of the name velocity chosen for the observable $\dot{Q}$.

The particle hamiltonian H_t is, as any operator, a function $H_t(P, Q)$ of the irreducible set of observables $\{Q, P\}$. Its form is restricted by (3.3) which can be rewritten

$$[Q, H_t] = iP/m. \tag{3.9}$$

To solve this equation for H_t we work in momentum space where by Fourier transformation one sees that $(\widetilde{Qf})(\eta) = i\tilde{f}'(\eta)$. In this representation the above commutation relation reads

$$\frac{d}{d\eta}(\widetilde{H_t f})(\eta) - (\widetilde{H_t f'})(\eta) = \frac{1}{m}\eta\tilde{f}(\eta).$$

Try a particular solution $H_t = H_0(P)$, a function of P only, so that $(\widetilde{H_0 f})(\eta) = H_0(\eta)\tilde{f}(\eta)$, and the above equation becomes

$$H_0'(\eta)\tilde{f}(\eta) = \frac{1}{m}\eta\tilde{f}(\eta)$$

which has the solution $H_0(\eta) = \eta^2/2m$; that is, (3.9) has a solution

$$H_0(P) = \frac{P^2}{2m}.$$

Write now the general solution as $H_0(P) + H_1(P,Q)$ and substitute in (3.9) to obtain $[Q, H_1] = 0$. Thus, H_1 is a function $V_t(Q)$, which may depend on t (that enters all equations as a parameter). We have then found that the hamiltonian of a spinless particle must be of the form

$$H_t = \frac{P^2}{2m} + V_t(Q),$$

in striking analogy with the hamiltonian of a single particle in classical mechanics. The analogy leads us to interpret $V_t(Q)$ as the potential energy, $P^2/2m$ as the kinetic energy and the parameter m as the mass. An interpretation whose plausibility is further reinforced by the relation $P = m\dot{Q}$ between the observables momentum and velocity.

Notice that m is an intrinsic parameter which galilean invariance requires, as we have seen, to be different from zero. *Zero-mass particles cannot exist in non-relativistic quantum mechanics.* The mass characterizes a system belonging to the type "one-dimensional spinless particle"; two such particles with the same mass are identical.

We stress that the considerations of kinematical galilean invariance that we have made refer to *the experimental instruments*. The galilean invariance of the results read in their dials, together with the postulates of quantum mechanics, determine the general form of the hamiltonian of a spinless particle in one dimension. It would be conceptually wrong to impose the invariance directly on the dynamical equations of the system, for the invariance principle is concerned with the results of experiments and those results have to do with the determination of the configurations of the experimental apparatus.

4. An example: the harmonic oscillator

Consider an experiment done on a particle in which field sources producing a time independent potential $V(Q)$ are present. As mentioned in Chapter IV, a knowledge of the spectral resolution of the hamiltonian

$$\frac{P^2}{2m} + V(Q)$$

yields through equation (IV.4.6) a complete solution of the dynamical problem. Suppose that the function $V(\xi)$ has a minimum, which without loss of generality we can take to be at $\xi = 0$, and that it can be expanded in a Taylor series

$$V(\xi) = V(0) + \tfrac{1}{2}\xi^2 V''(0) + \ldots$$

with $V''(0) > 0$. By a shift of the spectrum we can take $V(0) = 0$. Putting $V''(0) = m\omega^2$, we can write

$$H = \frac{1}{2m}\left(P^2 + m^2\omega^2 Q^2\right) \tag{4.1}$$

as an approximation of the original hamiltonian. Roughly speaking, the approximation will be valid if one is only interested in applying the operator H to vectors that lie in a subspace spanned by eigenvectors belonging to eigenvalues that are close to the minimum of $V(\xi)$.

In analogy with the classical definition, a particle with the hamiltonian (4.1) is called a harmonic oscillator. The harmonic oscillator is of great interest, not only because it constitutes an approximation to many systems, but also because, as we shall see, it is very important for the study of systems formed by many particles. It is for the latter reason that we shall use an algebraic method for the determination of the spectral resolution of H rather than working in a particular representation.

To simplify the writing it is convenient to introduce new operators

$$p = P/\sqrt{m\omega}, \qquad q = \sqrt{m\omega}\,Q, \qquad h = H/\omega$$

which amounts to a change to new units in which the hamiltonian is

$$h = \frac{1}{2}(p^2 + q^2). \tag{4.2}$$

Inspired by classical mechanics where a particle with this hamiltonian moves on a circle of phase space, we introduce a pair of mutually adjoint operators

$$a = \frac{1}{\sqrt{2}}(q + ip), \qquad a^\dagger = \frac{1}{\sqrt{2}}(q - ip) \tag{4.3}$$

that obey the commutation relation

$$[a, a^\dagger] = 1. \tag{4.4}$$

In terms of these operators (4.2) becomes

$$h = a^\dagger a + \tfrac{1}{2}.$$

Not to have to carry the constant, we introduce the operator

$$N = a^\dagger a = h - \tfrac{1}{2}. \tag{4.5}$$

Obviously N and h have the same eigenprojections and their spectra differ by a shift of $1/2$.

We now study the eigenvalue problem

$$N|\psi_n\rangle = n|\psi_n\rangle \tag{4.6}$$

and start with the assertion, to be proved later, that N has at least one eigenvalue. Since according to (4.2) h is positive, its spectrum is non-negative and therefore it follows from (4.5) that the spectrum of N is bounded from below by $-1/2$.

From the definition (4.5) of N and the commutation relation (4.4) we have

$$Na = (aa^\dagger - 1)a = a(N - 1), \qquad Na^\dagger = a^\dagger(a^\dagger a + 1) = a^\dagger(N + 1)$$

whence,

$$Na|\psi_n\rangle = a(N-1)|\psi_n\rangle = (n-1)a|\psi_n\rangle$$

$$Na^\dagger|\psi_n\rangle = a^\dagger(N+1)|\psi_n\rangle = (n+1)a^\dagger|\psi_n\rangle.$$

Thus if $|\psi_n\rangle$ is an eigenvector of N, $a|\psi_n\rangle$ and $a^\dagger|\psi_n\rangle$ are solutions of (4.6), that is, they are either eigenvectors of N or the zero vector. The norms of these vectors can also be obtained from (4.4) and (4.5)

$$\|a|\psi_n\rangle\|^2 = \langle\psi_n|a^\dagger a|\psi_n\rangle = n\langle\psi_n|\psi_n\rangle \tag{4.7}$$

$$\|a^\dagger|\psi_n\rangle\|^2 = \langle\psi_n|aa^\dagger|\psi_n\rangle = \langle\psi_n|a^\dagger a + 1|\psi_n\rangle = (n+1)\langle\psi_n|\psi_n\rangle. \tag{4.8}$$

Therefore, if $|\psi_n\rangle$ is an eigenvector of N, $a|\psi_n\rangle$ is the zero vector if and only if $n = 0$, while $a^\dagger|\psi_n\rangle$ is the zero vector if and only if $n = -1$. The latter case is not possible, for we know that the eigenvalues are bounded from below by $-1/2$. Thus, repeated application of $a^\dagger$ to an eigenvector $|\psi_n\rangle$ yields an infinite sequence of eigenvectors belonging to the increasing sequence of eigenvalues $n, n+1, n+2, \ldots$. Since the eigenvalues are bounded from below, repeated application of a to any eigenvector must eventually encounter the zero vector which implies that all the eigenvalues of N must be integers and that $n = 0$ must be the smallest of them. Therefore, the discrete spectrum of N consists of the non-negative integers, and the discrete spectrum of h of the numbers

$$n + \tfrac{1}{2}, \qquad n = 0, 1, 2, \ldots$$

We represent by $|n\rangle$ a normalized eigenvector of N belonging to the eigenvalue n. The expressions (4.7) and (4.8) for the norms show that a set of normalized eigenvectors obeys the relations

$$a|n\rangle = \sqrt{n}\,|n-1\rangle, \qquad a^\dagger|n\rangle = \sqrt{n+1}\,|n+1\rangle. \tag{4.9}$$

Starting with the normalized eigenvector $|0\rangle$ we obtain by induction a full sequence of normalized eigenvectors

$$|n\rangle = \frac{(a^\dagger)^n}{\sqrt{n!}}|0\rangle, \qquad n = 0, 1, 2, \ldots. \tag{4.10}$$

Next we show that, in the Hilbert space $\mathbb{H}$ of the spinless particle in one dimension, this sequence constitutes a basis and that therefore h has a pure discrete spectrum.

Let $\mathbb{M} \subset \mathbb{H}$ be the subspace spanned by the vectors (4.9). Any vector $|f\rangle \in \mathbb{M}$ can be written

$$|f\rangle = \sum_n f_n|n\rangle.$$

It is clear by (4.9) that application of a or $a^\dagger$ to $|f\rangle$ changes the coefficients f_n transforming $|f\rangle$ into another linear combination of the vectors $|n\rangle$. Thus both $a|f\rangle$ and $a^\dagger|f\rangle$ lie in $\mathbb{M}$. If $|g\rangle$ is orthogonal to $\mathbb{M}$ then $\langle n|g\rangle = 0$ for all $|n\rangle$ and, by (4.9), $\langle n|a|g\rangle = \sqrt{n+1}\langle n+1|g\rangle = 0$, $\langle n|a^\dagger|g\rangle = \sqrt{n}\langle n-1|g\rangle = 0$, that is, $a|g\rangle$ and $a^\dagger|g\rangle$ are also orthogonal to $\mathbb{M}$. Thus, $\mathbb{M}$ reduces a and $a^\dagger$ and, since by (4.3),

$$q = \frac{1}{\sqrt{2}}(a + a^\dagger), \qquad p = \frac{1}{i\sqrt{2}}(a - a^\dagger),$$

it also reduces q and p. Since, as we know, q and p form an irreducible set, it follows that $\mathbb{M} = \mathbb{H}$, and consequently the set of eigenvectors (4.10) constitutes a basis in $\mathbb{H}$.

It remains to show that N has an eigenvector. To see this we rewrite the eigenvalue equation (4.6) in the function space representation induced by $q = \int dx\, x|x\rangle\langle x|$. Putting $\langle x|n\rangle = \phi_n(x)$, we have for $n = 0$

$$\frac{d^2\phi_0}{dx^2} - (x^2 - 1)\phi_0 = 0$$

which has the normalized solution

$$\phi_0(x) = \pi^{-\frac{1}{4}} e^{-\frac{1}{2}x^2},$$

showing that $n = 0$ is an eigenvalue of N.

By use of (4.10) one can find the other eigenvectors in this representation. We have

$$\phi_n(x) = \frac{1}{\sqrt{n!}}\langle x|(a^\dagger)^n|n\rangle = \frac{1}{\sqrt{2^n n!}}\langle x|(q - ip)^n|n\rangle$$

or, replacing the operators by their representation in function space,

$$\phi_n(x) = \frac{1}{\sqrt{2^n \pi^{\frac{1}{2}} n!}}\left(x - \frac{d}{dx}\right)^n e^{-\frac{1}{2}x^2},$$

which after a little manipulation becomes

$$\phi_n(x) = \frac{1}{\sqrt{2^n \pi^{\frac{1}{2}} n!}} e^{-\frac{1}{2}x^2} H_n(x),$$

where $H_n(x)$ is the Hermite polynomial

$$H_n(x) = (-1)^n e^{x^2} \frac{d^n e^{-x^2}}{dx^n}.$$

Note that these eigenfunctions are normalized in the function space induced by q and not in the position space induced by Q.

Next we determine the time evolution of the pertinent Heisenberg operators which we denote by explicitly writing the argument t. We take their values at $t = 0$ to be the corresponding Schrödinger operators. Thus $Q(t)$ is a Heisenberg operator and $Q(0) = Q$. It is convenient to start with the operators a and $a^\dagger$ which, as would be expected from the classical analogy, have a very simple motion. First we note that by use of (4.9) the operator a can be written

$$a = \sum_n \sum_{n'} |n\rangle\langle n|a|n'\rangle\langle n'| = \sum_n \sqrt{n+1}\,|n\rangle\langle n+1|.$$

With the relation (4.5) between the hamiltonian ωh and the number operator N we have for the Heisenberg operator $a(t)$

$$a(t) = e^{i\omega ht} a e^{-i\omega ht} = \sum_n \sqrt{n+1}\, e^{i\omega Nt} |n\rangle\langle n+1| e^{-i\omega Nt}.$$

Since $e^{i\omega Nt}|n\rangle = e^{i\omega nt}|n\rangle$, the last expression simplifies to

$$\sum_n \sqrt{n+1}\, |n\rangle\langle n+1| e^{-i\omega t} = e^{-i\omega t} a$$

so that we get

$$a(t) = e^{-i\omega t} a.$$

One can obtain $a^\dagger(t)$ by a similar calculation or simply by noting that unitary operators preserve the adjoint operation and taking the adjoint of the last equation:

$$a^\dagger(t) = e^{i\omega t} a^\dagger.$$

Inverting the relations (4.3), which are obviously also valid for the Heisenberg operators, and using the last two equations, one obtains

$$q(t) = q \cos \omega t + p \sin \omega t, \qquad p(t) = -q \sin \omega t + p \cos \omega t$$

or in terms of the original observables

$$Q(t) = Q \cos \omega t + \frac{P}{m\omega} \sin \omega t, \qquad P(t) = -m\omega Q \sin \omega t + P \cos \omega t.$$

The time evolution of the coordinate and the momentum of the harmonic oscillator has the property, shared, as we shall see, with the free particle, of being formally identical with the classical one.

5. A particle in a constant force field

As another example we solve in this section the problem of a particle with hamiltonian

$$H = \frac{P^2}{2m} + mgQ \tag{5.1}$$

which is the quantum mechanical analogue of a classical particle in a constant field of force. This example not only has considerable intrinsic interest, but is also useful because it illustrates the use of distributions in dealing with operators with a continuous spectrum.

In determining the spectral resolution of the hamiltonian, it is convenient to work in momentum space. For the function representing the vector $|f\rangle$ we write $\langle\eta|f\rangle = f(\eta)$. From the commutation relation (2.4), or by Fourier transformation, it is easily seen that $\langle\eta|Q|f\rangle = (Qf)(\eta) = if'(\eta)$. The generalized eigenvalue equation for the hamiltonian becomes then

$$\left(\frac{\eta^2}{2m} + img\frac{d}{d\eta}\right)\psi_\epsilon(\eta) = \epsilon\psi_\epsilon(\eta).$$

This first order differential equation has general solution

$$\psi_\epsilon(\eta) = \kappa \exp\left[-\frac{i}{mg}\left(\epsilon\eta - \frac{\eta^3}{6m}\right)\right]$$

which for all real energies ϵ is a distribution. The spectrum is simple and consists of the whole real axis. To satisfy the normalization condition

$$\int_{-\infty}^{\infty} d\eta\ \psi_\epsilon(\eta)\psi^*_{\epsilon'}(\eta) = \delta(\epsilon - \epsilon'),$$

we can choose the integration constant $\kappa = 1/\sqrt{2\pi m|g|}$. Thus we take the generalized eigenvector corresponding to the energy ϵ to be given by

$$\psi_\epsilon(\eta) = \langle\eta|\epsilon\rangle = \frac{1}{\sqrt{2\pi m|g|}} \exp\left[-\frac{i}{mg}\left(\epsilon\eta - \frac{\eta^3}{6m}\right)\right]. \tag{5.2}$$

The spectral resolution of the hamiltonian is given by

$$H(\eta,\eta') = \langle\eta|H|\eta'\rangle = \int_{-\infty}^{\infty} d\epsilon\ \epsilon\psi_\epsilon(\eta)\psi^*_{\epsilon'}(\eta),$$

and the evolution operator by

$$U(\eta,\eta') = \langle\eta|e^{-iHt}|\eta'\rangle = \int_{-\infty}^{\infty} d\epsilon\ e^{-i\epsilon t}\psi_\epsilon(\eta)\psi^*_{\epsilon'}(\eta).$$

After substitution of the eigenfunction expression (5.2) into this equation and evaluation of the integral, one obtains

$$U(\eta,\eta') = \exp\left(i\frac{\eta^3 - \eta'^3}{6m^2 g}\right)\delta(\eta - \eta' + mgt). \tag{5.3}$$

Let the system be prepared at $t = 0$ in the state ρ_0 given in momentum space by $\rho_0(\eta,\eta') = \langle\eta|\rho_0|\eta'\rangle$. The state ρ_t at time t is then given by

$$\rho_t(\eta,\eta') = \int_{-\infty}^{\infty}\int_{-\infty}^{\infty} d\eta'' d\eta'''\ U(\eta,\eta'')\rho_0(\eta'',\eta''')U^*(\eta',\eta''').$$

Upon substitution of the expression (5.3) for the evolution operator one obtains without difficulty

$$\rho_t(\eta,\eta') = \exp\left[-\frac{it}{2m}(\eta+\eta'+mgt)(\eta-\eta')\right]\rho_0(\eta+mgt,\ \eta'+mgt). \tag{5.4}$$

The probability density $\pi_t(\eta)$ for obtaining the value η of the momentum at the measurement time t is

$$\pi_t(\eta) = \rho_t(\eta,\eta) = \rho_0(\eta+mgt,\ \eta+mgt) = \pi_0(\eta+mgt).$$

The shape of the momentum probability density is independent of t, it merely undergoes a translation by mgt, the amount that the momentum of a classical particle with hamiltonian (5.1) would change.

The probability density for the position, on the other hand, spreads with t. To see this take an initial state of the form

$$\rho_0(\eta,\eta') = \frac{1}{\beta\sqrt{\pi}}\exp\left[-\frac{1}{4\beta^2}(\eta+\eta')^2 - \frac{\alpha^2}{4}(\eta-\eta')^2\right]. \tag{5.5}$$

By using the relation

$$\langle\xi|\rho|\xi'\rangle = \int_{-\infty}^{\infty}\int_{-\infty}^{\infty} d\eta\ d\eta'\ \langle\xi|\eta\rangle\langle\eta|\rho|\eta'\rangle\langle\eta'|\xi'\rangle$$

or, equivalently,

$$\rho(\xi,\xi') = \frac{1}{2\pi}\int_{-\infty}^{\infty}\int_{-\infty}^{\infty} d\eta\ d\eta'\ e^{i(\xi\eta-\xi'\eta')}\rho(\eta,\eta'), \tag{5.6}$$

it is easy to verify that this is one state (among infinitely many) that yields initial gaussian probabilities for both the momentum and position:

$$\pi_0(\eta) = \frac{1}{\beta\sqrt{\pi}}e^{-\eta^2/\beta^2}, \qquad \pi_0(\xi) = \frac{1}{\alpha\sqrt{\pi}}e^{-\xi^2/\alpha^2}.$$

Substituting the initial state (5.5) into the expression (5.4) for the state at time t, we obtain with the help of (5.6)

$$\pi_t(\xi) = \frac{1}{2\pi^{3/2}\beta} \int_{-\infty}^{\infty} \int_{-\infty}^{\infty} dx dy \, \exp\left\{-ix\left[\xi + \frac{(2y+mgt)t}{2m}\right] - \frac{(y+mgt)^2}{\beta^2} - \frac{\alpha^2 x^2}{4}\right\},$$

where we have made the change of variables $x = \eta - \eta'$, $y = \frac{1}{2}(\eta + \eta')$. The integrations can be performed without difficulty and yield

$$\pi_t(\xi) = \frac{1}{\sqrt{\pi(\alpha^2 + \beta^2 t^2/m^2)}} \exp\left[-\frac{(\xi + \frac{1}{2}gt^2)^2}{\pi(\alpha^2 + \beta^2 t^2/m^2)}\right].$$

One can see that the probability density is still a gaussian but that its width spreads with time by an amount which depends on the initial spread of the momentum density.

It is of interest to note that a state of the form (5.5) is possible, that is, ρ_0 is a positive operator with unit trace, if and only if the widths α and β satisfy the condition $\alpha\beta \geq 1$. A condition that can be regarded as an expression of the uncertainty principle.

VII A SPINLESS PARTICLE IN THREE DIMENSIONS

The procedures applied to the one-dimensional particle in the previous chapter are generalized in this one to the more realistic case in which all three dimensions of space are important. After finding the restrictions imposed by galilean invariance on the hamiltonian of a particle without internal structure, we solve the dynamical problem for some simple cases. Phase space representations of the state are also introduced, and the framework to describe a system in arbitrary orthogonal coordinates is presented.

1. Position, momentum and the hamiltonian

A particle without internal degrees of freedom was defined at the beginning of the previous chapter as a system characterized by a Hilbert space $\mathbb{H}$ on which the position observable $\mathbf{Q}$ is a maximal observable. In that chapter we restricted the discussion to the case in which the experiments could be described in one dimension. Here we extend the discussion to the general case.

The position space induced by the vector observable $\mathbf{Q}$ is now the $\mathbb{L}^2$-space of functions $f(\boldsymbol{\xi}) = \langle \boldsymbol{\xi} | f \rangle$ of the vector variable $\boldsymbol{\xi}$ with scalar product defined by

$$\langle f | g \rangle = \iint d^3\xi \, f^*(\boldsymbol{\xi}) g(\boldsymbol{\xi}),$$

where the integral extends over all space.

The results of an experiment on a spinless particle depend on all three components of the position and velocity vectors $\mathbf{x}$ and $\dot{\mathbf{x}}$ of the parts of the experimental instruments. To find the restrictions imposed on the hamiltonian by galilean invariance, it is necessary to consider the full kinematical Galilei group $\mathfrak{G}_K$ studied in Sections V.5 and V.6. Except for some changes that will be noted in due time, we shall use here the notation of those sections for the transformations of the Galilei group and for its generators.

As in the case of one dimension we start by considering two galilean invariant experiments $\mathcal{E}$ and $\mathcal{E}'$ connected by a displacement $\mathfrak{p}(\mathbf{a}) = \{\mathbf{1}, \mathbf{a}, 0\}$. An argument identical to the one used in Section VI.2 shows that the position observable must satisfy the relation

$$\mathbf{Q} - \mathbf{a} = e^{-i\mathbf{P}\cdot\mathbf{a}}\mathbf{Q}e^{i\mathbf{P}\cdot\mathbf{a}},$$

where $\mathbf{P}$ is the generator of displacements (V.6.21). From this expression we get by the usual means the commutation relations

$$[Q_i, P_j] = i\delta_{ij}, \qquad i,j = 1,2,3 \tag{1.1}$$

which are often referred to as *canonical* commutation relations due to their analogy with the Poisson brackets between the canonical variables of classical mechanics. In tensor form they can be written

$$[\mathbf{Q}, \mathbf{P}] = i\mathbf{1}.$$

Here $[\mathbf{Q}, \mathbf{P}] = \mathbf{QP} - \mathbf{PQ}$, and $\mathbf{PQ}$ is the 3×3 matrix with elements P_iQ_j. The general solution of these commutation relations in position space is

$$(\mathbf{P}f)(\boldsymbol{\xi}) = -i\nabla f(\boldsymbol{\xi}) + g(\boldsymbol{\xi})f(\boldsymbol{\xi}),$$

where $g(\boldsymbol{\xi})$ is an arbitrary function which, to be consistent with the case of one dimension, we take equal to zero. Thus we have the two equivalent expressions

$$(\mathbf{P}f)(\boldsymbol{\xi}) = -i\nabla f(\boldsymbol{\xi}), \qquad \langle\boldsymbol{\xi}|\mathbf{P}|\boldsymbol{\xi}'\rangle = -i\nabla\delta(\boldsymbol{\xi} - \boldsymbol{\xi}').$$

As before, the momentum is also a maximal observable which induces a representation of the Hilbert space as the $\mathbb{L}^2$-space of functions $\langle\boldsymbol{\eta}|f\rangle = \tilde{f}(\boldsymbol{\eta})$ of the vector variable $\boldsymbol{\eta}$ related to position space by the Fourier transformation

$$\tilde{f}(\boldsymbol{\eta}) = (2\pi)^{-3/2}\int d^3\xi\; e^{-i\boldsymbol{\eta}\cdot\boldsymbol{\xi}} f(\boldsymbol{\xi}). \tag{1.2}$$

In Dirac's notation the Fourier kernel is

$$\langle\boldsymbol{\eta}|\boldsymbol{\xi}\rangle = (2\pi)^{-3/2}e^{-i\boldsymbol{\eta}\cdot\boldsymbol{\xi}}.$$

Next we consider two experiments $\mathcal{E}$ and $\mathcal{E}'$ linked by a homogeneous galilean transformation $\mathfrak{g}(\mathbf{v}) = \{\mathbf{1}, 0, \mathbf{v}\}$. A reasoning entirely analogous to the one of Section VI.3 shows that

$$\mathbf{Q} = e^{i\mathbf{G}\cdot\mathbf{v}}\mathbf{Q}e^{-i\mathbf{G}\cdot\mathbf{v}},$$

where $-\mathbf{G}$, given by equation (V.6.22), is the generator of homogeneous galilean transformations. This operator, therefore, commutes with $\mathbf{Q}$ and, since $\mathbf{Q}$ is maximal, one gets

$$\mathbf{G} = \mathbf{F}(\mathbf{Q}). \tag{1.3}$$

If we define the observable velocity by

$$\dot{\mathbf{Q}} = i[H_t, \mathbf{Q}], \tag{1.4}$$

the velocity transformation for the one-dimensional particle generalizes immediately to

$$\dot{\mathbf{Q}} - \mathbf{v} = e^{i\mathbf{G}\cdot\mathbf{v}}\dot{\mathbf{Q}}e^{-i\mathbf{G}\cdot\mathbf{v}}$$

which leads to the commutation relation

$$[\mathbf{G}, \dot{\mathbf{Q}}] = i\mathbf{1}. \tag{1.5}$$

To determine $\mathbf{G}$ we need its commutation relation with $\mathbf{P}$ which is given by (V.6.19) to be

$$[\mathbf{G}, \mathbf{P}] = im\mathbf{1},$$

where m is a real number. Using the result (1.3) and writing this commutation relation in position space we find that

$$\mathbf{G} = m\mathbf{Q}$$

and that, in order not to violate the velocity transformation formula, the *mass* m must be different from zero.

Substitution of the last equation into (1.5) shows that, as in the case of one dimension, $m\dot{\mathbf{Q}}$ obeys the same commutation relation with $\mathbf{Q}$ as $\mathbf{P}$ does, and that therefore we have

$$m\dot{\mathbf{Q}} = \mathbf{P} - \mathbf{A}_t(\mathbf{Q}), \tag{1.6}$$

where $\mathbf{A}_t(\mathbf{Q})$ is an arbitrary vector valued function of $\mathbf{Q}$ that may depend on t as a parameter. As we shall soon see, contrary to the situation in one dimension, we cannot in general eliminate this function by a unitary transformation that depends only on $\mathbf{Q}$.

We now regard (1.4) as an equation for the hamiltonian. The one-dimensional case suggests that we put

$$H_t = \tfrac{1}{2}m|\dot{\mathbf{Q}}|^2 + H_t'.$$

Using the commutation relation $[\mathbf{Q}, m\dot{\mathbf{Q}}] = i$ and substituting the expression for H_t into (1.4) we find that H_t' commutes with $\mathbf{Q}$, and must therefore be a scalar function $V_t(\mathbf{Q})$ of $\mathbf{Q}$ alone. Taking into account (1.6), we obtain the general form of the hamiltonian for a spinless particle in three dimensions

$$H_t = |\mathbf{P} - \mathbf{A}_t(\mathbf{Q})|^2 + V_t(\mathbf{Q}). \tag{1.7}$$

The analogy with the classical hamiltonian of a particle in an electromagnetic field is striking. However, it should be made clear that nowhere in the argument leading to (1.7) have we implied that the external field sources present in the experimental setups are of electromagnetic nature. The expression (1.7) is based solely on the galilean invariance of the experimental setup and apply no matter what type of field sources may be present.

In non-relativistic quantum mechanics the only external fields that are considered are electromagnetic or gravitational. In the electromagnetic case, it is experimentally known that the potentials $\mathbf{A}_t$ and V_t in the hamiltonian do not depend only on the field sources. They are of the form $e\mathbf{A}_{et}(\mathbf{Q})$ and $eV_{et}(\mathbf{Q})$ where the functions $\mathbf{A}_{et}(\cdot)$ and $V_{et}(\cdot)$ depend only on the field sources, while e the electric charge of the particle is an intrinsic parameter. In the gravitational case there is only a scalar potential of the form $mV_{gt}(\mathbf{Q})$, with the function $V_{gt}(\cdot)$ depending only on the sources. The spinless particle has no other intrinsic parameters besides the mass m and the electric charge e. Two such particles with the same mass and charge are therefore identical.

It was asserted above that the function $\mathbf{A}_t(\mathbf{Q})$ cannot be removed by a unitary transformation $S(\mathbf{Q})$ depending on $\mathbf{Q}$ alone. To show this we rewrite the equation

$$S(\mathbf{Q})[\mathbf{P} - \mathbf{A}_t(\mathbf{Q})]S^\dagger(\mathbf{Q}) = \mathbf{P}$$

in position space. After multiplying on the right by $S^\dagger(\mathbf{Q})$ and applying the result to a vector $|f\rangle$, we get

$$S(\boldsymbol{\xi})\mathbf{A}_t(\boldsymbol{\xi})f(\boldsymbol{\xi}) = if(\boldsymbol{\xi})\nabla S(\boldsymbol{\xi})$$

from which we conclude that

$$\mathbf{A}_t(\boldsymbol{\xi}) = i\nabla[\ln S(\boldsymbol{\xi})].$$

which shows that the unitary transformation $S(\mathbf{Q})$ does not exist unless $\mathbf{A}_t(\boldsymbol{\xi})$ is irrotational.

In the next section we will be considering the subgroup of rotations $\mathbf{j}(\mathbf{R})$. We have already used it implicitly because it is the galilean invariance under rotations that determined the vector nature of the generators $\mathbf{P}$ and $\mathbf{G}$, and, therefore, the scalar nature of the hamiltonian. We shall see that this invariance adds no further restrictions to the hamiltonian but we will uncover a very interesting structure in the Hilbert space of a particle.

2. Rotations, the orbital angular momentum

Imagine now two galilean invariant experimental setups $\mathcal{E}$ and $\mathcal{E}'$ connected by a rotation $\mathbf{j_e}(\varphi)$ of an angle φ about the axis $\mathbf{e}$. We denote the tensor that effects this rotation by $\mathbf{R_e}(\varphi)$, that is, $\mathbf{j_e}(\varphi) = \{\mathbf{R}, 0, 0\}$. In the case of a spinless particle we follow the common practice of denoting the generator of rotations (V.6.23) by $\mathbf{L}$ and its components by L_i. With this notation the unitary operator that represents the rotation $\mathbf{j_e}(\varphi)$ is,

$$e^{-i\mathbf{e}\cdot\mathbf{L}\varphi} = e^{-i(L_1\mathbf{e}_1+L_2\mathbf{e}_2+L_3\mathbf{e}_3)\cdot\mathbf{e}\varphi}.$$

As before the position observable of $\mathcal{E}'$ is measured by a set of counters labelled by their position vectors $\boldsymbol{\xi}$ with respect to the inertial frame associated with $\mathcal{E}'$. This set is identical to the set of counters of $\mathcal{E}$ when the labels of the latter are changed from $\boldsymbol{\xi}$ to $\mathbf{R_e}^{-1}(\varphi)\boldsymbol{\xi}$. Our familiar argument permits us to infer that

$$\mathbf{R_e}^{-1}(\varphi)\mathbf{Q} = \mathbf{e}^{-i\mathbf{e}\cdot\mathbf{L}\varphi}\mathbf{Q}\mathbf{e}^{i\mathbf{e}\cdot\mathbf{L}\varphi}. \tag{2.1}$$

To determine $\mathbf{L}$ it is convenient to restrict first one's attention to the particular case in which $\mathbf{R}$ is a rotation $\mathbf{R}_3(\varphi)$ of an angle φ about the 3-axis common to $\mathcal{E}$ and $\mathcal{E}'$:

$$\mathbf{R}_3(\varphi) = \begin{pmatrix} \cos\varphi & -\sin\varphi & 0 \\ \sin\varphi & \cos\varphi & 0 \\ 0 & 0 & 1 \end{pmatrix}$$

so that (2.1) becomes

$$\mathbf{R}_3^{-1}(\varphi)\mathbf{Q} = e^{-iL_3\varphi}\mathbf{Q}e^{iL_3\varphi}. \tag{2.2}$$

Before proceeding further we show that the spectrum of the operator L_3 is discrete. To see this imagine that experiment $\mathcal{E}'$ is rotated an angle of 2π with respect to $\mathcal{E}$. Let $|\phi\rangle$ be a basis vector of a basis associated with an apparatus of $\mathcal{E}$ and $|\phi'\rangle = e^{-2\pi i L_3}|\phi\rangle$ the corresponding basis vector in $\mathcal{E}'$. The readings of the apparatus in $\mathcal{E}$ and $\mathcal{E}'$ must be identical but, because a rotation of $\mathcal{E}'$ has been effected, $|\phi\rangle$ and $|\phi'\rangle$ may differ by a phase factor $e^{-2\pi i c}$ common to all basis vectors. Thus $e^{-2\pi i L_3}|\phi\rangle = e^{-2\pi i c}|\phi\rangle$. Since this equation holds for any basis vector we must have

$$e^{-2\pi i L_3} = e^{-2\pi i c} I,$$

or, writing $E(\lambda)$ for the spectral family of L_3, we get after multiplication by $e^{2\pi i c}$

$$\int_{-\infty}^{\infty} (e^{-2\pi i(\lambda - c)} - 1)\, dE(\lambda) = 0.$$

By applying this expression to a vector $|f\rangle$ in the range of the projection $E(\lambda'+\epsilon)-E(\lambda')$ with λ' arbitrary, and letting $\epsilon \to 0$ we see that the quantity in parenthesis must equal zero and that, therefore

$$\lambda = m + c, \qquad m \text{ integer.} \tag{2.3}$$

Thus the spectrum of L_3 is discrete as asserted and its eigenvalues differ by an integer.

We proceed now to determine the operator L_3. To this end rewrite (2.2) in components

$$\begin{aligned} \cos(\varphi)Q_1 + \sin(\varphi)Q_2 &= e^{-iL_3\varphi}Q_1 e^{iL_3\varphi}, \\ -\sin(\varphi)Q_1 + \cos(\varphi)Q_2 &= e^{-iL_3\varphi}Q_2 e^{iL_3\varphi}, \\ Q_3 &= e^{-iL_3\varphi}Q_3 e^{iL_3\varphi}. \end{aligned}$$

By the usual procedure we find the commutation relations

$$[L_3, Q_1] = iQ_2, \qquad [L_3, Q_2] = -iQ_1, \qquad [L_3, Q_3] = 0. \tag{2.4}$$

The commutation relations (V.6.18) among the generators of rotations and displacements give similar relations between L_3 and the components of $\mathbf{P}$,

$$[L_3, P_1] = iP_2, \qquad [L_3, P_2] = -iP_1, \qquad [L_3, P_3] = 0. \tag{2.5}$$

Multiplying on the right the first equations of (2.4) and (2.5) by P_2 and Q_2, respectively, subtracting the results and using the canonical commutation relations (1.1) we find $[L_3, Q_1P_2 - Q_2P_1] = 0$, which suggests that we try a solution of the form

$$L_3 = Q_1P_2 - Q_2P_1 + F(\mathbf{Q}, \mathbf{P})$$

for the equations (2.4) and (2.5). By substitution we find that F commutes with both $\mathbf{Q}$ and $\mathbf{P}$ and must therefore be a multiple of the identity that we may take equal to zero, and write

$$L_3 = Q_1P_2 - Q_2P_1. \tag{2.6}$$

Everything that we have said applies to a rotation about any of the coordinate axes. Calling L_1 and L_2 the generators of rotations about the 1- and 2- axes, we get for them expressions that can be obtained from (2.6) by circular permutation of the indices. We thus have

$$\mathbf{L} = \mathbf{Q} \times \mathbf{P} \tag{2.7}$$

which by analogy with classical mechanics is called the *orbital angular momentum.* The adjective "orbital" is used to distinguish it from the spin angular momentum, to be studied later, which has no classical analogue. The commutation relations among the components of $\mathbf{L}$ are given by (V.6.18) as

$$[L_i, L_j] = i\epsilon_{ijk} L_k \tag{2.8}$$

or more succinctly,

$$\mathbf{L} \times \mathbf{L} = i\mathbf{L}.$$

From the expression (2.7) for the orbital angular momentum we now show that the eigenvalues of its components are in fact integers. By (2.3) it is sufficient to show that zero is an eigenvalue of L_3. In position space the equation $L_3|\psi\rangle = 0$ reads after multiplication by i

$$\xi_1 \frac{\partial \psi}{\partial \xi_2} - \xi_2 \frac{\partial \psi}{\partial \xi_1} = 0$$

whose general solution is an arbitrary function of $\xi_1^2 + \xi_2^2$, and of ξ_3 which appears only as a parameter, thus the square integrable function $\exp(-|\boldsymbol{\xi}|^2)$ is an eigenfunction of L_3 (and of L_1 and L_2) belonging to the eigenvalue zero.

It would now be possible to determine the full spectral resolution of the orbital angular momentum but we shall instead use the more fruitful approach of studying in the next section the properties of operators that obey commutation relations similar to (2.8).

3. The angular momentum commutation relations

In this section we digress from the study of the single particle and consider a purely algebraic problem. Suppose that, on a space $\mathbb{V}$, three hermitean operators J_1, J_2, J_3, that satisfy commutation relations analogous to (2.8), are defined.

$$[J_1, J_2] = iJ_3, \qquad [J_2, J_3] = iJ_1, \qquad [J_3, J_1] = iJ_2. \tag{3.1}$$

As can be verified directly from the commutation relations (3.1), the operator

$$J^2 = J_1^2 + J_2^2 + J_3^2 \tag{3.2}$$

commutes with each J_k, $k = 1, 2, 3$. Suppose furthermore that that J^2 and J_3 form a maximal set of observables with a discrete spectrum. Our problem consists in determining the spectral resolution of these operators together with the structure of the of the space $\mathbb{V}$ on which they are defined.

Obviously J^2 is a positive operator; for reasons that will become apparent shortly, it is expedient to write its eigenvalues as $j(j+1)$, $j > 0$. We shall study now the spectral resolution of J_3 regarded as an operator on the eigensubspace $\mathbb{V}^{(j)}$ belonging to the eigenvalue $j(j+1)$ of J^2. On this space J^2 is a multiple $j(j+1)$ of the identity, and J_3 is maximal by hypothesis.

It is convenient to introduce a pair of mutually adjoint operators

$$J_+ = J_1 + iJ_2, \qquad J_- = J_1 - iJ_2 \tag{3.3}$$

which from (3.1) and (3.2) are seen to obey the commutation relations

$$[J_3, J_+] = J_+, \qquad [J_3, J_-] = -J_-, \qquad [J_+, J_-] = -2J_3.$$

Let $|\psi_m\rangle$ be an eigenvector of J_3 belonging to the eigenvalue m. Use of the last set of commutation relations yields

$$J_3 J_+|\psi_m\rangle = (J_+J_3 + J_+)|\psi_m\rangle = (m+1)J_+|\psi_m\rangle,$$
$$J_3 J_-|\psi_m\rangle = (J_-J_3 - J_-)|\psi_m\rangle = (m-1)J_-|\psi_m\rangle.$$

Thus, $J_+|\psi_m\rangle$ and $J_-|\psi_m\rangle$ are eigenvectors of J_3 belonging to the eigenvalues $m+1$ and $m-1$, respectively. Repeated application of J_+ (J_-) to an eigenvector of J_3 yields a sequence of eigenvectors belonging to an increasing (decreasing) sequence of eigenvalues that differ by an integer. It is clear that this sequence will terminate if and only if the application of J_+ (J_-) to an eigenvector $|\psi_\lambda\rangle$ yields the zero vector.

The sequences must, in fact, be finite because from the non-negativeness of

$$\begin{aligned}\|J_+|\psi_m\rangle\|^2 &= \langle\psi_m|J_-J_+|\psi_m\rangle = \langle\psi_m|J^2 - J_3^2 - J_3|\psi_m\rangle \\ &= (j-m)(j+m+1)\langle\psi_m|\psi_m\rangle\end{aligned} \tag{3.4}$$

and of

$$\begin{aligned}\|J_-|\psi_m\rangle\|^2 &= \langle\psi_m|J_+J_-|\psi_m\rangle = \langle\psi_m|J^2 - J_3^2 + J_3|\psi_m\rangle \\ &= (j+m)(j-m+1)\langle\psi_m|\psi_m\rangle,\end{aligned} \tag{3.5}$$

we conclude that both $(j-m)(j+m+1)$ and $(j+m)(j-m+1)$ must be non-negative or, equivalently, that

$$-j \le m \le j.$$

The finiteness of the sequences implies that, if m_{max} and m_{min} are the upper an lower bounds of the sequence of eigenvalues of J_3, then both $J_+|\psi_{m_{max}}\rangle$ and $J_-|\psi_{m_{min}}\rangle$ must be the zero vector. Thus, from the expressions for the norms we conclude that $m_{max} = j$ and that $m_{min} = -j$. Since these two numbers must differ by an integer, and j is positive we conclude that $2j$ must be a positive integer. If j itself is not an integer we shall call it, as is customary, a half-integer. The sequence of eigenvalues of J_3 obtained from m consists then of the $2j+1$ numbers

$$-j, -j+1, ..., j-1, j \tag{3.6}$$

and m itself must be one of these numbers so that, were we to start with a different m, we would obtain the same sequence. The operator J_3 has therefore the $2j+1$ eigenvalues (3.6) and, since it was assumed to be maximal, this number is precisely the dimension of the subspace $\mathbb{V}^{(j)}$.

The set of eigenvectors obtained by repeated application of J_+ or J_- to a normalized eigenvector will not, in general, be normalized. However, one can construct easily such a set as follows: let $|m\rangle$ be a normalized eigenvector belonging to m; then, if we put $|\psi_{m+1}\rangle = J_+|m\rangle$, we conclude from (3.4) that $\||\psi_m\rangle\|^2 = (j-m)(j+m+1)$ and that therefore

$$|m+1\rangle = [(j-m)(j+m+1)]^{-1/2} J_+|m\rangle \tag{3.7}$$

is a corresponding normalized eigenvector. Similarly, from (3.5) we see that

$$|m-1\rangle = [(j+m)(j-m+1)]^{-1/2} J_-|m\rangle \tag{3.8}$$

is also a normalized vector. Thus, if $|j\rangle$ is normalized, a set of normalized eigenvectors is given by

$$|m\rangle = \sqrt{\frac{(j+m)!}{(2j)!(j-m)!}}\, J_-^{j-m}|j\rangle, \qquad m = -j, -j+1, ..., j-1, j. \tag{3.9}$$

Similarly we can start with a normalized $|-j\rangle$ and obtain the normalized eigenvectors

$$|m\rangle = \sqrt{\frac{(j-m)!}{(2j)!(j+m)!}}\, J_+^{j+m}|-j\rangle, \qquad m = -j, -j+1, ..., j-1, j, \tag{3.10}$$

which may differ from (3.9) by a phase factor.

We now show that, if $j > 0$, the operators J_k do not have common eigenvectors, or equivalently that any pair of them is irreducible on $\mathbb{V}^{(j)}$. Suppose, on the contrary, that $|\psi\rangle$ were an eigenvector of both J_1 and J_2 with respective eigenvalues m_1, m_2; then, according to the first commutation relation (3.1),

$$iJ_3|\psi\rangle = (m_1m_2 - m_2m_1)|\psi\rangle = 0.$$

Thus, $|\psi\rangle$ would be an eigenvector of J_3 belonging to the eigenvalue zero, and, by the remaining relations (3.1), $J_1|\psi\rangle = 0$ and $J_2|\psi\rangle = 0$ (that is, $m_1 = m_2 = 0$). From the definition (3.2) of J^2 it would follow that $J^2|\psi\rangle = 0$, so that contrary to the hypothesis, $|\psi\rangle$ would be in the eigensubspace of J^2 corresponding to $j = 0$.

We summarize the results obtained up to now.

For each j such that $2j$ is a positive integer, there exists a Hilbert space $\mathbb{V}^{(j)}$ of dimension $2j+1$ on which hermitean operators $J_1^{(j)}$, $J_2^{(j)}$, $J_3^{(j)}$ that satisfy the commutation relations (3.1) are defined. In this space, each of the operators is maximal and any pair of them form an irreducible set. Their eigenvalues are the $2j+1$ numbers $-j, -j-1, ..., j-1, j$. The operator $(J^{(j)})^2$ defined by (3.2) is the multiple $j(j+1)I^{(j)}$, of the identity $I^{(j)}$ on $\mathbb{V}^{(j)}$

The space $\mathbb{V}^{(j)}$ is one of the spaces $\mathbb{V}$ that satisfies the conditions introduced at the beginning of this section. More general spaces $\mathbb{V}$ can be constructed by taking the direct sum of a (finite or infinite) sequence of spaces $\mathbb{V}^{(j)}$,

$$\mathbb{V} = \bigoplus_j \mathbb{V}^{(j)}. \tag{3.11}$$

In such a space, hermitean operators J_1, J_2, J_3 that obey the commutation relations (3.1) can be defined as the direct sums of the corresponding operators in the subspaces $\mathbb{V}^{(j)} \subset \mathbb{V}$

$$J_k = \bigoplus_j J_k^{(j)}, \qquad k = 1, 2, 3$$

Regarded as an operator on $\mathbb{V}$, the identity on $\mathbb{V}^{(j)}$ is the projection E_j on the subspace $\mathbb{V}^{(j)}$, and therefore, as an operator on $\mathbb{V}$, we have $(J^{(j)})^2 = j(j+1)E_j$. Consequently we have the spectral resolution of J^2 on $\mathbb{V}$

$$J^2 = \sum_j j(j+1)E_j.$$

Clearly it is not true that each of the observables J_k is maximal in $\mathbb{V}$. However, if all the numbers j are different, the pair formed by J^2 and one of them, say J_3, forms a maximal set of observables on $\mathbb{V}$. The eigenvectors $|j, m\rangle$ of $J_3^{(j)} = E_jJ_3E_j = E_jJ_3 = J_3E_j$ are eigenvectors of J^2 belonging to the eigenvalue $j(j+1)$

$$J^2|j, m\rangle = j(j+1)|j, m\rangle, \qquad J_3|j, m\rangle = m|j, m\rangle$$

and the set of all such eigenvectors constitutes a basis in $\mathbb{V}$. The operators J_k are said to be *reducible* with respect to $\mathbb{V}$ because this space can be decomposed into a direct sum of spaces $\mathbb{V}^{(j)}$ in which the corresponding operators form an irreducible set. Note that, if desired, one may include among the subspaces $\mathbb{V}^{(j)}$ the eigen-subspace $\mathbb{V}^{(0)}$ belonging to the zero eigenvalue of J^2, which may be regarded as a one-dimensional subspace spanned by a vector $|0,0\rangle$ such that, besides $J^2|0,0\rangle = 0$, we also have $J_k|0,0\rangle = 0$.

It may be that the operators J_k are defined on a space $\mathbb{H}$ which is larger than $\mathbb{V}$ in the sense that J^2 and J_3 form part of a maximal set containing other observables. Let the eigenvalues of the other observables be represented by λ, then the eigenvectors $|\lambda, j, m\rangle$ form a basis in $\mathbb{H}$. For each fixed λ the eigenvectors $|\lambda, j, m\rangle$ span a subspace $\mathbb{H}_\lambda \subset \mathbb{H}$ which is isomorphic to $\mathbb{V}$.

The results of this section show that the class of operators that obey the commutation relations (3.1) is more general than the class of orbital angular momenta, for the components of the latter were shown in the last section to admit only integer eigenvalues, while we have seen that the operators that obey (3.1) may have half-integer eigenvalues.

After this digression we return in the next section to the study of the orbital angular momentum of a particle.

4. Spectral resolution of the orbital angular momentum

To continue with the study of the orbital angular momentum $\mathbf{L} = \mathbf{Q} \times \mathbf{P}$ of a particle, we rewrite this equation in the position space of functions $f(\boldsymbol{\xi})$; we have

$$(\mathbf{L}f)(\boldsymbol{\xi}) = -i\boldsymbol{\xi} \times \nabla f(\boldsymbol{\xi}).$$

It is convenient to write the functions $f(\boldsymbol{\xi})$ in polar coordinates r, θ, φ with unit vectors $\mathbf{e}_r, \mathbf{e}_\theta, \mathbf{e}_\varphi$. In this representation, the scalar product $\langle f|g\rangle$ is given by

$$\langle f|g\rangle = \int_0^\infty r^2 dr \int d\Omega\, f^*(r,\theta,\varphi) g(r,\theta,\varphi), \tag{4.1}$$

where

$$\int d\Omega \cdot = \int_0^\pi d\theta \int_0^{2\pi} d\varphi\, \sin\theta \cdot$$

Denoting the operators in this function space with a circumflex, we write

$$\hat{\mathbf{L}} = -i\boldsymbol{\xi} \times \nabla = \hat{L}_r \mathbf{e}_r + \hat{L}_\theta \mathbf{e}_\theta + \hat{L}_\varphi \mathbf{e}_\varphi$$

and by writing the gradient in polar coordinates we conclude that

$$\hat{L}_r = 0, \qquad \hat{L}_\theta = -\frac{1}{\sin\theta}\frac{\partial}{\partial\varphi}, \qquad \hat{L}_\varphi = \frac{\partial}{\partial\theta}.$$

The corresponding cartesian components can now be obtained from these by the usual transformation formulae

$$\hat{L}_1 = i\cot\theta\cos\varphi\frac{\partial}{\partial\varphi} + i\sin\varphi\frac{\partial}{\partial\theta},$$

$$\hat{L}_2 = i\cot\theta\sin\varphi\frac{\partial}{\partial\varphi} - i\cos\varphi\frac{\partial}{\partial\theta}, \qquad \hat{L}_3 = -i\frac{\partial}{\partial\varphi}. \tag{4.2}$$

Since we shall be using the results of the previous section we also need the raising and lowering operators corresponding to (3.3)

$$\hat{L}_+ = e^{i\varphi}\left(\cot\theta\frac{\partial}{\partial\varphi} + \frac{\partial}{\partial\theta}\right), \qquad \hat{L}_- = e^{-i\varphi}\left(i\cot\theta\frac{\partial}{\partial\varphi} - \frac{\partial}{\partial\theta}\right). \tag{4.3}$$

It is clear that the operators $\hat{L}_k$ do not operate on the radial coordinate r. We can regard the position space of the functions $f(\boldsymbol{\xi})$ as the tensor product of the two spaces $\mathbb{L}^2_r$ and $\mathbb{L}^2_\Omega$ of functions $f_1(r)$ on the positive real axis and $f_2(\theta,\varphi)$ on the surface of the unit sphere, respectively, with scalar products defined by

$$\langle f_1|g_1\rangle = \int r^2 dr\ f_1^*(r)g_1(r), \qquad \langle f_2|g_2\rangle = \int d\Omega\ f_2^*(\theta,\varphi)g_2(\theta,\varphi) \tag{4.4}$$

The abstract Hilbert spaces of which $\mathbb{L}^2_r$ and $\mathbb{L}^2_\Omega$ are representations will be denoted by $\mathbb{H}_r$ and $\mathbb{H}_\Omega$ respectively, so that $\mathbb{H} = \mathbb{H}_r \otimes \mathbb{H}_\Omega$. We shall look in more detail at this decomposition in Section 9 below. For the moment it is sufficient to remark that the operators $\hat{L}_k$ can be thought of as been defined on $\mathbb{L}^2(\theta,\varphi)$. In $\mathbb{H}_\Omega$ we write the eigenvalue equations for L^2 and L_3 as

$$L^2|l,m\rangle = l(l+1)|l,m\rangle, \qquad L_3|l,m\rangle = m|l,m\rangle.$$

We know from the study of the previous section that l may have integer or half-integer values and that the values of m differ by integers and range from $-l$ to l. We also know that

$$L_+|l,l\rangle = 0.$$

In $\mathbb{L}^2(\theta,\varphi)$ the eigenvalue equation for $\hat{L}_3$ becomes by use of (4.2)

$$-i\frac{\partial}{\partial\varphi}\psi_l^m = m\psi_l^m,$$

where, as was shown in Section 2, the eigenvalue m is an integer. If, in a sense to be explained in Section 9, we write the Dirac eigenket of the coordinate operator on the unit sphere as $|\Omega\rangle$, we have $\psi_l^m = \psi_l^m(\theta,\varphi) = \langle\Omega|l,m\rangle$. The solution of the eigenvalue equation is

$$\psi_l^m(\theta,\varphi) = f_l^m(\theta)e^{im\varphi}. \tag{4.5}$$

We determine $|l,-l\rangle$ through the condition $L_-|l,-l\rangle = 0$, or by (4.3) and (4.5)

$$\left(l\cot\theta - \frac{\partial}{\partial\theta}\right)f_l^{-l}(\theta) = 0$$

whose solution is $f_l^{-l}(\theta) = c_l \sin^l\theta$. We have therefore

$$\psi_l^{-l}(\theta,\varphi) = c_l e^{il\varphi}\sin^l\theta.$$

Choosing a real constant c_l with the normalization condition $\int d\Omega\ |\psi_l^{-l}|^2 = 1$ we obtain a normalized solution

$$Y_l^{-l}(\theta,\varphi) = \frac{1}{2^l l!}\sqrt{\frac{(2l+1)!}{4\pi}}\, e^{-il\varphi}\sin^l\theta.$$

By repeated application of $\hat{L}_-$ and use of (3.10) and (4.3) we obtain

$$Y_l^m(\theta,\varphi) = \sqrt{\frac{(2l+1)!(l+m)!}{4\pi(2l)!(l-m)!}}\, e^{im\varphi}P_l^m(\cos\theta), \tag{4.6}$$

where $P_l^m(x)$ is the associated Legendre polynomial given for $|x| \leq 1$ and m an integer, $-l \leq m \leq l$, by

$$P_l^m(x) = \frac{(-1)^{l+m}}{2^l l!}(1-x^2)^{m/2}\frac{d^{(l+m)}}{dx^{l+m}}(1-x^2)^l, \qquad m \geq 0,$$

$$P_l^{-m}(x) = (-1)^m\frac{(l-m)!}{(l+m)!}P_l^m(x), \qquad m < 0.$$

Observe that the last formula leads to the relation

$$[Y_l^m(\theta,\varphi)]^* = (-1)^m Y_l^{-m}(\theta,\varphi). \tag{4.7}$$

The functions $Y_l^m(\theta,\varphi)$ are the well known spherical harmonics which for $l = 0, 1, \ldots$, $m = -l, -l+1, \ldots, l-1, l$, are known to constitute a complete set of eigenfunctions on the surface of the unit sphere. Thus the corresponding abstract space $\mathbb{H}_\Omega$ can be written in the notation of (3.11) as

$$\mathbb{H}_\Omega = \bigoplus_{l=0}^{\infty} \mathbb{H}_\Omega^{(l)}.$$

The components of $\mathbf{L}$ form an irreducible set on $\mathbb{H}_\Omega^{(l)}$, while the observables L^2 and L_3 (or any other component of $\mathbf{L}$) constitute a maximal set of observables on $\mathbb{H}_\Omega$. Since $|\mathbf{Q}|$ can be easily seen to be a maximal observable on the radial space $\mathbb{H}_r$, a maximal set of observables on the Hilbert space $\mathbb{H}$ of the spinless particle is formed by $|\mathbf{Q}|$, L^2 and L_3.

5. The free particle

The simplest example to consider is of course the free particle, namely one with a hamiltonian that contains no external fields

$$H = \frac{|\mathbf{P}|^2}{2m}.$$

We shall solve the dynamical problem that can be formulated as follows: knowing the state in which the particle has been prepared at $t = 0$, find the probability densities at time t for the momentum and the position which we denote by $\pi_t(\boldsymbol{\eta})$ and $\pi_t(\boldsymbol{\xi})$, respectively. The solution of this problem for the special case of a pure state is well covered in elementary texts. We shall be concerned here with a general initial state ρ_0.

Since the hamiltonian is time independent the evolution operator can be written immediately

$$U(t) = \exp\left(-i\frac{|\mathbf{P}|^2}{2m}t\right).$$

At time t the state is $\rho_t = U(t)\rho_0 U^\dagger(t)$. Because the hamiltonian depends only on $\mathbf{P}$ it is convenient to work in momentum space. For any operator A we use the notation $\tilde{A}(\boldsymbol{\eta}, \boldsymbol{\eta}') = \langle\boldsymbol{\eta}|A|\boldsymbol{\eta}'\rangle$. We have then

$$\tilde{\rho}_t(\boldsymbol{\eta}, \boldsymbol{\eta}') = \exp\left(-i\frac{\eta^2 - \eta'^2}{2m}t\right)\tilde{\rho}_0(\boldsymbol{\eta}, \boldsymbol{\eta}'). \tag{5.1}$$

The momentum probability density is

$$\pi_t(\boldsymbol{\eta}) = \langle\boldsymbol{\eta}|\rho_t|\boldsymbol{\eta}\rangle = \tilde{\rho}_t(\boldsymbol{\eta}, \boldsymbol{\eta}) = \tilde{\rho}_0(\boldsymbol{\eta}, \boldsymbol{\eta}) = \pi_0(\boldsymbol{\eta}) \tag{5.2}$$

and remains constant during the motion.

The probability density for the position is

$$\pi_t(\boldsymbol{\xi}) = \langle \boldsymbol{\xi}|\rho_t|\boldsymbol{\xi}\rangle = \iint d^3\eta\, d^3\eta'\, \langle \boldsymbol{\xi}|\boldsymbol{\eta}\rangle\langle \boldsymbol{\eta}|\rho_t|\boldsymbol{\eta}'\rangle\langle \boldsymbol{\eta}'|\boldsymbol{\xi}\rangle.$$

Using (5.1) and the expression following (1.2) for the Fourier kernel we get

$$\pi_t(\boldsymbol{\xi}) = (2\pi)^{-3} \iint d^3\eta\, d^3\eta'\, \exp\left[i(\boldsymbol{\eta} - \boldsymbol{\eta}')\cdot\boldsymbol{\xi} - i\,\frac{\eta^2 - \eta'^2}{2m}\,t\right] \tilde{\rho}_0(\boldsymbol{\eta}, \boldsymbol{\eta}').$$

A more concise expression results if we introduce new integration variables $\mathbf{K} = \frac{1}{2}(\boldsymbol{\eta}' + \boldsymbol{\eta})$, $\boldsymbol{\kappa} = \boldsymbol{\eta}' - \boldsymbol{\eta}$,

$$\pi_t(\boldsymbol{\xi}) = (2\pi)^{-3} \iint d^3K\, d^3\kappa\, \exp\left[-i\left(\xi - \frac{\mathbf{K}}{m}t\right)\cdot\boldsymbol{\kappa}\right] \tilde{\rho}_0\left(\mathbf{K} - \tfrac{1}{2}\boldsymbol{\kappa},\ \mathbf{K} + \tfrac{1}{2}\boldsymbol{\kappa}\right). \tag{5.3}$$

If we use the abbreviation

$$f_0(\boldsymbol{\eta}, \boldsymbol{\xi}) = (2\pi)^{-3} \int d^3\kappa\, e^{-i\boldsymbol{\kappa}\cdot\boldsymbol{\xi}} \tilde{\rho}_0\left(\boldsymbol{\eta} - \tfrac{1}{2}\boldsymbol{\kappa},\ \boldsymbol{\eta} + \tfrac{1}{2}\boldsymbol{\kappa}\right), \tag{5.4}$$

we can write the solution (5.3), (5.2) of our problem as

$$\pi_t(\boldsymbol{\xi}) = \int d^3\eta\, f_0\left(\boldsymbol{\eta},\ \boldsymbol{\xi} - \frac{\boldsymbol{\eta}}{m}t\right), \tag{5.5}$$

$$\pi_t(\boldsymbol{\eta}) = \pi_0(\boldsymbol{\eta}) = \int d^3\xi\, f_0(\boldsymbol{\eta}, \boldsymbol{\xi}).$$

The last equation tells us that

$$\iint d^3\eta\, d^3\xi\, f_0(\boldsymbol{\eta}, \boldsymbol{\xi}) = 1.$$

The average of $\mathbf{Q}$ at time t is found from (5.5) to be

$$\langle \mathbf{Q}\rangle_t = \int d^3\xi\, \boldsymbol{\xi} \int d^3\eta\, f_0\left(\boldsymbol{\eta},\ \boldsymbol{\xi} - \frac{\boldsymbol{\eta}}{m}t\right) = \iint d^3\eta\, d^3\xi\, \left(\boldsymbol{\xi} + \frac{\boldsymbol{\eta}}{m}t\right) f_0(\boldsymbol{\eta}, \boldsymbol{\xi}),$$

that is,

$$\langle \mathbf{Q}\rangle_t = \langle \mathbf{Q}\rangle_0 + \frac{\langle \mathbf{P}\rangle}{m}t.$$

The average value of the position moves classically. We shall show that, in a certain precise sense, the averages of all free particle observables move classically. But, before, we shall study in more detail in the next section the function introduced in (5.4).

6. The Wigner distribution function

We consider here a general spinless particle in a state ρ. and define the function

$$f(\boldsymbol{\eta},\boldsymbol{\xi}) = (2\pi)^{-6} \iint d^3\alpha\, d^3\beta\; e^{-i(\boldsymbol{\alpha}\cdot\boldsymbol{\eta}+\boldsymbol{\beta}\cdot\boldsymbol{\xi})}\mathrm{Tr}\,(e^{i(\boldsymbol{\alpha}\cdot\mathbf{P}+\boldsymbol{\beta}\cdot\mathbf{Q})}\rho) \tag{6.1}$$

which is called the *Wigner distribution function* for the state ρ. We shall show that this is precisely the function defined in the previous section by equation (5.4) and that its specification is equivalent to the specification of ρ.

The following lemma, known as Glauber's formula, will be useful in the calculations.

If A and B are two hermitean operators that commute with their commutator, then

$$e^{\lambda(A+B)} = e^{\lambda A}e^{\lambda B}e^{-\frac{1}{2}\lambda^2[A,B]}. \tag{6.2}$$

To prove the lemma call the left and right members $F(\lambda)$ and $G(\lambda)$ respectively. Since $F(0) = G(0) = I$, we only need to prove that both functions obey the same differential equation in λ. Clearly $F(\lambda)$ obeys the differential equation $F' = (A+B)F$. Differentiating $G(\lambda)$ we obtain

$$G' = \left(A + e^{\lambda A}Be^{-\lambda A} - \lambda[A,B]\right) G.$$

The derivative of the quantity in parentheses with respect to λ is $e^{\lambda A}[A,B]e^{-\lambda A} - [A,B] = 0$. This quantity is therefore independent of λ and its value can be obtained by putting $\lambda = 0$. We thus obtain $G' = (A+B)G$ and the lemma is proved.

Using Glauber's formula we evaluate the trace in (6.1) in momentum space

$$\mathrm{Tr}\,(e^{i(\boldsymbol{\alpha}\cdot\mathbf{P}+\boldsymbol{\beta}\cdot\mathbf{Q})}\rho) = \int d^3\eta'\; e^{-\frac{1}{2}i\boldsymbol{\alpha}\cdot\boldsymbol{\beta}}\langle\boldsymbol{\eta}'|e^{i\boldsymbol{\alpha}\cdot\mathbf{P}}e^{i\boldsymbol{\beta}\cdot\mathbf{Q}}\rho|\boldsymbol{\eta}'\rangle.$$

Since by the Fourier transform formula (1.2)

$$\langle\boldsymbol{\eta}|e^{i\boldsymbol{\beta}\cdot\mathbf{Q}} = (2\pi)^{-3/2}\int d^3\xi\; e^{-i(\eta-\beta)\cdot\xi}\langle\boldsymbol{\xi}| = \langle\boldsymbol{\eta}-\boldsymbol{\beta}|,$$

we have

$$\mathrm{Tr}\,(e^{i(\boldsymbol{\alpha}\cdot\mathbf{P}+\boldsymbol{\beta}\cdot\mathbf{Q})}\rho) = \int d^3\eta'\; e^{i\boldsymbol{\alpha}\cdot(\eta'-\frac{1}{2}i\boldsymbol{\beta})}\langle\boldsymbol{\eta}'-\boldsymbol{\beta}|\rho|\boldsymbol{\eta}'\rangle.$$

After substituting this expression in (6.1) one can perform the $\boldsymbol{\alpha}$ integration which yields $(2\pi)^{3/2}\delta(\boldsymbol{\eta}'-\boldsymbol{\eta}-\frac{1}{2}\boldsymbol{\beta})$. The $\boldsymbol{\eta}'$ integration then gives

$$f(\boldsymbol{\eta},\boldsymbol{\xi}) = (2\pi)^{-3}\int d^3\beta\; e^{-i\boldsymbol{\beta}\cdot\boldsymbol{\xi}}\rho(\boldsymbol{\eta}-\tfrac{1}{2}\boldsymbol{\beta},\ \boldsymbol{\eta}+\tfrac{1}{2}\boldsymbol{\beta}), \tag{6.3}$$

a formula identical to (5.4).

An identical calculation in position space yields the alternative expression

$$f(\boldsymbol{\eta}, \boldsymbol{\xi}) = (2\pi)^{-3} \int d^3\alpha \; e^{i\boldsymbol{\alpha}\cdot\boldsymbol{\eta}} \rho(\boldsymbol{\xi} - \tfrac{1}{2}\boldsymbol{\alpha}, \; \boldsymbol{\xi} + \tfrac{1}{2}\boldsymbol{\alpha}). \tag{6.4}$$

with $\rho(\boldsymbol{\xi}, \boldsymbol{\xi}') = \langle \boldsymbol{\xi} | \rho | \boldsymbol{\xi}' \rangle$. This formula tells us that $f(\boldsymbol{\eta}, \boldsymbol{\xi})$ is the Fourier transform of $\rho(\boldsymbol{\xi} - \frac{1}{2}\boldsymbol{\alpha}, \boldsymbol{\xi} + \frac{1}{2}\boldsymbol{\alpha})$ *qua* function of $\boldsymbol{\alpha}$ and is therefore a specification of the state equivalent to ρ, which can be obtained by inverting (6.4).

$$\rho(\boldsymbol{\xi}, \boldsymbol{\xi}') = \int d^3\eta \; e^{i(\boldsymbol{\xi} - \boldsymbol{\xi}')\cdot\boldsymbol{\eta}} f(\boldsymbol{\eta}, \tfrac{1}{2}(\boldsymbol{\xi} + \boldsymbol{\xi}')).$$

A similar formula for $\tilde{\rho}(\boldsymbol{\eta}, \boldsymbol{\eta}')$ is obtained by inversion of (6.3).

In the previous section we saw that the probability densities for the position and the momentum are obtained from $f(\boldsymbol{\eta}, \boldsymbol{\xi})$ by formally treating this function as if it were a joint probability density for $\boldsymbol{\eta}$ and $\boldsymbol{\xi}$. We will take farther the analogy between $f(\boldsymbol{\eta}, \boldsymbol{\xi})$ and a joint probability density: given an arbitrary classical dynamical variable $g(\boldsymbol{\eta}, \boldsymbol{\xi})$ for a particle with momentum $\boldsymbol{\eta}$ and position $\boldsymbol{\xi}$, we will construct a quantum mechanical observable G whose average can be obtained from the Wigner distribution function $f(\boldsymbol{\eta}, \boldsymbol{\xi})$ by the formula

$$\langle G \rangle = \iint d^3\eta \, d^3\xi \; g(\boldsymbol{\eta}, \boldsymbol{\xi}) f(\boldsymbol{\eta}, \boldsymbol{\xi}). \tag{6.5}$$

Such an observable G can always be written as a function $G(\mathbf{P}, \mathbf{Q})$ which must satisfy

$$\langle G \rangle = \mathrm{Tr}\, [G(\mathbf{P}, \mathbf{Q})\rho]. \tag{6.6}$$

Substituting the definition (6.1) into (6.5) we obtain

$$\langle G \rangle = \mathrm{Tr} \left(\iint d^3\alpha \, d^3\beta \; \bar{g}(\boldsymbol{\alpha}, \boldsymbol{\beta}) e^{i(\boldsymbol{\alpha}\cdot\mathbf{P} + \boldsymbol{\beta}\cdot\mathbf{Q})} \rho \right), \tag{6.7}$$

where $\bar{g}$ is the Fourier transform of g,

$$\bar{g}(\boldsymbol{\alpha}, \boldsymbol{\beta}) = (2\pi)^{-6} \iint d^3\eta \, d^3\xi \; g(\boldsymbol{\eta}, \boldsymbol{\xi}) e^{-i(\boldsymbol{\alpha}\cdot\boldsymbol{\eta} + \boldsymbol{\beta}\cdot\boldsymbol{\xi})}.$$

Equating the right members of equations (6.6) and (6.7) for arbitrary ρ shows that the desired observable is the hermitean operator

$$G = \iint d^3\alpha \, d^3\beta \; \bar{g}(\boldsymbol{\alpha}, \boldsymbol{\beta}) e^{i(\boldsymbol{\alpha}\cdot\mathbf{P} + \boldsymbol{\beta}\cdot\mathbf{Q})}. \tag{6.8}$$

In spite of its appearance, *the Wigner distribution function is not a probability density* because although it satisfies the conditions

$$\iint d^3\eta\, d^3\xi\; f(\boldsymbol{\eta},\boldsymbol{\xi}) = 1, \qquad f^*(\boldsymbol{\eta},\boldsymbol{\xi}) = f(\boldsymbol{\eta},\boldsymbol{\xi})$$

it does not satisfy, in general, the condition $f(\boldsymbol{\eta},\boldsymbol{\xi}) \geq 0$. In fact Wigner proved that if the latter condition were satisfied for any state ρ, the uncertainty relation between **P** and **Q** would be violated.

The expression (6.8) establishes a correspondence, known as the *Weyl correspondence*, between classical dynamical variables and quantum mechanical observables. The correspondence has the obviously desirable properties that to the dynamical variables $f(\boldsymbol{\eta})$, $g(\boldsymbol{\xi})$ there correspond respectively the observables $f(\mathbf{P})$, $g(\mathbf{Q})$, and that to $f(\boldsymbol{\eta}) + g(\boldsymbol{\xi})$ there corresponds $f(\mathbf{P}) + g(\mathbf{Q})$. One can also verify that to $\boldsymbol{\eta}\boldsymbol{\xi}$ there corresponds $\frac{1}{2}[\mathbf{PQ} + \mathbf{QP}]$.

The Weyl correspondence, however, is not the only one that has the properties mentioned. There are infinitely many such correspondences. They differ in the observable that they assign to dynamical variables of the form $\eta_i^n \xi_j^m$. There is no empirical evidence that allows us to choose among them.

Similarly the Wigner distribution function is not the only possible "phase space representation" of a state. To each correspondence there belongs a phase space representation. What characterizes the Wigner representation is the fact, derived in the previous section, that *for a free particle* its time evolution is identical to the time evolution of a classical ensemble density. It is this fact together with the expression (6.5) for the average of an observable that led us to state at the end of the previous section that in the case of a free particle the averages of all observables move classically.

7. The harmonic oscillator

In three dimensions the harmonic oscillator is defined as a particle in an attractive and time independent external scalar potential quadratic in the position **Q**. According to the expression (1.7) the general form for the hamiltonian of a harmonic oscillator is

$$H = \frac{|\mathbf{P}|^2}{2m} + \mathbf{Q}\cdot\mathbf{WQ}, \tag{7.1}$$

where **W** is a positive definite tensor whose components are complex numbers $w_{ij} = w_{ji}$, $i, j = 1, 2, 3$. We work in the coordinate system where **W** is diagonal with components $w_{ii} = \frac{1}{2}m\omega_i^2$. In this coordinate system the hamiltonian becomes

$$H = \frac{1}{2m}\sum_i (P_i^2 + m^2\omega_i^2 Q_i^2). \tag{7.2}$$

In view of the form of this expression it is useful to regard the Hilbert space $\mathbb{H}$ of the particle in three dimensions as the tensor product $\mathbb{H}_1 \otimes \mathbb{H}_2 \otimes \mathbb{H}_3$ of three Hilbert spaces with $\mathbb{H}_i$ belonging to a particle in one dimension with position $Q_i^{(i)}$ and momentum $P_i^{(i)}$ (we denote the operators on $\mathbb{H}_i$ with the superscript $^{(i)}$). From this point of view we have $Q_2 = I^{(1)} \otimes Q_2^{(2)} \otimes I^{(3)}$, where $I^{(i)}$ is the identity on $\mathbb{H}_i$, and similar expressions for the other components of $\mathbf{Q}$ and $\mathbf{P}$. If we define the hamiltonian $H^{(i)}$ on $\mathbb{H}_i$ by

$$H^{(i)} = \frac{1}{2m}(P_i^{(i)2} + m^2\omega_i^2 Q_i^{(i)2}),$$

then the hamiltonian (7.2) can be written

$$H = H^{(1)} \otimes I^{(2)} \otimes I^{(3)} + I^{(1)} \otimes H^{(2)} \otimes I^{(3)} + I^{(1)} \otimes I^{(2)} \otimes H^{(3)}.$$

Hereafter, unless necessary to avoid confusion, we shall omit the cumbersome identities. Thus the previous equation will be written more succinctly

$$H = H^{(1)} + H^{(2)} + H^{(3)}.$$

From Section VI.4 we can construct an orthonormal basis of eigenvectors $|n\rangle^{(i)} \in \mathbb{H}_i$ such that

$$H^{(i)}|n\rangle^{(i)} = \omega_i(n + \tfrac{1}{2})|n\rangle^{(i)}, \qquad n = 0, 1, 2, \ldots$$

The vectors of the product basis in $\mathbb{H}$

$$|n_1, n_2, n_3\rangle = |n_1\rangle^{(1)}|n_2\rangle^{(2)}|n_3\rangle^{(3)}$$

are eigenvectors of the hamiltonian H,

$$H|n_1, n_2, n_3\rangle = [\omega_1(n_1 + \tfrac{1}{2}) + \omega_2(n_2 + \tfrac{1}{2}) + \omega_3(n_3 + \tfrac{1}{2})]|n_1, n_2, n_3\rangle.$$

In case $\omega_1 = \omega_2 = \omega_3 = \omega$, we refer to the harmonic oscillator as isotropic. The hamiltonian (7.1) simplifies then to

$$H = \frac{1}{2m}(|\mathbf{P}|^2 + m^2\omega^2|\mathbf{Q}|^2) \tag{7.3}$$

and its eigenvalue equation becomes

$$H|n_1, n_2, n_3\rangle = \omega(n_1 + n_2 + n_3 + \tfrac{3}{2})|n_1, n_2, n_3\rangle.$$

Clearly in this case all the eigenvectors such that $n_1 + n_2 + n_3 = n$ belong to the same eigenvalue.

One can use the results of Section VI.4 to determine without much effort the time dependence of the Heisenberg operators $\mathbf{Q}(t)$ and $\mathbf{P}(t)$. For future reference we give the result in the case of the *isotropic* harmonic oscillator:

$$\mathbf{Q}(t) = \mathbf{Q}\cos\omega t + \frac{\mathbf{P}}{m\omega}\sin\omega t, \qquad \mathbf{P}(t) = -m\omega\mathbf{Q}\sin\omega t + \mathbf{P}\cos\omega t. \tag{7.4}$$

Because the hamiltonian (7.3) commutes with the angular momentum $\mathbf{L}$, the determination of its spectrum can also be performed in a simple manner by using the decomposition of $\mathbb{H}$ into a tensor product of the spaces $\mathbb{H}_r$ and $\mathbb{H}_\Omega$ of Section 4. We shall do this in the last section of this chapter but, before, we examine the decomposition of $\mathbb{H}$ into the tensor product of Hilbert spaces appropriate to deal with a problem in a general orthogonal coordinate system.

8. The spinless particle in a general coordinate system

In the previous section we have dealt with the harmonic oscillator by regarding the Hilbert space $\mathbb{H}$ of the particle as the tensor product of three Hilbert spaces in which a maximal observable consisted of one cartesian component of the position operator $\mathbf{Q}$. Depending on the hamiltonian for a particular experimental setup, it may be convenient to use a curvilinear coordinate system rather than a cartesian one. In this section we discuss how to decompose $\mathbb{H}$ into a tensor product of spaces which is suited to a general curvilinear orthogonal coordinate system with coordinates ζ_n, $n = 1, 2, 3$, and line element given by

$$ds^2 = \sum_n h_n^2(\zeta)d\zeta_n^2,$$

where we have denoted by ζ the triplet $\{\zeta_1, \zeta_2, \zeta_3\}$, as we shall often do.

In this coordinate system the instrument that measures the position observable

$$\mathbf{Q} = \int d^3\xi \; \boldsymbol{\xi}|\boldsymbol{\xi}\rangle\langle\boldsymbol{\xi}|$$

consists of a set of particle counters each of which is labelled by its position given in terms of the three coordinates $\zeta_1, \zeta_2, \zeta_3$. The Hilbert space $\mathbb{H}$ is characterized by the fact that the observable measured by the set of counters is maximal. We construct next a decomposition of $\mathbb{H}$ into a tensor product of three spaces by constructing on this tensor product a maximal observable measured by the set of counters. Needless to say that such an observable is equivalent to $\mathbf{Q}$.

Suppose that the labels corresponding to two of the coordinates are erased from each counter leaving only the label corresponding to the coordinate ζ_n. We take the resulting instrument to measure the maximal observable

$$Z_n^{(n)} = \int d\zeta_n \; \zeta_n |\zeta_n\rangle\langle\zeta_n|$$

on a Hilbert space $\mathbb{H}_n$ (unless otherwise specified the integrals with respect to a coordinate ζ_n are to be understood to extend over the range of the coordinate).

With the three Hilbert spaces thus defined we construct the tensor product $\mathbb{H}_1 \otimes \mathbb{H}_2 \otimes \mathbb{H}_3$ and on it the maximal observable $Z = \{Z_1, Z_2, Z_3\}$ formed by the set of operators Z_1, Z_2, and Z_3, defined by

$$Z_n = \iiint d\zeta_1 d\zeta_2 d\zeta_3 \; \zeta_n |\zeta\rangle\langle\zeta|,$$

where the ket $|\zeta\rangle = |\zeta_1, \zeta_2, \zeta_3\rangle$ is the tensor product $|\zeta_1\rangle|\zeta_2\rangle|\zeta_3\rangle$ (equivalently $Z_1 = Z_1^{(1)} \otimes I^{(2)} \otimes I^{(3)}$, etc.). It is clear that the set of counters measures Z, and that therefore $\mathbb{H} = \mathbb{H}_1 \otimes \mathbb{H}_2 \otimes \mathbb{H}_3$. Obviously Z and $\mathbf{Q}$ commute, for one measures $\mathbf{Q}$ by measuring Z. In the particular case in which the coordinate system is rectangular, the observables Z_1, Z_2, Z_3 are just the components Q_i along the three axes and the observable Z is identical to $\mathbf{Q}$. We then recover the decomposition of $\mathbb{H}$ of Section 7. In general, however, Z is different from $\mathbf{Q}$ and is not a vector observable.

It follows from the last remark that the representation of $\mathbb{H}$ as the function space induced by Z – the Z-representation – is in general different from, although equivalent to, its representation in position space – the $\mathbf{Q}$-representation. As before we denote the position representation of a vector $|f\rangle \in \mathbb{H}$ by $f(\boldsymbol{\xi}) = \langle\boldsymbol{\xi}|f\rangle$, while its Z-representation will be denoted by $f_Z(\zeta) = \langle\zeta|f\rangle$. The following remark is intended to make absolutely clear the meaning of the statement that the two representations are different.

Let $\xi_1(\zeta)$, $\xi_2(\zeta)$, $\xi_3(\zeta)$, be the expressions of the components of the position vector $\boldsymbol{\xi}$ along the unit vectors $\mathbf{e}_1(\zeta)$, $\mathbf{e}_2(\zeta)$, $\mathbf{e}_3(\zeta)$, of the curvilinear coordinate system. Substituting those expressions into $f(\boldsymbol{\xi}) = f(\xi_1, \xi_2, \xi_3)$ we obtain the function of ζ

$$f(\zeta) = f(\xi_1(\zeta), \xi_2(\zeta), \xi_3(\zeta)). \tag{8.1}$$

The precise meaning of the statement that the $\mathbf{Q}$- and Z- representations are different is that the function $f(\cdot)$ of (8.1) is different from $f_Z(\cdot)$.

Suppose the particle is at a certain time in a state $|f\rangle$, the probability $\mathcal{P}(\Delta|f)$ that at that time a counter in the region Δ of space be activated can be written in the Z-representation as

$$\mathcal{P}(\Delta|f) = \iiint\limits_{\Delta} d\zeta_1 d\zeta_2 d\zeta_3 \; |f_Z(\zeta)|^2 \tag{8.2}$$

or in the **Q**-representation as

$$\mathcal{P}(\Delta|f) = \int\limits_{\Delta} d^3\xi \; |f(\boldsymbol{\xi})|.^2$$

Using the expression for the volume element in terms of the coordinates ζ and expressing the functions of $\boldsymbol{\xi}$ as functions of ζ in accordance with (8.1), the last expression becomes

$$\mathcal{P}(\Delta|f) = \iiint\limits_{\Delta} d\zeta_1 d\zeta_2 d\zeta_3 \; h_1 h_2 h_3 |f(\zeta)|^2. \tag{8.3}$$

Comparing (8.2) with (8.3) we find the relation between the two representations (up to a phase factor)

$$f_Z(\zeta) = \sqrt{h_1(\zeta)h_2(\zeta)h_3(\zeta)}\; f(\zeta). \tag{8.4}$$

Since $f_Z(\zeta) = \langle\zeta|f\rangle$ and $f(\boldsymbol{\xi}) = \langle\boldsymbol{\xi}|f\rangle$, this relation can formally be transferred to the Dirac kets

$$|\zeta\rangle = \sqrt{h_1 h_2 h_3}\; |\boldsymbol{\xi}\rangle. \tag{8.5}$$

The scalar product $\langle f|g\rangle$ in the Z-representation is

$$\langle f|g\rangle = \iiint d\zeta_1 d\zeta_2 d\zeta_3 \; f_Z(\zeta) g_Z(\zeta), \tag{8.6}$$

while in the **Q**-representation it becomes in terms of the coordinates ζ_n

$$\langle f|g\rangle = \iiint d\zeta_1 d\zeta_2 d\zeta_3 \; h_1 h_2 h_3 f^*(\zeta) g(\zeta). \tag{8.7}$$

The maximal observable $\Pi = \{\Pi_1, \Pi_2, \Pi_3\}$ given in the Z-representation by

$$\langle\zeta|\Pi_n|f\rangle = -i\,\frac{\partial f_Z(\zeta)}{\partial\zeta_n}, \tag{8.8}$$

obeys a commutation relation with ζ analogous to (1.1),

$$[Z_n, \Pi_m] = i\delta_{nm}.$$

For this reason Π_n is called the momentum conjugate to the coordinate Z_n. Clearly the maximal observables Z and Π form an irreducible set of observables.

To complete the study of the connection between the Z- and $\mathbf{Q}$- representations we determine the relation between Π and $\mathbf{P}$. Denoting by ∇_n the component of the gradient along the unit vector $\mathbf{e}_n$, we have in the $\mathbf{Q}$-representation

$$\langle \boldsymbol{\xi}|P_n|f\rangle = -i\nabla_n f(\boldsymbol{\xi}) = -i\,\frac{1}{h_n}\frac{\partial f(\zeta)}{\partial \zeta_n}$$

or, by virtue of (8.4)

$$\langle \zeta|P_n|f\rangle = -i\sqrt{h_1h_2h_3}\,\frac{1}{h_n}\frac{\partial}{\partial \zeta_n}\frac{f_Z(\zeta)}{\sqrt{h_1h_2h_3}}.$$

Simplifying and using the definition (8.8) we obtain

$$\langle \zeta|P_n|f\rangle = \frac{1}{h_n}\Big(i f_Z(\zeta)\frac{\partial \ln\sqrt{h_1h_2h_3}}{\partial \zeta_n} + \langle \zeta|\Pi_n|f\rangle\Big).$$

If we set

$$G_n(\zeta) = \frac{\partial \ln\sqrt{h_1h_2h_3}}{\partial \zeta_n} \tag{8.9}$$

we have in abstract operator notation

$$P_n = \frac{1}{h_n(Z)}\,[\Pi_n + iG_n(Z)]. \tag{8.10}$$

Note that the components P_n of $\mathbf{P}$ are not hermitean unless Π_n commutes with $h_n(Z)$ or equivalently unless $h_n(Z)$ is independent of Z_n. The coordinates ζ_n can be regarded as functions of $\boldsymbol{\xi}$; if we put $G_n(\boldsymbol{\xi}) = G_n(\zeta(\boldsymbol{\xi}))$ and take into account that Z and $\mathbf{Q}$ commute, the above equation can also be written

$$\Pi_n = h_n(\mathbf{Q})P_n - iG_n(\mathbf{Q}).$$

As a simple example we look at the case of polar coordinates $\zeta_1 = r$, $\zeta_2 = \theta$, $\zeta_3 = \varphi$, already touched upon in Section 4. The line element is

$$ds^2 = dr^2 + r^2 d\theta^2 + r^2\sin^2\theta\, d\varphi^2$$

so that $h_1 = 1$, $h_2 = r$, $h_3 = r\sin\theta$. In obvious compressed notation the operator $Z = \{Z_r, Z_\theta, Z_\varphi\}$ is

$$Z = \int_0^\infty dr \int_0^\pi d\theta \int_0^{2\pi} d\varphi\ \{r,\theta,\varphi\}|r,\theta,\varphi\rangle\langle r,\theta,\varphi|$$

and the relation (8.4) between the Z- and $\mathbf{Q}$- representations gives

$$f_Z(r,\theta,\varphi) = r\sqrt{\sin\theta}\, f(r,\theta,\varphi). \tag{8.11}$$

By using the relation (8.5) the operator Z_r can be written

$$Z_r = \int_0^\infty dr\, r^2 \int_0^\pi d\theta\, \sin\theta \int_0^{2\pi} d\varphi\, r|\boldsymbol{\xi}\rangle\langle\boldsymbol{\xi}|$$

which, since $|\boldsymbol{\xi}| = r$, shows that $Z_r = |\mathbf{Q}|$.

The formulas (8.6) and (8.7) for the scalar product become

$$\langle f|g\rangle = \int_0^\infty dr \int_0^\pi d\theta \int_0^{2\pi} d\varphi\, f_Z(r,\theta,\varphi) g_Z(r,\theta,\varphi)$$

and

$$\langle f|g\rangle = \int_0^\infty dr\, r^2 \int_0^\pi d\theta\, \sin\theta \int_0^{2\pi} d\varphi f^*(r,\theta,\varphi) g(r,\theta,\varphi),$$

respectively, the latter being identical with (4.1).

In accordance with equation (8.9), we get $G_1 = 1/r$, $G_2 = \frac{1}{2}\cot\theta$, $G_3 = 0$ so that by (8.10) the momenta conjugate to the operators Z_r, Z_θ, Z_φ, are related to the components of the momentum $\mathbf{P}$ along the unit vectors $\mathbf{e}_r$, $\mathbf{e}_\theta$, $\mathbf{e}_\varphi$ by

$$\Pi_r = P_r - \frac{i}{Z_r}, \qquad \Pi_\theta = Z_r P_\theta - \tfrac{1}{2} i \cot Z_\theta, \qquad \Pi_\varphi = Z_r \sin Z_\theta P. \tag{8.12}$$

To abbreviate let us put for any operator A: $(A^{(Q)}f)(\boldsymbol{\xi}) = \langle\boldsymbol{\xi}|A|f\rangle$, $(A^{(Z)}f)(\zeta) = \langle\zeta|A|f\rangle$. By definition we have then

$$\Pi_r^{(Z)} = -i\frac{\partial}{\partial r}, \qquad \Pi_\theta^{(Z)} = -i\frac{\partial}{\partial\theta}, \qquad \Pi_\varphi^{(Z)} = -i\frac{\partial}{\partial\varphi},$$

while from the expression for the gradient in polar coordinates,

$$P_r^{(Q)} = -i\frac{\partial}{\partial r}, \qquad P_\theta^{(Q)} = -\frac{i}{r}\frac{\partial}{\partial\theta}, \qquad P_\varphi^{(Q)} = -\frac{i}{r\sin\theta}\frac{\partial}{\partial\varphi}.$$

From (8.12) we can find the corresponding expressions for Π_n in the function space induced by $\mathbf{Q}$:

$$\Pi_r^{(Q)} = -i\left(\frac{\partial}{\partial r} + \frac{1}{r}\right), \qquad \Pi_\theta^{(Q)} = -i\left(\frac{\partial}{\partial\theta} + \tfrac{1}{2}\cot\theta\right), \qquad \Pi_\varphi^{(Q)} = -i\frac{\partial}{\partial\varphi},$$

and of the components of $\mathbf{P}$ in the function space induced by Z

$$P_r^{(Z)} = -i\left(\frac{\partial}{\partial r} - \frac{1}{r}\right), \qquad P_\theta^{(Z)} = -\frac{i}{r}\left(\frac{\partial}{\partial\theta} - \tfrac{1}{2}\cot\theta\right), \qquad P_\varphi^{(Z)} = -\frac{i}{r\sin\theta}\frac{\partial}{\partial\varphi}.$$

On the Hilbert space $\mathbb{H} = \mathbb{H}_r \otimes \mathbb{H}_\theta \otimes \mathbb{H}_\varphi$ consider a product vector $|f_1\rangle|f_2\rangle|f_3\rangle$; in the Z-function space we denote it as the function $f_{1Z}(r)f_{2Z}(\theta)f_{3Z}(\varphi)$. By (8.11) we can write the $\mathbf{Q}$-representation of such a vector as $f_1(r)f_2(\theta)f_3(\varphi)$ where

$$f_1(r) = \frac{1}{r}f_{1Z}(r), \qquad f_2(\theta) = \frac{1}{\sqrt{\sin\theta}}f_{2Z}(\theta), \qquad f_3(\varphi) = f_{3Z}(\varphi), \tag{8.13}$$

As formula (8.4) shows, such a product decomposition is possible in a particular coordinate system if, as in the case at hand, the quantity $h_1h_2h_3$ is a product of functions of a single coordinate.

The first expression (8.13) shows that in the space $\mathbb{H}_r$, already introduced in Section 4, the scalar product can be written

$$\langle f_1|g_1\rangle = \int_0^\infty drr^2 f_1(r)g_1(r)$$

in agreement with the first equation (4.2).

The other space introduced in Section 4 was the space $\mathbb{H}_\Omega = \mathbb{H}_\theta \otimes \mathbb{H}_\varphi$. The notation $|\Omega\rangle$ introduced there is explained next. We write $|\theta,\varphi\rangle = |\theta\rangle|\varphi\rangle$ for the eigenket of the maximal observable $Z_\theta \otimes Z_\varphi$, on $\mathbb{H}_\theta \otimes \mathbb{H}_\varphi$ and $f_{\Omega Z}(\theta,\varphi) = \langle\theta,\varphi|f_\Omega\rangle$, $|f_\Omega\rangle \in \mathbb{H}_\Omega$. Guided by the last two equations (8.13) we define $|\Omega\rangle = (\sin\theta)^{-1/2}|\theta,\varphi\rangle$ and, correspondingly, the function space of the functions

$$\langle\Omega|f_\Omega\rangle = f_\Omega(\theta,\varphi) = \frac{1}{\sqrt{\sin\theta}}f_{\Omega Z}(\theta,\varphi).$$

In this function space we have

$$\langle f_\Omega|g_\Omega\rangle = \int_0^\pi d\theta \int_0^{2\pi} d\varphi \sin\theta f_\Omega^*(\theta,\varphi)g_\Omega(\theta,\varphi)$$

in agreement with the second expression (4.4)

9. The particle in a central field

The example at the end of the last section is of great importance for it permits the simplification of the problem of determining the spectral resolution of the hamiltonian of a particle in a central field,

$$H = \frac{P^2}{2m} + V(Q), \tag{9.1}$$

where, since there is no danger of confusion with a one-dimensional problem we write $P = |\mathbf{P}|, Q = |\mathbf{Q}|$. The simplification results because the hamiltonian can be written as a sum of operators with each term involving the tensor product of an operator on $\mathbb{H}_r$ and an operator on $\mathbb{H}_\Omega$. To see this we find an expression for L^2, the square of the magnitude of the orbital angular momentum. Applying the formula $(\mathbf{a}\times\mathbf{b})\cdot(\mathbf{c}\times\mathbf{d}) = (\mathbf{a}\cdot\mathbf{c})(\mathbf{b}\cdot\mathbf{d}) - (\mathbf{a}\cdot\mathbf{d})(\mathbf{b}\cdot\mathbf{c})$ to the vector operators $\mathbf{Q}$ and $\mathbf{P}$ and exercising care to keep the correct order we get

$$L^2 = |\mathbf{Q}\times\mathbf{P}|^2 = \sum_i\sum_j (Q_iP_jQ_iP_j - Q_iP_jQ_jP_i),$$

where the subscripts $i, j = 1, 2, 3$, denote components along rectangular axes (the components along the unit vectors of the polar coordinate system to be introduced shortly will be denoted by subscripts r, θ, φ). Repeated use of the commutation relation $[Q_i, P_j] = i\delta_{ij}$ yields

$$L^2 = \sum_i\sum_j (Q_iQ_iP_jP_j - Q_iP_iQ_jP_j) + i\sum_i Q_iP_i$$

or in vector form

$$L^2 = Q^2P^2 - (\mathbf{Q}\cdot\mathbf{P})(\mathbf{Q}\cdot\mathbf{P}) + i\mathbf{Q}\cdot\mathbf{P}.$$

If we introduce polar coordinates r, θ , φ, the preceding equation becomes

$$L^2 = Q^2P^2 - QP_rQP_r + iQP_r$$

Introducing the momentum Π_r conjugate to $Z_r = Q$ defined in the first expression (8.12) we obtain

$$L^2 = Q^2P^2 - (Q\Pi_r + i)(Q\Pi_r + i) + i(Q\Pi_r + i)$$

which on account of the commutation relation $[Q, \Pi_r] = i$ simplifies to

$$L^2 = Q^2(P^2 - \Pi_r^2).$$

Multiplying this equation on the left by Q^{-2} and substituting in the hamiltonian (9.1) we obtain

$$H = \frac{\Pi_r^2}{2m} + \frac{Q^{-2}L^2}{2m} + V(Q)$$

which is of the desired form.

As we know from Section 4, the observables L^2 and L_3 constitute a maximal set on $\mathbb{H}_\Omega$ with a basis of eigenvectors $|l, m\rangle$. A basis of eigenvectors of H will be of the form $|\psi_n, l, m\rangle = |\psi_{nl}\rangle|l, m\rangle$. Upon substitution in the eigenvalue equation for H,

$$\left[\frac{\Pi_r^2}{2m} + \frac{Q^{-2}L^2}{2m} + V(Q)\right] |\psi_n, l, m\rangle = E_{nl}|\psi_n, l, m\rangle$$

we see that the eigenvalues E_{nl} are independent of m so that E_{nl} will be at least $(2l+1)$-degenerate. The vectors $|\psi_{nl}\rangle \in \mathbb{H}_r$ obey the eigenvalue equation

$$\left[\frac{\Pi_r^2}{2m} + \frac{l(l+1)}{2m} Q^{-2} + V(Q)\right] |\psi_n\rangle = E_{nl}|\psi_{nl}\rangle. \tag{9.2}$$

To use the customary notation we write the position space representation of a vector $|\psi\rangle \in \mathbb{H}_r$ as $\psi(r)$ and its Z-function space representation as $\chi(r) = r\psi(r)$. In the latter representation (9.2) reads

$$\left[-\frac{1}{2m}\frac{d^2}{dr^2} + \frac{l(l+1)}{2mr^2} + V(r)\right] \chi_{nl}(r) = E_{nl}\chi_{nl}(r). \tag{9.3}$$

In solving this equation we must ensure that $|\psi\rangle$ is in the domain of the hermitean operator on the left and in particular of $\Pi_r^{(Z)} = -d^2/dr^2$, which requires that $\chi(r)$ tend to zero with r.

The most important example of a central field is, of course, the one for which $V(Q) = -e^2Q^{-1}$. Since the solution of (9.2) for this problem is discussed in detail in every textbook we shall refrain from repeating it here.

VIII DYNAMICAL SYMMETRIES AND CONSERVATION LAWS

In previous chapters we have studied and made extensive use of the kinematical symmetries of quantum mechanics. In this chapter we introduce the concept of dynamical symmetry and study some of its applications. In particular we study space inversion and reversal of motion. We end the the chapter by introducing the full Galilei group and the restrictions that galilean relativity imposes on the hamiltonian of an *isolated* system.

1. The dynamical symmetry transformations of quantum mechanics

As we know, kinematical symmetries link equivalent descriptions of quantum systems, they are the quantum equivalent of the classical transformations of the phase space onto itself. As briefly discussed in Appendix D, the dynamical transformations of classical mechanics include not only phase space transformations, but also time transformations. In a similar vein we consider here *dynamical transformations* $\mathbf{T} = \{\mathbf{V}, \tau\}$ that consist of a kinematical symmetry, which may depend on time, and an invertible change $t \mapsto \bar{t}$ of the time variable given by $\bar{t} = \tau(t)$ with derivative $\tau'(t)$. The results in this section are slight generalizations of the results in Section V.4 to which the reader is invited to refer.

Let $X = \{X_r\}$, $r = 1, \ldots, n$, be the Schrödinger picture of the fundamental set of observables of the system under consideration, V_t the unitary or anti-unitary operator that represents the kinematical symmetry $\mathbf{V}$, and $A_t = A(X,t)$ a Schrödinger observable. The transform $\bar{A}_{\bar{t}}$ of A_t under $\mathbf{T}$ is defined by

$$\bar{A}_{\bar{t}} = V_{\tau(\bar{t})} A_{\tau(\bar{t})} V^{\dagger}_{\tau(\bar{t})}.$$

in the special case that $\bar{A}_t = A_t$ or, more explicitly,

$$V_{\tau(t)} A_{\tau(t)} V^{\dagger}_{\tau(t)} = A_t,$$

we say that A is *invariant* under the transformation.

In the following, whenever $\bar{t}$ and $\tau(\bar{t})$ appear in an equation, we shall write for brevity $t = \tau(\bar{t})$. Thus, the first equation will be written

$$\bar{A}_{\bar{t}} = V_t A_t V_t^\dagger.$$

In this and other equations, it will be evident that the above abbreviation has been used because both $\bar{t}$ and t appear in the same equation.

How does the dynamical equation

$$\frac{\partial U}{\partial t} = -iH_t U \tag{1.1}$$

transform under the dynamical transformation $\mathbf{T}$? An obvious generalization of the discussion leading to equation (V.4.3) shows that the transformed evolution operator is

$$\bar{U}'(\bar{t}, \bar{t}_0) = V_t U(t, t_0) V_{t_0}^\dagger, \tag{1.2}$$

with $t_0 = \tau(\bar{t}_0)$, and that it obeys the equation

$$\frac{\partial \bar{U}'}{\partial \bar{t}} = \mp i \bar{H}_{\bar{t}}' \bar{U}'. \tag{1.3}$$

where, with $\partial V_t = \partial V_t / \partial t$, the changed hamiltonian is

$$\bar{H}_{\bar{t}}' = \tau'(\bar{t})(\bar{H}_t \pm i\partial V_t V_t^\dagger), \qquad \bar{H}_t = V_t H_t V_t^\dagger \tag{1.4}$$

and the upper (lower) sign applies to unitary (antiunitary) V_t.

We are now ready to give a definition of dynamical symmetry which, *mutatis mutandis*, is the same as the classical definition of D.1.4.3.

A dynamical symmetry of a system is a transformation $\mathbf{T}$ *that preserves the dynamical mapping.*

More explicitly, $\mathbf{T}$ must be such that the mapping $\bar{\rho}(t_0) \mapsto \bar{\rho}(t)$, given by

$$\bar{\rho}(t) = \bar{U}'(t, t_0)\bar{\rho}(t_0)\bar{U}'^\dagger(t, t_0)$$

be identical to the mapping $\rho(t_0) \mapsto \rho(t)$ given by

$$\rho(t) = U(t, t_0)\rho(t_0)U^\dagger(t, t_0).$$

Comparing these two equations one concludes that $U(t, t_0)$ and $\bar{U}'(t, t_0)$ are equal up to a phase factor. By (1.1) and (1.3) this implies that

$$\pm \bar{H}_t' = H_t + \alpha(t) I \tag{1.5}$$

where $\alpha(t)$ is an arbitrary real function and $\bar{H}'_t$ is defined by (1.4). Particular cases of dynamical symmetries are unitary transformations for which $\bar{H}'_t = H_t$, they are said to *preserve* the hamiltonian. Time independent unitary transformations that leave the hamiltonian invariant constitute an important class of transformations that preserve the hamiltonian.

As in Section V.4, the transform under **T** of a Heisenberg operator can be found to be

$$\bar{A}_H(\bar{t}) = V_{t_o} A_H(t) V_{t_o}^\dagger.$$

The changed Heisenberg hamiltonian is defined by

$$\bar{H}'_H(t) = \bar{U}'^\dagger(t, t_0)\bar{H}'_t\bar{U}'(t, t_0) \tag{1.6}$$

Substituting in this equation the expressions (1.2) for $\bar{U}'$ and (1.4) for $\bar{H}'_t$ we obtain

$$\bar{H}'_H(\bar{t}) = \tau'(\bar{t})V_{t_o}[H_H(t) + i(V_t^\dagger \partial V_t)_H]V_{t_o}^\dagger.$$

This equation can also be written

$$\bar{H}'_H(\bar{t}) = \tau'(\bar{t})\left[\bar{H}_H(t) \pm i\overline{(V_t^\dagger \partial V_t)_H}\right]. \tag{1.7}$$

Note the formal difference between this hamiltonian and (1.4).

In case **T** is a dynamical symmetry, we can substitute the condition (1.5) for $\bar{H}'_t$ into (1.6), taking into account that, in this case $\bar{U}'$ and U differ at most by a phase factor, we find a relation similar to (1.5)

$$\pm\bar{H}'_H(t) = H_H(t) + \alpha(t)I. \tag{1.8}$$

In case the kinematical symmetry **V** is time independent, the expressions in the Heisenberg and in the Schrödinger pictures become formally identical. In the general case we shall work whenever possible in the Schrödinger picture where, with the definition of **V** that we have adopted, the transformation expressions are substantially simpler.

The most important transformations **T** considered in the applications are those for which V_t is unitary and $\tau' = 1$, and those for which V_t is antiunitary and $\tau' = -1$. In the following section we study one class of transformations of the former type.

2. Unitary transformations and constants of the motion

In this section we consider transformations **T** which are just unitary transformations, that is, they consist of a unitary kinematical symmetry **V**, and $\bar{t} = \tau(t) = t$, the formulae (1.2) and (1.7) for the evolution operator and the Schrödinger hamiltonian in the transformed description become

$$\bar{U}'(t,t_0) = V_t U(t,t_0) V_{t_o}^{\dagger},$$
$$\bar{H}_t' = \bar{H}_t + i\partial V_t V_t^{\dagger}. \tag{2.1}$$

Unitary transformations play, in quantum mechanics, a role analogous to the canonical transformations of classical mechanics (D.1.5). In particular they preserve the form of Heisenberg's equations of motion. That is, the system of equations for the fundamental observables $X_{Hr} = X_{Hr}(t)$ in the Heisenberg picture

$$\dot{X}_{Hr} = i[H_H, X_{Hr}]$$

is transformed by $\mathbf{T}$ into

$$\dot{\bar{X}}_{Hr} = i[\bar{H}_H', \bar{X}_{Hr}],$$

where the dot denotes the Heisenberg derivative. In case $\mathbf{V}$ is a dynamical symmetry the condition (1.8), which now reads

$$\bar{H}_H' = H_H + \alpha(t)I,$$

implies that both systems of equations are identical because the commutators are not affected by the multiple of the identity.

We are interested in a family of unitary transformations $\mathbf{V}_{\epsilon t}$, depending continuously on the parameter ϵ, that are dynamical symmetries. By Stone's theorem we can write the corresponding unitary operator on $\mathbb{H}$ as

$$V_{\epsilon t} = e^{-iG_t\epsilon} \tag{2.2}$$

with the generator G_t hermitean. The dynamical symmetry condition (1.5) yields by (2.1)

$$e^{-iG_t\epsilon} H_t e^{iG_t\epsilon} + \left(\partial e^{-iG_t\epsilon}\right) e^{iG_t\epsilon} = H_t + \alpha_\epsilon(t)I.$$

Evaluating the ϵ-derivative of this expression at $\epsilon = 0$ one gets

$$[H_t, G_t] + \partial G_t = \alpha_1(t)I \tag{2.3}$$

where $\alpha_1 = \partial\alpha_\epsilon/\partial\epsilon|_{\epsilon=0}$. Remembering the definition of the Heisenberg derivative $\dot{G}_H = [H_H, G_H] + (\partial G_t)_H$, one can write (2.3) in the Heisenberg picture as

$$\dot{G}_H = \alpha_1(t)I$$

In particular, if $\alpha_1(t) = 0$, that is, if the hamiltonian is preserved by the dynamical symmetry, $\bar{H}_t' = H_t$, the above equation shows that G_H is a constant of the motion.

Conversely, the time independence of G_H implies the invariance of the hamiltonian under $\mathbf{T}$. We can, therefore, assert:

The constants of the motion are the generators of the unitary transformations that preserve the hamiltonian.

In particular the constants of the motion represented by time-independent Schrödinger operators are the generators of unitary transformations under which the hamiltonian is invariant.

The generator of translations is the total momentum and the total angular momentum generates rotations. We already know from Chapter VII this to be true for a particle and we will see in future chapters that it is also the case for composite systems. We therefore can state the following two conservation laws.

If the hamiltonian H_t of a system is invariant under translations, its total momentum is conserved. If it is invariant under rotations, the total angular momentum is conserved.

If H_t is only invariant under translations along or rotations about a fixed axis then the respective components along that axis are conserved.

If a system is prepared in an eigenstate of a constant of the motion, its dynamical evolution takes place in the corresponding eigensubspace which often simplifies greatly the solution of the dynamical problem. For this reason the eigenvalues of constants of the motion are often referred to as *good quantum numbers.*

Under the transformation $\mathbf{V}_{\epsilon t}$, a Schrödinger state $\rho(t)$ becomes

$$\bar{\rho}(t) = e^{-iG_t\epsilon}\rho(t)^{iG_t\epsilon}.$$

If the system has a time independent hamiltonian H and we take $G_t = H$, the preceding equation tells us that $\bar{\rho}(t) = \rho(t+\epsilon)$, that is, that H generates the dynamical mapping. By analogy with classical mechanics we say, in a rather loose manner, that H generates the motion. We shall see that, in an infinitesimal sense, this is also true if the hamiltonian is time dependent.

To this end, and to make a connection with classical mechanics, we study next the infinitesimal transformations corresponding to (2.2). Following a notation similar to the one in Section D.2, we denote by $\bar{A}_{\epsilon t}$ the approximation to first order in ϵ to the transform

$$\bar{A}_t = e^{-iG_t\epsilon}A_t e^{iG_t\epsilon}$$

of A_t. With the definition $\delta_\epsilon A_t = \bar{A}_{\epsilon t} - A_t$, we obtain

$$\delta_\epsilon A_t = i\epsilon[A_t, G_t] \tag{2.4}$$

an equation that can be formally obtained from the classical one in D.2.2.2. by replacing the Poisson bracket with i times the commutator. In the case of the hamiltonian we write $\bar{H}'_{\epsilon t}$ for the approximation to $\bar{H}'_t$ and $\Delta_\epsilon H_t = \bar{H}'_{\epsilon t} - \bar{H}_t$. From the expression (2.1) for $\bar{H}'_t$ and the definition (2.4) we obtain

$$\Delta_\epsilon H_t = \delta_\epsilon H_t + \partial G_t$$

which is to be compared with the corresponding classical expression in D.2.2.3.

In equation (2.4) let us replace the generator G_t by H_t and the operator A_t by the state $\rho(t)$, we get

$$\delta_\epsilon \rho(t) = -i\epsilon[H_t, \rho(t)].$$

Differentiating this expression with respect to ϵ one obtains

$$\left.\frac{\partial \rho_\epsilon(t)}{\partial \epsilon}\right|_{\epsilon=0} = -i[H_t, \rho(t)].$$

Now, by von Neumann's equation (V.5.2) one also has

$$\left.\frac{\partial \rho_\epsilon(t+\epsilon)}{\partial \epsilon}\right|_{\epsilon=0} = \frac{\partial \rho(t)}{\partial t} = -i[H_t, \rho(t)]$$

which shows that H_t is the infinitesimal generator of the motion.

An aside into the transformation of Heisenberg operators. – Except for brief excursions, we have been working in the Schrödinger picture because our definition of the time dependent transformation $\mathbf{V}$ is well suited to it. If one wants to work in the Heisenberg picture, it is more convenient to define instead of $\mathbf{V}$ a different unitary transformation $\mathbf{V}'$ to which corresponds instead of the unitary operator V_t of equation (2.2) a Heisenberg unitary operator

$$V'_{H\epsilon}(t) = e^{iG_H(t)\epsilon}$$

and such that the transform under $\mathbf{V}'$ of the fundamental observable X_{Hr} is

$$X'_{Hr} = e^{iG_H\epsilon} X_{Hr} e^{-iG_H\epsilon}.$$

The corresponding infinitesimal transformation is

$$\delta'_\epsilon X_{Hr} = i\epsilon[G_H, X_{Hr}].$$

in complete formal agreement with (D.2.1).

If (in analogy with the classical definition D.1.4) we define the transform A'_H of $A_H = A_H(X_H, t)$ as [1]

$$A'_H(X'_H, t) = A_H\left(e^{-iG_H\epsilon} X'_H e^{iG_H\epsilon}, t\right),$$

and define the infinitesimal transformation $\delta'_\epsilon A_H$ as the first order in ϵ of $A'_H(X_H, t) - A_H(X_H, t)$ we obtain

$$\delta'_\epsilon A_H = i\epsilon[A_H, G_H],$$

all the results in Section 2 of Appendix D can then be transcribed into relations between Heisenberg operators if, as usual, one replaces the Poisson bracket by i times the commutator.

3. Space inversion, the parity observable

Consider again a particle without internal degrees of freedom with Hilbert space $\mathbb{H}_0$. As we know any linear operator A on $\mathbb{H}_0$ can be written as a function of the fundamental set of observables $\{\mathbf{P}, \mathbf{Q}\}$. We define *space inversion* as a transformation from the set of linear operators on $\mathbb{H}_0$ to itself such that

$$A(\mathbf{P}, \mathbf{Q}) \mapsto A(-\mathbf{P}, -\mathbf{Q}).$$

It is necessary and sufficient to require $\mathbf{P} \mapsto -\mathbf{P}, \mathbf{Q} \mapsto -\mathbf{Q}$.

The hamiltonian of a particle in the presence of a central external field

$$H = \frac{|\mathbf{P}|^2}{2m} + V(|\mathbf{Q}|). \tag{3.1}$$

is invariant under the space inversion, and the form of Heisenberg's equations

$$\frac{d\mathbf{Q}}{dt} = i[H, \mathbf{Q}], \qquad \frac{d\mathbf{P}}{dt} = i[H, \mathbf{P}]$$

is therefore preserved by it (since we are dealing here with a time independent transformation, the transformation properties of operators in the Schrödinger and in the Heisenberg pictures are identical and we are free to regard the observables as been given by operators in either picture). Whether the space inversion is a dynamical symmetry of such a particle or not, depends on whether it is also a

[1] This somewhat unnatural definition is necessary because in the corresponding classical definitions of D.2.1., p and q are treated as coordinates and not as dynamical variables. They correspond to the Schrödinger representation of the fundamental observables.

kinematical symmetry. We will show that it is by showing that it induces a unitary transformation Λ of $\mathbb{H}_0$ to itself such that

$$-\mathbf{P} = \Lambda \mathbf{P} \Lambda^\dagger, \qquad -\mathbf{Q} = \Lambda \mathbf{Q} \Lambda^\dagger \tag{3.2}$$

which will be explicitly constructed shortly. Suppose that such a Λ exists, any operator of the form $e^{i\alpha}\Lambda$ with α real will satisfy the same relations. If we multiply equations (3.2) by Λ on the left and by $\Lambda^\dagger$ on the right we find, after making use of (3.2) itself, that $\mathbf{P} = \Lambda^2 \mathbf{P} \Lambda^{2\dagger}$ and $\mathbf{Q} = \Lambda^2 \mathbf{Q} \Lambda^{2\dagger}$, that is, Λ^2 commutes with both $\mathbf{P}$ and $\mathbf{Q}$ and is therefore a multiple of the identity. By appropriately choosing the phase factor we can then choose an unitary operator $\Pi = e^{i\alpha}\Lambda$ that satisfies (3.2) and such that $\Pi^2 = I$. Multiplying this equation on the left by $\Pi^\dagger$ we find that $\Pi = \Pi^\dagger$, i.e., Π is hermitean.

Next we construct explicitly a unitary operator Λ that obeys (3.2). In Section (VII.7) we discussed the three-dimensional harmonic oscillator and obtained the time evolution of the momentum and position, equation (VII.7.4). For an oscillator with hamiltonian $\frac{1}{2}(|\mathbf{P}|^2 + |\mathbf{Q}|^2)$, after putting $m = \omega = 1$ and $t = \pi$ in that equation, we obtain immediately

$$\Lambda \mathbf{P} \Lambda^\dagger = -\mathbf{P}, \qquad \Lambda \mathbf{Q} \Lambda^\dagger = -\mathbf{Q} \tag{3.3}$$

where we have written

$$\Lambda = e^{\frac{1}{2} i\pi(|\mathbf{P}|^2 + |\mathbf{Q}|^2)}$$

In view of (3.3) we can see that this unitary operator satisfies the relations (3.2) and is therefore one operator of the type sought.

To summarize: in the Hilbert space of the spinless particle there exists a unitary, hermitean operator Π, the *parity operator*, such that

$$\Pi \mathbf{P} \Pi = -\mathbf{P}, \qquad \Pi \mathbf{Q} \Pi = -\mathbf{Q}.$$

Not only is space inversion a dynamical symmetry of a particle with hamiltonian (3.1) but it is performed by the unitary operator Π which, being also hermitean, is a conserved observable. Since $\Pi^2 = I$, the eigenvalues of Π are $+1$ and -1. Eigenstates of Π belonging to the eigenvalue $+1$ (-1) are said to be of *positive (negative) parity.*

To determine the action of Π on a vector of $\mathbb{H}_0$, we choose arbitrary vectors $|f\rangle$, $|g\rangle$ and an arbitrary operator function $F(\mathbf{Q})$. We, then have $\langle g|\Pi F(\mathbf{Q})\Pi|f\rangle = \langle g|F(-\mathbf{Q})|f\rangle$ or, in position space, with $(\Pi f)(\boldsymbol{\xi}) = \langle \boldsymbol{\xi}|\Pi|f\rangle$ and $f(\boldsymbol{\xi}) = \langle \boldsymbol{\xi}|f\rangle$,

$$\int d^3\xi\, (\Pi g)^*(\boldsymbol{\xi}) F(\boldsymbol{\xi}) (\Pi f)(\boldsymbol{\xi}) = \int d^3\xi\, g^*(\boldsymbol{\xi}) F(-\boldsymbol{\xi}) f(\boldsymbol{\xi}) = \int d^3\xi\, g^*(-\boldsymbol{\xi}) F(\boldsymbol{\xi}) f(-\boldsymbol{\xi});$$

the last equality follows by a change of the integration variable. Since the equality between the first and last expressions holds for arbitrary $F(\boldsymbol{\xi})$ we conclude that $(\Pi g)^*(\boldsymbol{\xi})(\Pi f)(\boldsymbol{\xi}) = g^*(-\boldsymbol{\xi})f(-\boldsymbol{\xi})$. That is, for any $f(\boldsymbol{\xi})$,

$$(\Pi f)(\boldsymbol{\xi}) = f(-\boldsymbol{\xi}).$$

An entirely analogous calculation yields in momentum space

$$(\widetilde{\Pi f})(\boldsymbol{\eta}) = \tilde{f}(-\boldsymbol{\eta}) \tag{3.4}$$

with $\tilde{f}(\boldsymbol{\eta}) = \langle \boldsymbol{\eta} | f \rangle$.

In view of future generalizations we discuss now some properties that apply not only to the parity operator just defined but to any linear operator R on a general Hilbert space $\mathbb{H}$, which is a square root of the identity in the sense that it is a solution of the equation $R^2 = I$.

It is apparent that the eigenvalues of R are ± 1. Therefore R is unitary. Any arbitrarily chosen subspace $\mathbb{E} \subset \mathbb{H}$ determines a solution of the equation $R^2 = I$ given by $R = E - (I - E) = 2E - I$, where E is the projection on $\mathbb{E}$; moreover, the general solution is of this form. It follows that R is also hermitean. The subspace $\mathbb{E}$ is the eigensubspace of R belonging to the eigenvalue 1, while the eigensubspace belonging to -1 is the orthogonal complement $\mathbb{E}^{\perp}$ of $\mathbb{E}$.

Consider now an operator A that commutes with a given R and let $|\phi_\lambda\rangle$ be a normalized eigenket of A belonging to the eigenvalue λ : $A|\phi_\lambda\rangle = \lambda|\phi_\lambda\rangle$. By multiplying this equation by R we get $AR|\phi_\lambda\rangle = \lambda R|\phi_\lambda\rangle$, so that $R|\phi_\lambda\rangle$ is also an eigenket of A belonging to λ. If $|\phi_\lambda\rangle$ and $R|\phi_\lambda\rangle$ are linearly dependent then $|\phi_\lambda\rangle$ is also an eigenket of R and therefore either $R|\phi_\lambda\rangle = |\phi_\lambda\rangle$ or $R|\phi_\lambda\rangle = -|\phi_\lambda\rangle$. If not, we can form two kets $|\psi_{\lambda+}\rangle = |\phi_\lambda\rangle + R|\phi_\lambda\rangle$ and $|\psi_{\lambda-}\rangle = |\phi_\lambda\rangle - R|\phi_\lambda\rangle$ which by the hermiticity of R and the property $R^2 = I$ can be verified to be orthogonal eigenkets of both A and R belonging to the eigenvalue λ of A and to the eigenvalues $+1$ and -1, respectively, of R.

Returning to the particle in three dimensions consider as a first example a free particle. Its hamiltonian $|\mathbf{P}|^2/2m$ has a doubly degenerate spectrum and commutes with the parity operator Π. The momentum eigenkets $|\boldsymbol{\eta}\rangle$, given in position space by

$$\psi_{\boldsymbol{\eta}}(\boldsymbol{\xi}) = (2\pi)^{3/2} e^{-i\boldsymbol{\eta}\cdot\boldsymbol{\xi}},$$

are eigenstates of the hamiltonian but not of Π – in fact $(\Pi\psi_{\boldsymbol{\eta}})(\boldsymbol{\xi}) = \psi_{-\boldsymbol{\eta}}(\boldsymbol{\xi}) = \psi^*_{\boldsymbol{\eta}}(\boldsymbol{\xi})$. A pair of orthonormal stationary states of definite parity belonging to the eigenvalue $|\boldsymbol{\eta}|^2/2m$ of the hamiltonian are $(|\boldsymbol{\eta}\rangle \pm \Pi|\boldsymbol{\eta}\rangle)/\sqrt{2}$ or in position space,

$$\frac{1}{4\pi^3}\cos(\boldsymbol{\eta}\cdot\boldsymbol{\xi}), \qquad -\frac{1}{4\pi^3}\sin(\boldsymbol{\eta}\cdot\boldsymbol{\xi}).$$

The first one has positive, the second one negative parity.

The eigenvectors $|l,m\rangle$ of the orbital angular momentum $\mathbf{L} = \mathbf{Q}\times\mathbf{P}$ are also parity eigenstates because under space inversion $\mathbf{Q}\times\mathbf{P} \mapsto (-\mathbf{Q})\times(-\mathbf{P}) = \mathbf{Q}\times\mathbf{P}$. To determine their parity note that in spherical coordinates the transformation $\boldsymbol{\xi} \mapsto -\boldsymbol{\xi}$ becomes $r \mapsto r$, $\varphi \mapsto \varphi + \pi$, $\theta \mapsto \pi - \theta$, and that by the expression of the spherical harmonics in Section VII.4, $Y_{lm}(\varphi,\theta) \mapsto (-1)^l Y_{lm}(\varphi,\theta)$, therefore

$$\Pi|l,m\rangle = (-1)^l|l,m\rangle.$$

The states with even (odd) l have positive (negative) parity.

The general hamiltonian (VII.1.7) that we obtained for the spinless particle from considerations of galilean invariance does not commute with the parity operator unless the potentials are such that $\mathbf{A}_t(-\mathbf{Q}) = -\mathbf{A}_t(\mathbf{Q})$ and $V_t(-\mathbf{Q}) = V_t(\mathbf{Q})$. These relations are obeyed by the electromagnetic potentials, as is apparent from Maxwell's equations. Thus galilean invariance implies that the hamiltonian of a charged particle in a magnetic field is invariant under space inversion.

4. Time reversal

A classical particle with a hamiltonian analogous to (3.1) possesses time reversal symmetry which can be traced to the fact that the hamiltonian is invariant under a transformation that changes the sign of the momentum which in this case is proportional to the velocity. As mentioned in D.2.3, classical time reversal is not a canonical transformation, which leads us to expect that the corresponding quantum mechanical transformation might not be unitary.

In analogy with the classical case we define in quantum mechanics the *time reversal transformation* of a particle without internal degrees of freedom as a transformation of the set of operators A on the Hilbert space $\mathbb{H}_0$ to itself and of the time t,

$$A(\mathbf{P},\mathbf{Q}) \mapsto A(-\mathbf{P},\mathbf{Q}), \qquad t \mapsto -t. \tag{4.1}$$

The first relation is equivalent to $\mathbf{P} \mapsto -\mathbf{P}, \mathbf{Q} \mapsto \mathbf{Q}$.

Next we shall construct a time independent kinematical symmetry transformation that satisfies the first of (4.1) by constructing an invertible *time reversal operator* K on $\mathbb{H}_0$ such that

$$K\mathbf{P}K^{-1} = -\mathbf{P}, \qquad K\mathbf{Q}K^{-1} = \mathbf{Q}. \tag{4.2}$$

By Wigner's theorem we know that, if K exists, it must be either unitary or antiunitary.

It is easy to see that the transformation cannot be unitary or even linear, for if K were linear we would have by (4.1) and (4.2), $K[\mathbf{Q},\mathbf{P}]K^{-1} = -[\mathbf{Q},\mathbf{P}]$, which implies $KiK^{-1} = -i$ in contradiction with the definition of a linear operator. The last relation is however consistent with the transformation being antiunitary.

We seek therefore an antiunitary operator K on $\mathbb{H}_0$ that satisfies (4.2). If K is antiunitary then K^2 is a unitary operator that by (4.2) commutes with both $\mathbf{P}$ and $\mathbf{Q}$ and must be a multiple of the identity: $K^2 = cI$, or $K = cK^{-1}$, with $|c| = 1$. Multiplying by K and using its antilinear property we get $K^2 = c^*I$, so that $c = c^*$. Therefore we must have $c = \pm 1$ and

$$K^2 = \pm I.$$

We will now construct an operator K that has the desired properties. Let $|f\rangle$ and $|g\rangle$ be arbitrary vectors of $\mathbb{H}_0$ and $F(\mathbf{Q})$ an observable function of $\mathbf{Q}$. We write $|\bar{f}\rangle = K|f\rangle, |\bar{g}\rangle = K|g\rangle$. The antiunitarity of K gives by A.5.4.1

$$\langle \bar{f}|\bar{F}(\mathbf{Q})|\bar{g}\rangle = \langle f|F(\mathbf{Q})|g\rangle^*$$

but, since $\bar{F}(\mathbf{Q}) = F(K\mathbf{Q}K^{-1}) = F(\mathbf{Q})$, this expression becomes

$$\langle \bar{f}|F(\mathbf{Q})|\bar{g}\rangle = \langle f|F(\mathbf{Q})|g\rangle^*.$$

Or, in position space, with $f(\boldsymbol{\xi}) = \langle \boldsymbol{\xi}|f\rangle$, $\bar{f}(\boldsymbol{\xi}) = \langle \xi|K|f\rangle = \hat{K}f(\boldsymbol{\xi})$,

$$\int d^3\xi\ \bar{f}^*(\boldsymbol{\xi})F(\boldsymbol{\xi})\bar{g}(\boldsymbol{\xi}) = \int d^3\xi\ f(\boldsymbol{\xi})F(\boldsymbol{\xi})g^*(\boldsymbol{\xi}).$$

Since this equality must hold for any $F(\boldsymbol{\xi})$ one must have $\bar{f}^*(\boldsymbol{\xi})\bar{g}(\boldsymbol{\xi}) = f(\boldsymbol{\xi}))g^*(\boldsymbol{\xi})$, which gives

$$\bar{f}(\boldsymbol{\xi}) = \hat{K}f(\boldsymbol{\xi}) = e^{i\theta(\xi)}f^*(\boldsymbol{\xi})$$

with the arbitrary phase $\theta(\boldsymbol{\xi})$ independent of the vector $|f\rangle$. One can verify immediately that $K^2 = I$ and that $K^{-1} = K^\dagger = K$. The phase is further restricted by the required transformation of the momentum. We must have

$$\langle \boldsymbol{\xi}|K\mathbf{P}K^{-1}|f\rangle = \langle \boldsymbol{\xi}| - \mathbf{P}|f\rangle,$$

or, in position space,

$$\hat{K}\left[-i\nabla\left(e^{i\theta(\boldsymbol{\xi})}f^*(\boldsymbol{\xi})\right)\right] = i\nabla f(\boldsymbol{\xi}),$$

which shows that $\theta(\boldsymbol{\xi}) = \alpha$, a constant.

We have thus shown that for the spinless particle there exists an antiunitary time reversal operator of the form

$$K = e^{i\alpha}K_0,$$

where K_0 is the operator of complex conjugation in position space

$$\hat{K}_0 f(\boldsymbol{\xi}) = f^*(\boldsymbol{\xi})$$

or in Dirac's notation

$$\langle\boldsymbol{\xi}|K_0|f\rangle = \langle\boldsymbol{\xi}|f\rangle^*. \tag{4.3}$$

As is customary we shall take the phase equal to zero. The operator just defined is usually called the complex conjugation operator, omitting the reference to position space. One should not be misled by this name into thinking that K_0 represents complex conjugation in any function space representation of $\mathbb{H}_0$. Indeed if we use momentum space we have with $\tilde{f}(\boldsymbol{\eta}) = \langle\boldsymbol{\eta}|f\rangle$ and $\langle\boldsymbol{\eta}|K_0|f\rangle = \tilde{K}_0\tilde{f}(\boldsymbol{\eta})$

$$\tilde{K}_0\tilde{f}(\boldsymbol{\eta}) = (2\pi)^{-3/2}\int d^3\xi\; e^{-i\boldsymbol{\eta}\cdot\boldsymbol{\xi}}f^*(\boldsymbol{\xi}) = \tilde{f}^*(-\boldsymbol{\eta}). \tag{4.4}$$

It is easy to see that that the transformations (4.3) and (4.4) can be written in terms of the Dirac kets as

$$K_0|\boldsymbol{\xi}\rangle = |\boldsymbol{\xi}\rangle, \qquad K_0|\boldsymbol{\eta}\rangle = |-\boldsymbol{\eta}\rangle.$$

Having found the time reversal operator it remains to verify under what conditions the time reversal transformation (4.1) is a dynamical symmetry. Since the antiunitary kinematical transformation represented by K is time independent, and since, in the notation of Section 1, the time transformation is given by $\tau(\bar{t}) = -\bar{t}$, which gives $\tau'(\bar{t}) = -1$, the transformed hamiltonian is according to (1.4)

$$\bar{H}'_t = -\bar{H}_{-t}, \qquad \bar{H}_t = KH_tK^\dagger$$

so that the condition (1.5) for dynamical symmetry becomes

$$KH_{-t}K^\dagger = H_t$$

or more explicitly

$$H(-\mathbf{P},\mathbf{Q},-t) = H(\mathbf{P},\mathbf{Q},t).$$

It is interesting to note that for time reversal to be a dynamical symmetry of the spinless particle with hamiltonian (VII.1.7) then the vector potential $\mathbf{A}_t$ must be such that $K\mathbf{A}_{-t}K^{-1} = -\mathbf{A}_t$. Such is the case with the classical electromagnetic vector potential when one thinks of time reversal as reversing the motion of the electric charges.

In concluding this section we examine for future reference how does the orbital angular momentum transform under time reversal. From its definition, $\mathbf{L} = \mathbf{Q} \times \mathbf{P}$, we have

$$K\mathbf{L}K^\dagger = -\mathbf{L}, \qquad KL^2K^\dagger = L^2.$$

From these relations we obtain for the eigenvectors $|l,m\rangle$ of the angular momentum defined in the space $\mathbb{H}_\Omega$ of Section VII.4

$$KL^2|l,m\rangle = L^2K|l,m\rangle = l(l+1)K|l,m\rangle,$$

$$KL_3|l,m\rangle = -L_3K|l,m\rangle = m|l,m\rangle,$$

which show that $K|l,m\rangle$ is an eigenvector of L^2 belonging to the eigenvalue $l(l+1)$ and an eigenvector of L_3 belonging to the eigenvalue $-m$. With the phases chosen for the eigenfunctions (VII.4.6) we see from (VII.4.7) that

$$\hat{K}_0 Y_l^m(\varphi,\theta) = (-1)^m Y_l^{-m}(\varphi,\theta).$$

5. The galilean transformations as dynamical symmetries

In this section we consider again galilean transformations but from a point of view essentially different from the one taken in Chapter V. There we studied two experimental setups that differed only in that one was rotated, displaced and in uniform motion with respect to the other and we compared their depictions at corresponding stages in the progress of the two experiments. Here, instead, we shall compare their depictions at the same instant. From the present point of view, it is equivalent, and more convenient, to consider a single experimental setup $\mathcal{E}$ described with respect to different galilean frames. As we saw in the last two chapters, the results of Chapter V are useful when one seeks to find restrictions on the hamiltonian of a general system. The results of this section lead to stronger restrictions in the important case of *isolated* systems.

Let $\mathbf{x}$, $\dot{\mathbf{x}}$ be the position and velocity of a feature of the experimental setup, and t the laboratory time with respect to a galilean frame $\mathcal{F}$. Let $\bar{\mathcal{F}}$ be another frame which, with respect to $\mathcal{F}$, is rotated by an amount represented by the rotation matrix $\mathbf{R}^{-1}$, displaced by $-\mathbf{a}$ at $t = 0$, moves with velocity $-\mathbf{v}$, an its time is shifted by $-s$. When measured with respect to $\bar{\mathcal{F}}$ we denote by $\bar{\mathbf{x}}$ and $\dot{\bar{\mathbf{x}}}$ the position and velocity of the feature, and by $\bar{t}$ the time of the laboratory clock. We have then

$$\bar{\mathbf{x}} = \mathbf{R}\mathbf{x} + \mathbf{v}t + \mathbf{a}, \qquad \dot{\bar{\mathbf{x}}} = \dot{\mathbf{x}} + \mathbf{v}, \qquad \bar{t} = t + s. \tag{5.1}$$

These equations define a ten parameter Lie group of transformations

$$\mathfrak{t} = \{\mathbf{R}, \mathbf{a}, \mathbf{v}, s\} \tag{5.2}$$

called the *Galilei group* $\mathfrak{G}$, with product operation

$$\{\mathbf{R}_2, \mathbf{a}_2, \mathbf{v}_2, s_2\}\{\mathbf{R}_1, \mathbf{a}_1, \mathbf{v}_1, s_1\} = \{\mathbf{R}_2\mathbf{R}_1, \mathbf{a}_2 + \mathbf{R}_2\mathbf{a}_1 + \mathbf{v}_2 s_1, \mathbf{v}_2 + \mathbf{R}_2\mathbf{v}_1, s_1 + s_2\}$$

and inverse

$$\mathfrak{t}^{-1} = \{\mathbf{R}, \mathbf{a}, \mathbf{v}, s\}^{-1} = \{\mathbf{R}^{-1}, \mathbf{R}^{-1}\mathbf{v}s - \mathbf{R}^{-1}\mathbf{a}, -\mathbf{R}^{-1}\mathbf{v}, -s\}$$

Examining the group operations of the kinematical Galilei group $\mathfrak{G}_K$ that we studied in Chapter V one can see that $\mathfrak{G}_K$ is a subgroup of $\mathfrak{G}$. We denote an element of $\mathfrak{G}_K$ by $\mathfrak{t}_K$,

$$\mathfrak{t}_K = \{\mathbf{R}, \mathbf{a}, \mathbf{v}, 0\}. \tag{5.3}$$

The transformation (5.2) can therefore be decomposed as follows

$$\mathfrak{t} = \mathfrak{k}(s)\mathfrak{t}_K,$$

where

$$\mathfrak{k}(s) = \{\mathbf{1}, 0, 0, s\}. \tag{5.4}$$

is an element of the one-parameter subgroup of time translations. As an element of $\mathfrak{G}_K$, the transformation $\mathfrak{t}_K$ can be decomposed in a product of transformations belonging to the set of nine one-parameter subgroups consisting of the three subgroups of displacements

$$\mathfrak{p}_i(a) = \{\mathbf{1}, a\mathbf{e}_i, 0, 0\}, \tag{5.5}$$

the three subgroups of rotations

$$\mathfrak{j}_i(\varphi) = \{\mathbf{R}_i(\varphi), 0, 0, 0\} \tag{5.6}$$

and the three subgroups of homogeneous galilean transformations

$$\mathfrak{g}_i(v) = \{\mathbf{1}, 0, v\mathbf{e}_i, 0\}. \tag{5.7}$$

Let A be a measuring instrument and let A be the operator that describes it when its configuration is referred to the inertial frame $\mathcal{F}$ (we work here in the Schrödinger picture). When its configuration is referred to the frame $\bar{\mathcal{F}}$, it will, in general, be described by a different operator $\bar{A}$ (for example, if the instrument is a set of counters that measures position, a given counter will be labelled by $\mathbf{x}$ in the first case and by $\bar{\mathbf{x}}$ [equation (5.1)] in the second). Clearly A and $\bar{A}$ are related by a kinematical symmetry to which there corresponds a unitary operator $V(\mathbf{t})$ that represents an element $\mathbf{t}$ of $\mathfrak{G}$ (the unitarity is implied by the continuity of the group). The same kinematical symmetry will relate, of course, the two descriptions ρ and $\bar{\rho}$ of the state.

As we did in Chapter V for $\mathfrak{G}_K$ we need to determine the structure of the infinitesimal transformations of $\mathfrak{G}$. Most of the work was already done there. In the notation of Section V.6 all that remains is to determine the new structure constants $c_{10,n}^{l}$ and the local phases $\beta_{10,n}$ introduced by the tenth one-parameter subgroup, namely the subgroup of time translations, whose generator we shall call K.

The multiplication law tells us that the time translations commute with the displacements (5.5) and the rotations (5.6) but not with the homogeneous galilean transformations (5.7). The corresponding structure constants are determined by looking at the transformation $\mathbf{t}_i = \mathfrak{g}_i(\lambda)\mathfrak{k}(\lambda)\mathfrak{g}_i^{-1}(\lambda)\mathfrak{k}^{-1}(\lambda)$ which, to lowest order in λ, is

$$\mathbf{x} \mapsto \mathbf{x} - \lambda^2 \mathbf{e}_i, \qquad t \mapsto t,$$

a displacement $of - \lambda^2$ along $\mathbf{e}_i$. Therefore, $\delta\mathbf{t}_i = -\delta\mathfrak{p}_i$. Remembering that we denoted by $-G_i$ the generator of $\mathfrak{g}_i$ we obtain the commutation relation

$$[G_i, K] = iP_i + ik_i I.$$

Following the procedure used in Section V.6, we find that the k_i can be taken equal to zero as well as all the other local phases except, as before, those corresponding to the pairs of homogeneous galilean transformations and displacements along the same direction. We thus obtain the following commutation relations among the generators of the Galilei group

$$[P_i, P_j] = 0, \qquad [G_i, G_j] = 0,$$

$$[J_i, J_j] = i\epsilon_{ijk} J_k, \qquad [J_i, P_j] = i\epsilon_{ijk} P_k, \qquad [J_i, G_j] = i\epsilon_{ijk} G_k,$$

$$[G_i, P_j] = i\mu\delta_{ij} I, \tag{5.8}$$

$$[P_i, K] = 0, \qquad [J_i, K] = 0, \qquad [G_i, K] = iP_i. \tag{5.9}$$

It is worth remarking that, because the first two transformation equations (5.1) depend on t, the generators P_i, J_i and G_i may be time dependent (in fact we shall soon exhibit an example in which G_i does depend on t).

Having determined the structure of the Galilei group, we turn our attention to the case in which the experiment $\mathcal{E}$ deals with an isolated system, and our first task is to attempt to find the meaning of this term.

Intuitively one would say that the system under investigation in $\mathcal{E}$ is isolated if the experimental setup does not contain any instruments that are field sources in the sense explained in Section IV.2. Suppose that this is the case and let t_0 be the preparation time, and t the measurement time. If there are no field sources, the same experimental instruments form part of the preparation apparatus for any choice of the reference time τ such that $t_0 < \tau < t$. Similarly, for this range of reference times, the same instruments form part of the measuring apparatus. Let $\rho(\tau)$ be the Schrödinger state at time τ [recall that in Section IV.2 $\rho(\tau)$ was defined to be the prepared state $\rho^{(\tau)}$ with respect to the reference time τ], a shift in the reference time from τ to $\tau + s$ ($t_0 < \tau + s < t$) corresponds to a new depiction of the experiment in which the state is $\rho(t+s)$, where, in terms of the evolution operator $U(t, t_0)$,

$$\rho(t+s) = U(t+s, t)\rho(t)U^{\dagger}(t+s, t).$$

Since, as just explained, in the absence of field sources, the shift in reference time is achieved without a redefinition of the preparation and measuring instruments, the new depiction is identical with the description of the system with respect to an inertial frame $\bar{\mathcal{F}}$ obtained from $\mathcal{F}$ by a time translation $\mathbf{t}(s)$. We therefore have

$$\rho(t+s) = e^{-iKs}\rho(t)e^{iKs}.$$

Comparing the last two equations we see that, except for an unimportant phase factor

$$U(t+s, t) = e^{-iKs}$$

from which we can see that K is precisely the hamiltonian of the isolated system. One may, therefore, define an isolated system as follows:

An isolated system is a system whose hamiltonian H is time independent and equal to the generator of time translations.

Thus, for an isolated system, the equations (5.9) become equations for the commutators of the hamiltonian with the generators of the kinematical subgroup $\mathfrak{G}_K$.

$$[P_i, H] = 0, \qquad [J_i, H] = 0, \qquad [G_i, H] = iP_i. \tag{5.10}$$

We shall show next that, always in the case of an isolated system, the transformations $\mathbf{t}_K \in \mathfrak{G}_K$ are, in fact, dynamical symmetries.

Let W_0 be the unitary operator that represents $\mathbf{t}_K$ at $t = t_0$.The transformed state at this time is

$$\bar{\rho}(t_0) = W_0 \rho(t_0) W_0^\dagger$$

and, if we put $\tau = t - t_0$, we have

$$\bar{\rho}(t) = e^{-iH\tau} W_0 \rho(t_0) W_0^\dagger e^{iH\tau}$$

but, since $\rho(t) = e^{-iH\tau}\rho(t_0)e^{iH\tau}$, we can write

$$\bar{\rho}(t) = e^{-iH\tau} W_0 e^{iH\tau} \rho(t) e^{-iH\tau} W_0^\dagger e^{iH\tau}.$$

Therefore, at time t, the transformation $\mathbf{t}_K$ is represented by

$$W_t = e^{-iH\tau} W_0 e^{iH\tau}. \tag{5.11}$$

By (5.10) the generators $\mathbf{P}$ and $\mathbf{J}$ of displacements and rotations commute with H. Therefore if $\mathbf{t}_K$ is a translation or a rotation (or a product thereof), $W_t = W_0 = W$ is time independent and so are the generators $\mathbf{P}$ and $\mathbf{J}$.

The situation is different if $\mathbf{t}_K$ is a homogeneous galilean transformation $\mathfrak{g}(\mathbf{v})$, for the generator $-$ $\mathbf{G}$ does not commute with H and, as (5.11) shows, W_t is time dependent and so is $\mathbf{G}$. If $\mathbf{G}_0$ and $\mathbf{G}_t$ denote the values of $\mathbf{G}$ at t_0 and t, equation (5.11) reads

$$W_t = e^{i\mathbf{G}_t\cdot\mathbf{v}} = e^{-iH\tau} e^{i\mathbf{G}_0\cdot\mathbf{v}} e^{iH\tau}$$

or

$$\mathbf{G}_t = e^{-iH\tau} \mathbf{G}_0 e^{iH\tau}.$$

Taking into account the last commutation relation (5.10) and using the same procedure that led to equation (V.6.24) one obtains

$$\mathbf{G}_t = \mathbf{G}_0 - \mathbf{P}(t - t_0)$$

and

$$W_t = e^{i(\mathbf{G}_0 - \mathbf{P}\tau\cdot\mathbf{v})} = e^{i\mathbf{G}_0\cdot\mathbf{v}} e^{-i\mathbf{P}\tau\cdot\mathbf{v}} e^{\frac{1}{2}i\mu|\mathbf{v}|^2\tau} \tag{5.12}$$

where Glauber's formula (VII.6.2) and the commutation relation (5.8) have been used.

We show next that $\mathfrak{g}(\mathbf{v})$ is a dynamical symmetry. To this end we calculate the transformed hamiltonian, which, according to (1.4) is

$$\bar{H}' = W_t H W_t^\dagger + i\partial W_t W_t^\dagger.$$

Substituting (5.12) into this equation one gets

$$\bar{H}' = e^{i\mathbf{G}_0\cdot\mathbf{v}}He^{-i\mathbf{G}_0\cdot\mathbf{v}} + e^{i\mathbf{G}_0\cdot\mathbf{v}}\mathbf{P}\cdot\mathbf{v}e^{-i\mathbf{G}_0\cdot\mathbf{v}} - \tfrac{1}{2}\mu|\mathbf{v}|^2,$$

where, as we often do, we have omitted the identity symbol I. By (5.10) and (5.12) the first term on the right is $\bar{H}$ and the second $\bar{\mathbf{P}}\cdot\mathbf{v}$, where the bar denotes the transform under $\mathfrak{g}(\mathbf{v})$. The above expression can therefore be written

$$\bar{H}' = \bar{H} + \bar{\mathbf{P}}\cdot\mathbf{v} - \tfrac{1}{2}\mu|\mathbf{v}|^2,$$

a result formally identical with the corresponding classical result. Now, by our familiar procedure, $\bar{H}$ is found to equal $H - \bar{\mathbf{P}}\cdot\mathbf{v}$ so that we get

$$\bar{H}' = H - \tfrac{1}{2}\mu|\mathbf{v}|^2$$

which satisfies the condition of Section 1 for $\mathfrak{g}(\mathbf{v})$ to be a dynamical symmetry. It follows that all transformations $\mathbf{t}_K$ are dynamical symmetries; a result which is an expression of the principle of galilean relativity.

The question arises: does the introduction of the time translations alter in any way the physical interpretation of $\mathbf{P}$, $\mathbf{J}$ and $\mathbf{G}$ obtained in Chapter VI and VII?

The results of these chapters are obviously applicable to our present case in the absence of time translations. There we found that $\mathbf{P}$ and $\mathbf{J}$ were the momentum and the angular momentum, respectively, while $\mathbf{G}$ was related to the particle position $\mathbf{Q}$ by $\mathbf{G} = m\mathbf{Q}$. Therefore, given a time t, $\mathbf{P}$ and $\mathbf{J}$ are now the t-depictions of the momentum and angular momentum, and if $\mathbf{Q}$ is the t-depiction of the observable position, the relation $\mathbf{G}_t = m\mathbf{Q}$ holds.

Suppose we work in the Schrödinger picture and we choose a particular time t_0 to define it, then, because, as we have seen, the generators $\mathbf{P}$ and $\mathbf{J}$ are time independent, they will be the momentum and angular momentum in the chosen picture at any time time t. However, because $\mathbf{G}$ depends on t, the relation $\mathbf{G} = m\mathbf{Q}$ will only hold if we choose for $\mathbf{G}$ its value $\mathbf{G}_{t_0}$ at time t_0.

Our next task is to find what restrictions are imposed by the structure of $\mathfrak{G}$ on the hamiltonian of a free (i.e., isolated) spinless particle.

We found in Chapter VII that the hamiltonian of a spinless particle has to be of the form

$$H_t = \frac{1}{2m}|\mathbf{P} - \mathbf{A}_t(\mathbf{Q})|^2 + V_t(\mathbf{Q})$$

In case the particle is free, we already found out that the hamiltonian must be time independent. Moreover, the first of the commutation relations (5.10) tells us that it has to be invariant under displacements and, therefore, it cannot depend on $\mathbf{Q}$.

We, thus, conclude that the hamiltonian of a spinless free particle has to be of the form

$$H = \frac{1}{2m}|\mathbf{P}|^2.$$

It is a simple matter to verify that this hamiltonian satisfies all the commutation relations (5.10).

In future chapters we shall apply the procedure used here to determine what conditions does galilean relativity impose on the hamiltonian of more complicated isolated systems.

IX A PARTICLE WITH SPIN

As is well known there are many physical systems that we call particles which possess "internal" observables that are commensurable with both the position and the momentum and cannot be regarded as functions of them. It follows that in these systems neither the position nor the momentum are maximal nor do they constitute an irreducible set of observables. Such systems are referred to as particles with internal degrees of freedom. We are concerned here with the case in which the internal observable is the spin. We first discuss the description of such a particle and, later, the general form that its hamiltonian may have in the light of galilean invariance. We end the chapter with a simple example in which the dynamical problem for neutral particle with spin in an inhomogeneous magnetic field is solved.

1. Internal degrees of freedom and galilean invariance

Consider an experimental setup with a measuring instrument consisting of the set of particle counters described in Section VI.1. As in that section, we assume that in each experimental run exactly one counter of the set is activated. In the idealized limit in which each counter occupies a vanishingly small region of space, to the set of counters there corresponds the vector observable position $\mathbf{Q}$ whose spectrum, in the sense of B.2.10, is the whole euclidean space $\mathbb{E}_3$ and in Dirac's notation has spectral resolution given by

$$\mathbf{Q} = \int d^3\xi\ \boldsymbol{\xi}\Gamma_Q(\boldsymbol{\xi}), \tag{1.1}$$

where the integration extends over all $\mathbb{E}_3$. In Chapters VI and VII we studied the case in which the observable $\mathbf{Q}$ was maximal and consequently the Dirac operator $\Gamma_Q(\boldsymbol{\xi})$ was of the form $|\boldsymbol{\xi}\rangle\langle\boldsymbol{\xi}|$. In this chapter we remove that assumption.

If $\mathbf{Q}$ is not maximal its spectrum is degenerate and may consist of regions of different multiplicities as explained in C.3.5. We shall show next that, if the experimental setup obeys the principle of galilean invariance of Section V.5, every region of the spectrum has the same multiplicity.

To this end, in the notation of Section VII.1, consider two identical experiments $\mathcal{E}$ and $\mathcal{E}'$ linked by a displacement $\mathfrak{p}(\mathbf{a})$, and let $\mathbf{P}$ be the generator of displacements. A by now familiar argument tells us that the position $\mathbf{Q}$ must obey the relation

$$\mathbf{Q} - \mathbf{a} = e^{-i\mathbf{P}\cdot\mathbf{a}}\mathbf{Q}e^{i\mathbf{P}\cdot\mathbf{a}}. \tag{1.2}$$

Since a unitary transformation preserves the spectrum of an operator, it follows that $\mathbf{Q}$ and $\mathbf{Q} - \mathbf{a}$, for arbitrary $\mathbf{a}$, have the same spectrum. Thus, if a region Σ of the spectrum has degeneracy k, so does the region obtained from Σ by an arbitrary displacement $\mathbf{a}$, which proves the desired result.

Assume that the degeneracy of the spectrum of $\mathbf{Q}$ is k. Then, according to the discussion in C.3.6.3, the Hilbert space $\mathbb{H}$ of the system can be written as a direct sum

$$\mathbb{H} = \bigoplus_{n=1}^{k} \mathbb{H}_n,$$

and correspondingly the operator $\mathbf{Q}$ can be written as

$$\mathbf{Q} = \bigoplus_{n=1}^{k} \mathbf{Q}_n, \tag{1.3}$$

where $\mathbf{Q}_n$ is an operator on $\mathbb{H}_n$ with a pure continuous simple spectrum:

$$\mathbf{Q}_n = \int d^3\xi \; \boldsymbol{\xi}|\boldsymbol{\xi}, n\rangle\langle\boldsymbol{\xi}, n|.$$

This operator induces a function space representation of $\mathbb{H}_n$. If $|f_n\rangle \in \mathbb{H}_n$ we write its representation as $f_n(\boldsymbol{\xi}) = \langle\boldsymbol{\xi}, n|f_n\rangle$.

From now on we shall regard $\mathbf{Q}_n$ as an operator on $\mathbb{H}$. Accordingly we shall write the direct sum (1.3) as a sum of operators on $\mathbb{H}$,

$$\mathbf{Q} = \sum_{n=1}^{k} \mathbf{Q}_n.$$

Similarly we shall view $|\boldsymbol{\xi}, n\rangle$ as a generalized vector of $\mathbb{H}$. The projection operator E_n on $\mathbb{H}$ that projects on the subspace $\mathbb{H}_n \subset \mathbb{H}$ equals the identity on $\mathbb{H}_n$. Therefore the completeness relation on $\mathbb{H}_n$ of the family $|\boldsymbol{\xi}, n\rangle$ becomes

$$E_n = \int d^3\xi \; |\boldsymbol{\xi}, n\rangle\langle\boldsymbol{\xi}, n|. \tag{1.4}$$

The Dirac operators $\Gamma_Q(\boldsymbol{\xi})$ of the spectral resolution (1.1) can then be written

$$\Gamma_Q(\boldsymbol{\xi}) = \sum_{n=1}^{k} |\boldsymbol{\xi}, n\rangle\langle\boldsymbol{\xi}, n|. \tag{1.5}$$

The operator $\mathbf{Q}$ together with the set of projections E_n constitutes a maximal set of observables with the complete set of generalized eigenvectors $|\boldsymbol{\xi}, n\rangle$, $\boldsymbol{\xi} \in \mathbb{E}_3$, $n = 1, \ldots, k$. The set of observables induces a function space representation of $\mathbb{H}$. In this function space a vector $|f\rangle \in \mathbb{H}$ is represented by the direct sum $\langle\boldsymbol{\xi}, 1|f\rangle \oplus \ldots \oplus \langle\boldsymbol{\xi}, n|f\rangle$, which is often written as a column matrix

$$\begin{pmatrix} \langle\boldsymbol{\xi}, 1|f\rangle \\ \langle\boldsymbol{\xi}, 2|f\rangle \\ \cdots \\ \langle\boldsymbol{\xi}, k|f\rangle \end{pmatrix} = \begin{pmatrix} f_1(\boldsymbol{\xi}) \\ f_2(\boldsymbol{\xi}) \\ \cdots \\ f_k(\boldsymbol{\xi}) \end{pmatrix}.$$

It is more convenient, and we shall do so, to regard the Hilbert space $\mathbb{H}$ as the tensor product $\mathbb{H}^0 \otimes \mathbb{H}^I$ of an infinite dimensional space $\mathbb{H}^0$ in which the position observable is maximal, and a k-dimensional space $\mathbb{H}^I$. For each value of $\boldsymbol{\xi}$, the above column matrix can be thought of as a representation of a vector of $\mathbb{H}^I$ with respect to a basis $\{|n\rangle\}, n = 1, \ldots, k$. Similarly, for each value of n, $f_n(\boldsymbol{\xi})$ is the position representation of a vector of $\mathbb{H}^0$. We will write $\mathbf{Q}^0$ for the position operator on $\mathbb{H}^0$,

$$\mathbf{Q}^0 = \int d^3\xi\ \boldsymbol{\xi}|\boldsymbol{\xi}\rangle\langle\boldsymbol{\xi}|,$$

so that in tensor product notation

$$\mathbf{Q} = \mathbf{Q}^0 \otimes I^I, \qquad |\boldsymbol{\xi}, n\rangle = |\boldsymbol{\xi}\rangle|n\rangle, \tag{1.6}$$

where I^I is the identity on $\mathbb{H}^I$. In this notation, the projection operators (1.4) become

$$E_n = I^0 \otimes |n\rangle\langle n|.$$

If we use (1.5) and the completeness relation in $\mathbb{H}^I$,

$$I^I = \sum_{n=1}^{k} |n\rangle\langle n|,$$

we obtain a corresponding expression for the operators $\Gamma_Q(\boldsymbol{\xi})$ of (1.1)

$$\Gamma_Q(\boldsymbol{\xi}) = |\boldsymbol{\xi}\rangle\langle\boldsymbol{\xi}| \otimes I^I,$$

which is but another way of expressing the first equation (1.6). The family of generalized operators $\Gamma_Q(\boldsymbol{\xi})$ is uniquely determined by the operator $\mathbf{Q}$ but, as shown clearly by the last expression, the basis $\{|n\rangle\}$ in $\mathbb{H}^I$ is arbitrary.

Contrary to the case studied in Chapter VII, the transformation equation (1.2) is not sufficient to determine the generator $\mathbf{P}$ because the observable $\mathbf{Q}$ is not maximal. To determine $\mathbf{P}$ we need to know the behaviour of a maximal observable on $\mathbb{H}$ under the displacement $\mathfrak{p}(\mathbf{a})$.

If the operator

$$A_1^I = \sum_{n=1}^{k} \alpha_n |n\rangle\langle n|$$

on $\mathbb{H}^I$ is maximal, then the pair $\{\mathbf{Q},\ A_1\}, A_1 = I^0 \otimes A_1^I$ is a composite observable in the sense of B.2.10, which is maximal on $\mathbb{H}$ and has a complete set of generalized eigenvectors $|\boldsymbol{\xi}, n\rangle$. We call $f_n(\boldsymbol{\xi}) = \langle \boldsymbol{\xi}, n|f\rangle$ the position-A_1 representation of the vector $|f\rangle = \mathbb{H}$.

We shall now assume that we are dealing with a system in which A_1 is a member of a set $\{A_p\}$ of observables of the form $A_p = I^0 \otimes A_p^I$ such that $\{A_p^I\}$ is a fundamental set of observables on $\mathbb{H}^I$. We shall furthermore assume that the observables A_p are preserved by a galilean transformation. The term "preserved" is used here to mean that the operators remain unchanged under a displacement or a homogeneous galilean transformation, and that under a rotation they change in accordance with their tensorial character.

Under a displacement $\mathbf{a}$, the composite operator $\{\mathbf{Q}, A_p\}$ satisfies the relation

$$\{\mathbf{Q} - \mathbf{a}, A_p\} = e^{-i\mathbf{P}\cdot\mathbf{a}}\{\mathbf{Q}, A_p\}e^{i\mathbf{P}\cdot\mathbf{a}}.$$

Since, by the assumption just made, $\mathbf{P}$ commutes with all the A_p, it has the form $\mathbf{P}^0 \otimes I^I$, where $\mathbf{P}^0$ obeys the relation

$$\mathbf{Q}^0 - \mathbf{a}I^0 = e^{-i\mathbf{P}^0\cdot\mathbf{a}}\mathbf{Q}^0 e^{i\mathbf{P}\cdot\mathbf{a}}.$$

The problem of determining $\mathbf{P}^0$ becomes identical to the problem studied Section VII.1 for the case of the spinless particle. Thus $\mathbf{P}^0$ and $\mathbf{Q}^0$ obey the commutation relations

$$\left[Q_i^0, P_j^0\right] = i\delta_{ij}I^0,$$

which, as in Section VII.1, yields in the $\mathbf{Q}^0$-representation of the space $\mathbb{H}^0$

$$(\mathbf{P}^0 f^0)(\boldsymbol{\xi}) = -i\nabla f^0(\boldsymbol{\xi}), \qquad \langle\boldsymbol{\xi}|\mathbf{P}^0|\boldsymbol{\xi}'\rangle = -i\nabla\delta(\boldsymbol{\xi} - \boldsymbol{\xi}')$$

with $|f^0\rangle \in \mathbb{H}^0$ and $f^0(\boldsymbol{\xi}) = \langle \boldsymbol{\xi}|f^0\rangle$. The unitary operator that represents $\mathfrak{p}(\mathbf{a})$ is then determined to be

$$e^{-i\mathbf{P}\cdot\mathbf{a}} = e^{i\mathbf{P}^0\cdot\mathbf{a}} \otimes I^I.$$

As in Chapter VII we call the generator of displacements,

$$\mathbf{P} = \mathbf{P}^0 \otimes I^I, \tag{1.7}$$

the momentum of the particle. Evidently $\mathbf{Q}$ and $\mathbf{P}$ commute with $A_p = I^0 \otimes A_p^I$, and obey the same commutation relations as $\mathbf{Q}^0$ and $\mathbf{P}^0$. In vector notation

$$[\mathbf{Q}, \mathbf{P}] = i\mathbf{1}.$$

In the position-A_1 representation we have the two equivalent relations

$$(\mathbf{P}f)_n(\boldsymbol{\xi}) = -i\nabla f_n(\boldsymbol{\xi}),$$
$$\langle \boldsymbol{\xi}, n|\mathbf{P}^0|\boldsymbol{\xi}', n'\rangle = -i\delta_{nn'}\nabla\delta(\boldsymbol{\xi} - \boldsymbol{\xi}').$$

A *particle with internal degrees of freedom* is a system that possesses a fundamental set of maximal observables consisting of the position $\mathbf{Q}$, the momentum $\mathbf{P}$, and of a set $\{A_p\}$ of "internal" observables of the form $A_p = I^0 \otimes A_p^I$ which are preserved by a galilean transformation.

As is well known, early in the development of quantum mechanics, it was discovered experimentally that in particles such as the electron and the proton, the spin is an internal observable. In those particles, two components of the spin along different directions constitute an irreducible set of operators on the "internal" Hilbert space $\mathbb{H}^I$. Systems with these properties are the elementary particles of (non-relativistic) quantum mechanics. We shall refer to them as particles with spin. Their study will be the object of the following sections.

Besides elementary particles, there are systems that in many cases can be treated approximately as particles with internal degrees of freedom. Such is the case, for example, of an atom when the detailed electronic structure can be ignored. In this case the hamiltonian of the atom (at rest) can be approximated by an operator on $\mathbb{H}^I$, and its angular momentum can be treated as a spin. The resulting structure of the space $\mathbb{H}^I$ is, of course, more complicated that in the case of an elementary particle.

2. Spin and rotations

In the introductory chapters of this book the spin 1/2 example was used extensively. It was pointed out in Section II.2 that, in the experiments there discussed, it was possible to consider the spin of a particle independently of the other observables. This possibility arises because, as already mentioned, the spin components are internal observables that are maximal on the internal subspace introduced in the previous section which shall now be called spin space. In preparation for the discussion of the next section, it is convenient to regard the observable spin as an operator on $\mathbb{H}^I$, and generalize some of the results obtained in Chapter II for spin 1/2 to the case of an arbitrary spin.

As in that chapter, a Stern-Gerlach measuring instrument with the magnetic field gradient in the direction given by the unit vector $\mathbf{e}$ is taken to define an observable $\Sigma_{\mathbf{e}}$ on $\mathbb{H}^I$. We will label the possible outcomes of the measurement of $\Sigma_{\mathbf{e}}$ by arbitrarily chosen numbers λ_n, $n = 1, \ldots, k$.

Following a procedure analogous to the one in Section VII.2, we consider two galilean invariant, identical experimental setups $\mathcal{E}$ and $\mathcal{E}'$ linked by a rotation $\mathbf{j}_{\mathbf{u}}(\varphi)$ by an angle φ about the axis determined by the unit vector $\mathbf{u}$. For now we call $\mathbf{K} = \mathbf{e}_1 K_1 + \mathbf{e}_2 K_2 + \mathbf{e}_3 K_3$ the generator of rotations on $\mathbb{H}^I$. Its components satisfy the commutation relations

$$[K_i, K_j] = i\epsilon_{ijk} K_k. \tag{2.1}$$

The unitary operator on $\mathbb{H}^I$ that represents $\mathbf{j}_{\mathbf{u}}(\varphi)$ is, therefore,

$$V(\mathbf{u}, \varphi) = e^{-i\mathbf{u}\cdot\mathbf{K}\varphi}.$$

The tensor that effects the rotation in space will be called $\mathbf{R} = \mathbf{R}_{\mathbf{u}}(\varphi)$.

Measurements are performed with Stern-Gerlach instruments. One such instrument is entirely defined by the direction $\mathbf{e}$ of its field gradient. A Stern-Gerlach instrument associated with the direction $\mathbf{e}$ with respect to the inertial frame associated with $\mathcal{E}'$ is identical to an instrument of $\mathcal{E}$ associated with the direction $\mathbf{R}^{-1}\mathbf{e}$ in its own frame. If, using the notation of Section VII.2 we call $\Sigma_{\mathbf{e}}(\mathbf{R})$ and $\Sigma_{\mathbf{e}}$ the observables corresponding to the instruments of $\mathcal{E}'$ and $\mathcal{E}$, respectively, we have

$$\Sigma_{\mathbf{e}}(\mathbf{R}) = \Sigma_{\mathbf{R}^{-1}\mathbf{e}}.$$

By an argument that would be tedious to repeat, one concludes that

$$\Sigma_{\mathbf{R}^{-1}\mathbf{e}} = e^{-i\mathbf{u}\cdot\mathbf{K}\varphi} \Sigma_{\mathbf{e}} e^{i\mathbf{u}\cdot\mathbf{K}\varphi}. \tag{2.2}$$

We saw in Section VII.3 that there are finite-dimensional Hilbert spaces in which operators K_i, $i = 1, 2, 3$, that obey the commutation relations (2.1) are maximal and any pair of them form an irreducible set of hermitean operators. We shall now assume that the space $\mathbb{H}^I$ is such that the generators of the rotations about the three coordinate axes have the above properties. To stress that we a considering now a special case, we will use the symbol $\mathbf{s}$, instead of $\mathbf{K}$, to denote the generator of rotations. According to Section VII.3, the space $\mathbb{H}^I$ is finite-dimensional. If we call its dimension $2s+1$, the operators s_i have discrete spectra consisting of integer or half integer values ranging from $-s$ to $+s$. The operator $s^2 = s_1^2 + s_2^2 + s_3^2$ is then a multiple of the identity that we write as $s^2 = s(s+1)I^I$ (this widely used notation is somewhat ambiguous; the operator s^2 should not be confused with the square of the positive number s). We shall denote the eigenvalues and eigenvectors of s_3 by μ and $|\mu\rangle$, respectively. Thus

$$s_3|\mu\rangle = \mu|\mu\rangle, \qquad |\mu\rangle \in \mathbb{H}^I, \qquad \mu = -s, -s+1, \ldots, s-1, s.$$

Particularize equation (2.2) for a Stern-Gerlach instrument with the field gradient along $\mathbf{e}_3$ and a rotation about the same axis. Since $\mathbf{R}_3(\varphi)\mathbf{e}_3 = \mathbf{e}_3$, we have

$$\Sigma_3 = e^{-is_3}\Sigma_3 e^{is_3},$$

where, for brevity we put $\Sigma_i = \Sigma_{\mathbf{e}_i}$. Evidently, the above relation tells us that Σ_3, and s_3 commute. Since, by assumption, s_3 is maximal, it follows that Σ_3 is a function of s_3.

Experience shows that in the case under consideration, Σ_3 is also a maximal observable on $\mathbb{H}^I$. Therefore, s_3 and Σ_3 are equivalent observables in the sense, used in Chapter II, that the basis $\{|\mu\rangle\}$ is a set of non-degenerate eigenvectors of both s_3 and Σ_3. The number of possible outcomes of the Stern-Gerlach measurement is then $k = 2s+1$, and the outcomes λ_n can be relabelled with the eigenvalues μ of s_3. With these labels, the Stern-Gerlach instrument with field gradient along $\mathbf{e}_3$ defines the observable s_3, the 3-*component of the spin* (of a system of spin s). More generally, an instrument with its field gradient along $\mathbf{e}$ can have its possible outcome configurations relabelled by the eigenvalues of the component $\mathbf{e}\cdot\mathbf{s}$ of the generator $\mathbf{s}$, which in this case we call the *spin*. We call the spin an angular momentum because, as the orbital angular momentum, it is a vector operator that obeys the commutation relations (2.1) or,

$$[s_i, s_j] = \epsilon_{ijk}s_k. \tag{2.3}$$

The expression for the unitary operator on H^I that represents the rotation $\mathbf{j}_{\mathbf{e}}(\varphi)$ becomes now

$$V(\mathbf{e},\varphi) = e^{-i\mathbf{e}\cdot\mathbf{s}\varphi}. \tag{2.4}$$

A widely used matrix representation of the spin operator on $\mathbb{H}^I$ is the representation with respect to a basis of eigenvectors of one of the components, say s_3, along the coordinate axes. We did use it in the spin 1/2 example of Chapter II. The results of Section VII.3 provide a simple method to construct such a representation in the case of an arbitrary spin.

As an example let us find the s_3-representation of the spin components of a particle of spin 3/2. In the four-dimensional space $\mathbb{H}^I$ we take a basis of eigenvectors $|\mu\rangle$, $\mu = 3/2, 1/2, -1/2, 3/2$, of s_3. Obviously, we have

$$s_3 = \frac{1}{2}\begin{pmatrix} 3 & 0 & 0 & 0 \\ 0 & 1 & 0 & 0 \\ 0 & 0 & -1 & 0 \\ 0 & 0 & 0 & -3 \end{pmatrix}.$$

To find the matrices representing the components s_1 and s_2, we use the formulae (VII.3.7-3.8) to determine the representations of $s_+ = s_1 + is_2$ and $s_- = s_1 - is_2$. With $j = s = 3/2$ and $m = \mu$, we have by virtue of the orthogonality of the $|\mu\rangle$,

$$\langle\nu|s_+|\mu\rangle = \sqrt{(3/2 - \mu)(5/2 + \mu)}\, \delta_{\nu,\mu+1}.$$

$$\langle\nu|s_-|\mu\rangle = \sqrt{(3/2 + \mu)(5/2 - \mu)}\, \delta_{\nu,\mu-1},$$

from which we obtain the representations

$$s_+ = \begin{pmatrix} 0 & \sqrt{3} & 0 & 0 \\ 0 & 0 & \sqrt{4} & 0 \\ 0 & 0 & 0 & \sqrt{3} \\ 0 & 0 & 0 & 0 \end{pmatrix}, \qquad s_- = \begin{pmatrix} 0 & 0 & 0 & 0 \\ \sqrt{3} & 0 & 0 & 0 \\ 0 & \sqrt{4} & 0 & 0 \\ 0 & 0 & \sqrt{3} & 0 \end{pmatrix}.$$

The matrices representing $s_1 = \frac{1}{2}(s_- + s_+)$ and $s_2 = \frac{1}{2}i(s_- - s_+)$ can now be found immediately,

$$s_1 = \begin{pmatrix} 0 & \sqrt{3} & 0 & 0 \\ \sqrt{3} & 0 & \sqrt{4} & 0 \\ 0 & \sqrt{4} & 0 & \sqrt{3} \\ 0 & 0 & \sqrt{3} & 0 \end{pmatrix}, \qquad s_2 = \begin{pmatrix} 0 & -i\sqrt{3} & 0 & 0 \\ i\sqrt{3} & 0 & -i\sqrt{4} & 0 \\ 0 & i\sqrt{4} & 0 & -i\sqrt{3} \\ 0 & 0 & i\sqrt{3} & 0 \end{pmatrix}.$$

In the case of spin one, the above procedure yields the following s_3-representation of the spin components on the three-dimensional space $\mathbb{H}^I$

$$s_1 = \frac{1}{\sqrt{2}}\begin{pmatrix} 0 & 1 & 0 \\ 1 & 0 & 1 \\ 0 & 1 & 0 \end{pmatrix}, \qquad s_2 = \frac{1}{\sqrt{2}}\begin{pmatrix} 0 & -i & 0 \\ i & 0 & -1 \\ 0 & i & 0 \end{pmatrix}, \qquad s_3 = \begin{pmatrix} 1 & 0 & 0 \\ 0 & 0 & 0 \\ 0 & 0 & -1 \end{pmatrix}.$$

In this particular case, however, a more symmetrical representation is often used. It is the representation with respect to a basis formed by the eigenvectors $|x_0\rangle, |y_0\rangle, |z_0\rangle$, of s_1, s_2, s_3, respectively, belonging to the zero eigenvalues of these operators. One can easily verify that, in the s_3-representation, we may take

$$|x_0\rangle = \frac{1}{\sqrt{2}}\begin{pmatrix} 1 \\ 0 \\ -1 \end{pmatrix}, \qquad |y_0\rangle = \frac{1}{\sqrt{2}}\begin{pmatrix} i \\ 0 \\ i \end{pmatrix}, \qquad |z_0\rangle = \frac{1}{\sqrt{2}}\begin{pmatrix} 0 \\ 1 \\ 0 \end{pmatrix},$$

and that these vectors[1] form an orthonormal basis on $\mathbb{H}^I$. The unitary matrix that changes from the s_3-representation to the one with respect to the basis $|x_0\rangle, |y_0\rangle, |z_0\rangle$, is

$$S = \frac{1}{\sqrt{2}}\begin{pmatrix} 1 & 0 & -1 \\ i & 0 & i \\ 0 & \sqrt{2} & 0 \end{pmatrix}.$$

If we call $\bar{s}_j$ the matrices of the new representation of the spin components, we have $\bar{s}_j = S s_j S^\dagger$, that is,

$$\bar{s}_1 = \begin{pmatrix} 0 & 0 & 0 \\ 0 & 0 & i \\ 0 & -i & 0 \end{pmatrix}, \qquad \bar{s}_2 = \begin{pmatrix} 0 & 0 & -i \\ 0 & 0 & 0 \\ i & 0 & 0 \end{pmatrix}, \qquad \bar{s}_3 = \begin{pmatrix} 0 & -i & 0 \\ i & 0 & 0 \\ 0 & 0 & 0 \end{pmatrix}.$$

In this representation the operator $e^{-is_j\varphi}$ corresponding to the a rotation by an angle φ about the i-axis in euclidean space can be found, with the help of the spectral resolution of s_j, to be given by matrices like the rotation matrices of Section 5.4 formed by the components of the euclidean tensors $\mathbf{R}_j(\varphi)$. A result which is to be expected in view of the remark that follows.

The family of unitary operators (2.4) on the space $\mathbb{H}^I$ constitutes, in the language of group theory, a $(2s+1)$-dimensional representation of the group of rotations on the three-dimensional euclidean space (or, more precisely, of its covering group). The representation is said to be irreducible because from the irreducibility of the spin components on $\mathbb{H}^I$ it follows that no subspace of $\mathbb{H}^I$ reduces the family $V(\mathbf{e}, \varphi)$.

3. The particle with spin

A particle with spin is a particle with internal degrees of freedom in which the irreducible set of internal observables on $\mathbb{H}^I$ (the A_p of Section 1) is formed by the spin components. A maximal observable on the Hilbert space $\mathbb{H}^0 \otimes \mathbb{H}^I$ is, then, the composite observable $\{\mathbf{Q}, s_3\}$ formed by the position and a component of the spin

[1] The arbitrary phase factors have been chosen in such a way that $|x_0\rangle = e^{i\pi s_2/2}|z_0\rangle$ and $|y_0\rangle = e^{i\pi s_3/2}|x_0\rangle$.

$s_3 = I^0 \otimes s_3^I$, where s_3^I is the spin component operator on $\mathbb{H}^I$ of the previous section (as is customary we have arbitrarily taken the 3-component of s). Its spectral resolution is written, in the notation of B.2.10, as

$$\{\mathbf{Q}, s_3\} = \sum_{\mu=-s}^{s} \int d^3\xi \; \{\boldsymbol{\xi}, \mu\} |\boldsymbol{\xi}, \mu\rangle\langle\boldsymbol{\xi}, \mu|$$

One can imagine this observable being measured by an idealized Stern-Gerlach device of vanishingly small dimensions, labelled by its position vector $\boldsymbol{\xi}$ as well as the eigenvalues $\mu = -s, -s+1, \ldots, s-1, s$, of s_3. As assumed in Section 1, the outcome of a spin measurement performed with this device is unaffected by displacements and (because we are neglecting relativistic effects) by homogeneous galilean transformations.

It follows that the generator of displacements is, as in (1.7), the momentum $\mathbf{P} = \mathbf{P}^0 \otimes I^I$, and that in position-spin space it is given by the two equivalent relations

$$(\mathbf{P}f)_\mu(\boldsymbol{\xi}) = -i\nabla f_\mu(\boldsymbol{\xi}),$$
$$\langle\boldsymbol{\xi}, \mu|\mathbf{P}^0|\boldsymbol{\xi}', \mu'\rangle = -i\delta_{\mu\mu'}\nabla\delta(\boldsymbol{\xi} - \boldsymbol{\xi}').$$

Similarly the generator $\mathbf{G}$ of the homogeneous galilean transformations is of the form $\mathbf{G} = \mathbf{G}^0 \otimes I^I = m\mathbf{Q}^0 \otimes I^I = m\mathbf{Q}$.

Let $\mathbf{L}^0 = \mathbf{Q}^0 \times P^0$ be the orbital angular momentum on $\mathbb{H}^0$. Evidently it obeys the same commutation relations on $\mathbb{H}^0$ as the orbital angular momentum of a spinless particle does

$$\left[L_i^0, L_j^0\right] = i\epsilon_{ijk}L_k^0, \qquad \left[L_i^0, P_j^0\right] = i\epsilon_{ijk}P_k^0, \qquad \left[L_i^0, G_j^0\right] = i\epsilon_{ijk}G_k^0.$$

We show that the generator of rotations on $\mathbb{H}$ is the operator $\mathbf{J} = \mathbf{L}^0 \otimes I^I + I^0 \otimes \mathbf{s}^I$. In order to do so, we need only verify that the commutation relations (V.6.18) among $\mathbf{J}$, $\mathbf{P}$, and $\mathbf{G}$ are obeyed. This follows by an easy calculation from the definition of the operators and the commutation relations for the components of $\mathbf{L}^0$ and of $\mathbf{s}^I$. We thus find that the unitary operator $W(\mathbf{e}, \varphi)$ on $\mathbb{H}$ that represents the rotation $\mathbf{j}_\mathbf{e}(\varphi)$ is

$$W(\mathbf{e}, \varphi) = e^{-i\mathbf{e}\cdot\mathbf{J}\varphi}, \qquad \mathbf{J} = \mathbf{L} + \mathbf{s}.$$

The generator $\mathbf{J}$ is called the (*total*) *angular momentum* of the particle.

In the notation of Section VII.4, consider the "angular" subspace $\mathbb{H}_\Omega^0 \subset \mathbb{H}^0$ in which the operators $(L^0)^2$ and L_3^0 form a maximal set of observables, and let $\mathbb{H}_\Omega^{0(l)} \subset \mathbb{H}_\Omega^0$ be the eigensubspace of $(L^0)^2$ belonging to its eigenvalue $l(l+1)$. In this subspace the eigenvectors $|m\rangle$, $m = -l, -l+1 \ldots, l-1, l$, form a basis. The components of $\mathbf{L}$ together with the components of $\mathbf{s}$ are irreducible on $\mathbb{H}_\Omega^{(l)} = \mathbb{H}_\Omega^{0(l)} \otimes \mathbb{H}^I \subset \mathbb{H}$. A basis in $\mathbb{H}_\Omega^{(l)}$ is the product basis $|m,\mu\rangle = |m\rangle|\mu\rangle$ of the eigenvectors of L_3 and those of s_3. It is clear that the subspaces $\mathbb{H}_\Omega^{(l)}$, $l = 0, 1, \ldots$, reduce $\mathbf{J}$. Therefore the problem of finding the spectral resolution of J^2 and J_3 can be studied separately in each of these subspaces.

In the following paragraphs we shall prove the so-called *angular momentum addition theorem*. According to it [see equation (3.2) below] the operators J^2 and J_3 constitute a maximal set on $\mathbb{H}_\Omega^{(l)}$. There exists a basis of their common eigenvectors which we denote by $|j,k\rangle$ such that

$$J^2|j,k\rangle = j(j+1)|j,k\rangle, \qquad J_3|j,k\rangle = k|j,k\rangle,$$

$$j = |l-\mu|, |l-\mu|+1, \ldots, l+\mu, \qquad k = -j, -j+1, \ldots, j.$$

Thus, the number of eigenvalues of J^2 is the lesser of $2l+1$ and $2\mu+1$.

In a notation similar to the one of Section VII.3, let $\mathbf{J}_a = \{J_{a1}, J_{a2}, J_{a3}\}$ be a triplet of operators[2] that obey commutation relations of the form (VII.3.1)

$$[J_{a1}, J_{a2}] = iJ_{a3}, \qquad [J_{a2}, J_{a3}] = iJ_{a1}, \qquad [J_{a3}, J_{a1}] = iJ_{a2}. \tag{3.1}$$

We know that the three operators determine a $(2j_a+1)$-dimensional space $\mathbb{V}_a$ on which any of them is maximal, any pair is irreducible, and $J_a^2 = J_{a1}^2 + J_{a2}^2 + J_{a3}^2$ equals the identity multiplied by $j_a(j_a+1)$. Let $\mathbf{J}_b = \{J_{b1}, J_{b2}, J_{b3}\}$ be another such triplet and let the corresponding $(2j_b+1)$-dimensional space be $\mathbb{V}_b$. As is customary, we call $|j_a, m_a\rangle \in \mathbb{V}_a$ the eigenvector of J_{a3} belonging to the eigenvalue $m_a = -j_a, \ldots, j_a$ (the index j_a in the ket serves only to label the dimension of the space $\mathbb{V}_a$). Similarly $|j_b, m_b\rangle \in \mathbb{V}_b, m_b = -j_b, \ldots, j_b$, denotes an eigenvector of J_{b3}.

Our problem is to study the spectral resolution of the operator $\mathbf{J} = \mathbf{J}_a + \mathbf{J}_b = \{J_1, J_2, J_3\}$, $J_k = J_{ak} + J_{bk}$, $k = 1,2,3$, on the tensor product space $\mathbb{V} = \mathbb{V}_a \otimes \mathbb{V}_b$ of dimension $(2j_a+1)(2j_b+1)$ (we use the same notation for an operator on one of the factor spaces and for its extension to the tensor product). A basis in $\mathbb{V}$ is the product basis formed by the eigenvectors $|j_a, j_b, m_a, m_b\rangle = |j_a, m_a\rangle|j_b, m_b\rangle$. Since

$$J_3|j_a, j_b, m_a, m_b\rangle = (J_{a3} + J_{b3})|j_a, j_b, m_a, m_b\rangle = (m_a + m_b)|j_a, j_b, m_a, m_b\rangle,$$

the vector $|j_a, j_b, m_a, m_b\rangle$ is an eigenvector of J_3 belonging to the eigenvalue $m =$

[2] We do not call the triplet a *vector* because no assumption is made here about the behaviour of the operators under a rotation.

$m_a + m_b$, but is not an eigenvector of J^2. Evidently we have

$$m = -(j_a + j_b), -(j_a + j_b) + 1, \ldots, (j_a + j_b) - 1, (j_a + j_b).$$

Thus, there are $2(j_a + j_b) + 1$ different eigenvalues of J_3. This number is less than the dimension of $\mathbb{V}$ (unless j_a or j_b equal zero); some eigenvalues m are therefore degenerate, and J_3 is not maximal on $\mathbb{V}$.

Since $\mathbf{J}$ obeys commutation relations of the form (3.1) on $\mathbb{V}$, it follows from Section VII.3 that this space can be decomposed into a direct sum of subspaces $\mathbb{V}^{(j)}$ which are eigensubspaces of J^2 belonging to the eigenvalue $j(j+1)$ and in which there exists a basis $|j, m\rangle$, of eigenvectors of J_3 belonging to the eigenvalues $m = -j, \ldots, j$. We shall now find the numbers j and verify that they are all different.

The degeneracy of the eigenvalue m of J_3 on $\mathbb{V}$ equals the number of ways in which it can be written in the form $m_a + m_b$ with $m_a = -j_a, \ldots, j_a$, and $m_b = -j_b, \ldots, j_b$. This number equals the number of decompositions of the non-negative integer $n = m + j_b + j_a \le 2j_b + 2j_a$ into the sum of two non-negative integers $p = m_a + j_a \le 2j_a$ and $q = m_b + j_b \le 2j_b$. Assume with no loss of generality that $j_a \ge j_b$. It is obvious that if $n \le 2j_b$, i.e., $m \le -(j_a - j_b)$, there are $n + 1 = m + j_a + j_b + 1$ decompositions corresponding to the values $0, 1, \ldots, n$ of q (a little thought will show that there are $2j_b + 1$ decompositions for $2j_b < n \le 2j_a$, and $2(j_a + j_b) - n + 1$ for $n \ge 2j_a$, but these results are not required).

The smallest eigenvalue m of J_3 is $-(j_a + j_b)$ and the above count shows that it is simple. Its eigenvector must lie in an eigensubspace $\mathbb{V}^{(j_a+j_b)}$ of J^2 which belongs to the eigenvalue $j = j_a + j_b$, and has $2(j_a + j_b) + 1$ dimensions. Evidently this subspace, as well as its orthogonal complement $(\mathbb{V}^{(j_a+j_b)})^\perp$, reduce both J^2 and J_3. In $\mathbb{V}^{(j_a+j_b)}$ there is a basis of eigenvectors $|j_a + j_b, m\rangle$ of J_3 belonging to the eigenvalues $m = -(j_a + j_b), \ldots, j_a + j_b$.

According to our count, the eigenvalue $-(j_a + j_b) + 1$ of J_3 has multiplicity 2. Hence, there is an eigenvector of J_3 which lies in $(\mathbb{V}^{(j_a+j_b)})^\perp$ and belongs also to that eigenvalue. It is possible to regard J^2 and J_3 as operators on $(\mathbb{V}^{(j_a+j_b)})^\perp$ because it is a subspace that reduces both operators. The smallest eigenvalue of J_3 on this subspace is $-(j_a + j_b) + 1$, and is simple. We can repeat, *mutatis mutandis*, what was said in the previous paragraph: there exists an eigensubspace $\mathbb{V}^{(j_a+j_b-1)} \subset \mathbb{V}$ of J^2, belonging to the eigenvalue $j = j_a + j_b - 1$, of dimension $2(j_a + j_b - 1) + 1$, in which there is a basis of eigenvectors $|j_a + j_b - 1, m\rangle$ of J_3 belonging to the eigenvalues $m = -(j_a + j_b - 1), \ldots, j_a + j_b - 1$. Clearly, the subspaces $\mathbb{V}^{(j_a+j_b)}$ and $\mathbb{V}^{(j_a+j_b-1)}$ are orthogonal.

We can now proceed recursively until we reach the eigenvalue $m = j_a - j_b$

of J_3. We will find a sequence of orthogonal subspaces $\mathbb{V}^{(j)} \subset \mathbb{V}$, $j = j_a - j_b, j_a - j_b + 1, \ldots, j_a + j_b$, in each of which there exists a basis of eigenvectors $|j,m\rangle$, $m = -j, \ldots, j$, of J_3. The direct sum of the subspaces $\mathbb{V}^{(j)}$ is, in fact, the space $\mathbb{V}$. Indeed the dimension of the direct sum is

$$[2(j_a + j_b) + 1] + [2(j_a + j_b - 1) + 1] + \ldots + [2(j_a - j_b) + 1] = (2j_a + 1)(2j_b + 1)$$

which is precisely the dimension of $\mathbb{V}$.

The above results are incorporated into the following theorem.

Let $\mathbf{J}_a$ and $\mathbf{J}_b$ be two triplets of operators which are irreducible on the spaces $\mathbb{V}_a$, of dimension $2j_a + 1$, and $\mathbb{V}_b$, of dimension $2j_b + 1$, respectively, and that obey commutation relations of the form (3.1). Let $\mathbf{J} = \mathbf{J}_a + \mathbf{J}_b$ be a triplet defined on the space $\mathbb{V} = \mathbb{V}_a \otimes \mathbb{V}_b$. Then there exists in $\mathbb{V}$ a basis of common eigenvectors $|j,m\rangle$ of J^2 and J_3 such that

$$\begin{gathered} J^2|j,m\rangle = j(j+1)|j,m\rangle, \qquad J_3|j,m\rangle = m|j,m\rangle, \\ j = |j_a - j_b|, |j_a - j_b| + 1, \ldots, j_a + j_b, \qquad m = -j, -j+1, \ldots, j. \end{gathered} \tag{3.2}$$

Because the components of angular momentum are triplets that satisfy commutation relations of the form (3.1), this theorem is known as the *addition of angular momentum* theorem. We remark again that the validity of the theorem rests only on the commutation relations and is independent of the vector nature of the angular momentum.

We note for future reference the following easy to prove relation among operators on $\mathbb{V}$,

$$J_{a1}J_{b1} + J_{a2}J_{b2} + J_{a3}J_{b3} = \tfrac{1}{2}\left[J^2 - j_a(j_a+1) - j_b(j_b+1)\right]. \tag{3.3}$$

The scalar products $\langle j_a, j_b, m_a, m_b|j,m\rangle$ are the elements of the unitary matrix that changes a representation of $\mathbb{V}$ with respect to the basis $\{|j,m\rangle\}$ into one with respect to the product basis $\{|j_a, j_b, m_a, m_b\rangle\}$. They are known as the Wigner or Clebsch-Gordan coefficients. Because of their importance in spectroscopy they have been extensively studied and tabulated. A closed formula for their evaluation was obtained by Wigner from which numerous simple relations between the coefficients can be found. We shall be content with quoting here the relation

$$\langle j_a, j_b, m_a, m_b|j,m\rangle = (-1)^{j_a+j_b-j}\langle j_b, j_a, m_b, m_a|j,m\rangle \tag{3.4}$$

which will be needed in Chapter XI.

4. The hamiltonian of a particle with spin

In this section we examine the restrictions that galilean invariance imposes on the hamiltonian of a particle with spin subject to the influence of an external electromagnetic field. Because, as shown by Dirac's theory of the electron, the relativistic nature of the electromagnetic field is closely linked with the concept of spin, we can only expect the results to be valid for low energies. In the course of the discussion we will have to make additional assumptions that are consistent with this limited range of validity. Consequently, the results will not be as compelling as in the case of the spinless particle, treated in Section VII.5.

Matters are considerably simplified if we restrict ourselves to particles for which space inversion and time reversal are dynamical symmetries. Since it is a result of experience that this is the case for all known particles that can be treated non-relativistically, we shall not discuss a more general case.

Since the spin has the properties of angular momentum, it is reasonable to assume that it has the same transformation properties as the latter. Under space inversion the orbital angular momentum remains unchanged; we therefore define the space inversion of a particle with spin as the transformation from the set of linear operators of $\mathbb{H}$ to itself such that

$$A(\mathbf{P},\mathbf{Q},\mathrm{s}) \mapsto A(-\mathbf{P},-\mathbf{Q},\mathrm{s}).$$

Or equivalently

$$\mathbf{P} \mapsto -\mathbf{P}, \qquad \mathbf{Q} \mapsto -\mathbf{Q}, \qquad \mathrm{s} \mapsto \mathrm{s}.$$

If we call Π^0 the parity operator introduced in Section VIII.3 on the spinless particle space $\mathbb{H}^0$, it follows that the operator $\Pi = \Pi^0 \otimes I^I$ on $\mathbb{H}$ is unitary and hermitean and such that

$$\Pi\mathbf{P}\Pi = -\mathbf{P}, \qquad \Pi\mathbf{Q}\Pi = -\mathbf{Q}, \qquad \Pi\mathrm{s}\Pi = \mathrm{s}.$$

The operator Π is the parity operator for the particle with spin.

Because the orbital angular momentum changes sign under time reversal, we define now the time reversal as the transformation of the operators of $\mathbb{H}$ onto themselves and of the time t such that

$$A(\mathbf{P},\mathbf{Q},\mathrm{s}) \mapsto A(-\mathbf{P},\mathbf{Q},-\mathrm{s}), \qquad t \mapsto -t.$$

Instead of the first relation we can write the equivalent one

$$\mathbf{P} \mapsto -\mathbf{P}, \qquad \mathbf{Q} \mapsto \mathbf{Q}, \qquad \mathrm{s} \mapsto -\mathrm{s}.$$

As in Section VIII.4, we seek an antiunitary operator K that satisfies

$$K\mathbf{P}K^{-1} = -\mathbf{P}, \qquad K\mathbf{Q}K^{-1} = \mathbf{Q}, \qquad K\mathbf{s}K^{-1} = -\mathbf{s}.$$

If we find an antiunitary operator K^I on the spin space $\mathbb{H}^I$ that satisfies the third relation above (when s is regarded as an operator on $\mathbb{H}^I$), then the operator $K_0^0 \otimes K^I$, where K_0^0 is the complex conjugation operator on $\mathbb{H}^0$, will have the desired properties.

We shall now construct an operator K^I on $\mathbb{H}^I$ (the superscript I on s will be omitted for brevity). A rotation of π about the 2-axis changes the sign of s_1 and s_3, but not of s_2. Since this rotation is represented by the unitary operator $W = e^{-i\pi s_2}$, we have

$$Ws_1W^\dagger = -s_1, \qquad Ws_2W^\dagger = s_2, \qquad Ws_3W^\dagger = -s_3.$$

If an antiunitary transformation Λ which changes the sign of s_2 alone can be found, then, the antiunitary operator $W\Lambda$ will have the required properties. Such a Λ is compatible with the commutation relations of s. Indeed in the commutation relation $[s_3, s_1] = is_2$ both sides would be left unchanged by the transformation, while both sides of the other two commutation relations will change sign. The operator Λ is defined by its action on a basis of $\mathbb{H}^I$. Since we want s_3 to be unaffected, we define Λ by requiring that it leaves unchanged a given basis of eigenvectors $|\mu\rangle$ of s_3. We take the basis to be one of the form (VII.3.10), that is, one in which eigenvectors belonging to neighbouring eigenvalues are related by

$$s_+|\mu\rangle = a_\mu|\mu + 1\rangle, \qquad s_-|\mu\rangle = b_\mu|\mu - 1\rangle. \tag{4.1}$$

where $s_+ = s_1 + is_2$ and $s_- = s_1 - is_2$ are the raising and lowering operators, and a_μ and b_μ are real. The antiunitary operator Λ is then defined by $\Lambda|\mu\rangle = |\mu\rangle$. Multiplying this equation on the left by Λ, we get $\Lambda^2|\mu\rangle = |\mu\rangle$ from which, since Λ^2 is linear, we conclude that $\Lambda^2 = I^I$. Therefore, $\Lambda^{-1} = \Lambda$, and $\Lambda^{-1}|\mu\rangle = |\mu\rangle$. Next we verify that the Λ thus defined leaves s_3 unchanged. For every basis vector $|\mu\rangle$ we have

$$\Lambda s_3 \Lambda^{-1}|\mu\rangle = \Lambda s_3|\mu\rangle = \Lambda\mu|\mu\rangle = \mu\Lambda|\mu\rangle = \mu|\mu\rangle = s_3|\mu\rangle,$$

that is,

$$\Lambda s_3 \Lambda^{-1} = s_3. \tag{4.2}$$

Since the coefficients a_μ and b_μ are real, it is easy to show from (4.1) that $\Lambda s_+ \Lambda^{-1} = s_+$, and $\Lambda s_- \Lambda^{-1} = s_-$. From which, noting that $s_1 = (s_+ + s_-)/2$, and $s_1 = (s_+ - s_-)/2i$, we obtain immediately

$$\Lambda s_1 \Lambda^{-1} = s_1, \qquad \Lambda s_2 \Lambda^{-1} = -s_2. \tag{4.3}$$

Having found an antiunitary operator Λ with the required properties (4.2) and (4.3), the time reversal operator on the spin space $\mathbb{H}^I$ can be taken to be

$$K^I = e^{-i\pi s_2}\Lambda.$$

An argument similar to the one in Section VIII.4 tells us that $K^{I^2} = \pm I^I$. We verify that this is the case. First we observe that $\Lambda e^{-i\pi s_2}\Lambda = \Lambda e^{-i\pi s_2}\Lambda^{-1} = e^{-i\pi s_2}$, as can be seen by expanding the exponential in a Taylor series and using the transformation equation (4.3) for s_2. We have, then

$$K^{I^2} = e^{-i\pi s_2}\Lambda e^{-i\pi s_2}\Lambda = e^{-i2\pi s_2}.$$

The action of this linear operator on an eigenvector $|\nu\rangle$ of s_2 belonging to the eigenvalue ν is

$$K^{I^2}|\nu\rangle = e^{-i2\pi s_2}|\nu\rangle = e^{-i2\pi\nu}|\nu\rangle = (-1)^{2s}|\nu\rangle$$

because $e^{-i2\pi\nu} = +1, (-1)$ if the spin s is integer (half-integer). Since the eigenvectors $|\nu\rangle$ constitute a basis on $\mathbb{H}^I$, it follows that

$$K^{I^2} = (-1)^{2s} I^I.$$

Having chosen K^I as the time reversal operator on $\mathbb{H}^I$, we take accordingly as time reversal operator on the full space $\mathbb{H}$ the tensor product of the complex conjugation in position space times K^I

$$K = K_0^0 \otimes K^I.$$

We have, thus, shown that for the particle with spin the mapping of the set of operators $A(\mathbf{Q}, \mathbf{P}, \mathbf{s})$ on $\mathbb{H}$ onto itself, defined by $A(\mathbf{P}, \mathbf{Q}, \mathbf{s}) \mapsto A(-\mathbf{P}, \mathbf{Q}, -\mathbf{s})$ constitutes a kinematical symmetry and we have found a corresponding antiunitary operator K.

We can now turn our attention to the problem of determining the general form of the hamiltonian of a particle with spin that satisfies galilean invariance and for which space inversion and time reversal are dynamical symmetries.

Since, as discussed in Section I, the momentum $\mathbf{P}$ and the generator of homogeneous galilean transformations $\mathbf{G}$ are of the form $\mathbf{P} = \mathbf{P}^0 \otimes I^I$ and $\mathbf{G} = \mathbf{G}^0 \otimes I^I$, the discussion in Section VII.1 leading to the relations

$$[\dot{\mathbf{Q}}, \mathbf{G}] = -i\mathbf{1}, \qquad [\mathbf{G}, \mathbf{P}] = im\mathbf{1}, \qquad \mathbf{G} = m\mathbf{Q}$$

can be repeated almost *verbatim*. As there, we find that $m\dot{\mathbf{Q}} - \mathbf{P}$ commutes with $\mathbf{Q}$. Now, the general form of an operator that commutes with $\mathbf{Q}$ is a function $F(\mathbf{Q}, \mathbf{s})$.

To see this, remember that every operator on $\mathbb{H}$ is of the form $F(\mathbf{Q},\mathbf{P},\mathbf{s})$, that is, a polynomial (or a limit thereof) in the components of $\mathbf{Q}$, $\mathbf{P}$, and $\mathbf{s}$. Since $\mathbf{s}$ commutes with $\mathbf{Q}$ and $\mathbf{P}$ every monomial term can be written as a tensor product $M(\mathbf{Q}^0,\mathbf{P}^0)\otimes N(\mathbf{s}^I)$, and, for this term to commute with $\mathbf{Q}$, we must have $M(\mathbf{Q}^0,\mathbf{P}^0) = M(\mathbf{Q}^0)$, or, $F(\mathbf{Q},\mathbf{P},\mathbf{s}) = F(\mathbf{Q},\mathbf{s})$. Therefore, we have

$$m\dot{\mathbf{Q}} = \mathbf{P} - \mathbf{A}_t(\mathbf{Q},\mathbf{s}). \tag{4.4}$$

where $\mathbf{A}_t$ is obviously a vector operator.

To determine the form of the hamiltonian H_t, we proceed as in Section VII.3. From the commutation relation $[\dot{\mathbf{Q}},\mathbf{Q}] = -i/m$ we can verify that a particular solution of $\dot{\mathbf{Q}} = i[H_t,\mathbf{Q}]$, regarded as an equation for H_t, is $\frac{1}{2}m|\dot{\mathbf{Q}}|^2$. If we now put $H_t = \frac{1}{2}m|\dot{\mathbf{Q}}|^2 + H_t'$ we find that H_t' commutes with $\mathbf{Q}$ and must therefore be a scalar function $V_t(\mathbf{Q},\mathbf{s})$. Thus, taking into account the expression (4.4) for $\dot{\mathbf{Q}}$, we have

$$H_t = \frac{1}{2m}|\mathbf{P} - \mathbf{A}_t(\mathbf{Q},\mathbf{s})|^2 + V_t(\mathbf{Q},\mathbf{s}). \tag{4.5}$$

We shall now assume that we are dealing with an experimental setup in which the field sources are electromagnetic, which is the most interesting case. Furthermore, for simplicity, the hamiltonian is taken to be time independent, and, as mentioned at the beginning of this section, will be assumed to be invariant under space inversion and time reversal. Without loss of generality we can write

$$\begin{aligned} \mathbf{A}_t(\mathbf{Q},\mathbf{s}) &= e\mathbf{A}(\mathbf{Q}) + \mathbf{A}_s(\mathbf{Q},\mathbf{s}), \\ V_t(\mathbf{Q},\mathbf{s}) &= eV(\mathbf{Q}) + V_s(\mathbf{Q},\mathbf{s}), \end{aligned} \tag{4.6}$$

where $\mathbf{A}(\mathbf{Q})$ and $V(\mathbf{Q})$ are the vector and scalar electromagnetic potentials produced by the field sources. In accordance with the correspondence mentioned in Section VII.6 we take them to be the same function of $\mathbf{Q}$ as the classical potentials are of position.

As remarked in Chapter VIII, $\mathbf{A}$ changes sign under space inversion and time reversal, while V preserves its sign under these two transformations. For the hamiltonian to be invariant under both of them, $\mathbf{A}_s$ and V_s must present similar behaviour. These quantities should vanish in the absence of applied fields or in the case of zero spin. We shall only consider them to the lowest order of the spin-field interaction because, for the reasons mentioned at the beginning of this section, we can only expect our results to be valid in the low energy limit. The operator $\mathbf{A}_s$ must be a vector operator that changes sign under both space inversion and time reversal. In lowest order, there are two vectors, $\mathbf{A}\times\mathbf{s}$ and $V\mathbf{s}$, that can be formed with the potentials and the spin. None of them has the required transformation properties. Therefore $\mathbf{A}_s$ cannot be linear in the potentials. A similar argument shows that neither can the scalar V_s. We shall see next that if, instead of the potentials, one considers the electric and magnetic fields, then, quantities linear in the fields with the correct transformation properties can be constructed.

Classically the electric and magnetic field at a point $\mathbf{x}$ are given by $-\nabla V(\mathbf{x})$, and $\nabla \times \mathbf{A}(\mathbf{x})$, respectively. The corresponding operators are $\mathbf{E}(\mathbf{Q}) = -(\nabla V)(\mathbf{Q})$ and $\mathbf{B}(\mathbf{Q}) = (\nabla \times V)(\mathbf{Q})$, where the notation $(\nabla F)(\cdot)$ means the function obtained when the gradient of F is calculated, and $(\nabla \times \mathbf{F})(\cdot)$ has a similar meaning. Naturally, when position-spin representation is used, if

$$\langle \boldsymbol{\xi}, \mu | f \rangle = f_\mu(\boldsymbol{\xi}),$$

then,

$$\langle \boldsymbol{\xi}, \mu | \mathbf{E}(\mathbf{Q}) | f \rangle = -\left[\nabla V(\boldsymbol{\xi})\right] f_\mu(\boldsymbol{\xi}),$$
$$\langle \boldsymbol{\xi}, \mu | \mathbf{B}(\mathbf{Q}) | f \rangle = \left[\nabla \times V(\boldsymbol{\xi})\right] f_\mu(\boldsymbol{\xi}).$$

From which one can verify the following identities

$$\begin{aligned} \mathbf{E}(\mathbf{Q}) &= -i\left[\mathbf{P}V(\mathbf{Q}) - V(\mathbf{Q})\mathbf{P}\right], \\ \mathbf{B}(\mathbf{Q}) &= i\left[\mathbf{P} \times \mathbf{A}(\mathbf{Q}) + \mathbf{A}(\mathbf{Q}) \times \mathbf{P}\right]. \end{aligned} \tag{4.7}$$

From these expressions one can see that, under space inversion, $\mathbf{E}$ changes sign and $\mathbf{B}$ is invariant, while, under time reversal (which being antiunitary changes the sign of i), $\mathbf{E}$ is invariant and $\mathbf{B}$ changes sign. With the fields $\mathbf{E}$ and $\mathbf{B}$ and the spin we can form only one vector, $\mathbf{E} \times \mathbf{s}$, with the same transformation properties as $\mathbf{A}$, and one scalar, $\mathbf{B} \cdot \mathbf{s}$, with those of V. We thus write

$$\mathbf{A}_s(\mathbf{Q}, \mathbf{s}) = -a\mathbf{E} \times \mathbf{s}, \qquad V_s(\mathbf{Q}, \mathbf{s}) = -b\mathbf{B} \cdot \mathbf{s},$$

where a and b are undetermined constants. When one uses this result in the expressions (4.7) for the potentials, the hamiltonian (4.5) becomes

$$H = \frac{1}{2} m |\dot{\mathbf{Q}}|^2 + eV(\mathbf{Q}) - b\mathbf{B}(\mathbf{Q}) \cdot \mathbf{s}, \tag{4.8}$$

$$\dot{\mathbf{Q}} = \frac{1}{m}\left[\mathbf{P} - e\mathbf{A}(\mathbf{Q}) - a\mathbf{E}(\mathbf{Q}) \times \mathbf{s}\right]. \tag{4.9}$$

Since we are working to lowest order in the spin-field interaction, we must calculate $|\dot{\mathbf{Q}}|^2$ to this order only. To do this, note first that by the identity (4.7) we have

$$\mathbf{P} \times \mathbf{E} = i\mathbf{P} \times V\mathbf{P} = i\mathbf{P}V \times \mathbf{P} = i(\mathbf{P}V - V\mathbf{P}) \times \mathbf{P} = -\mathbf{E} \times \mathbf{P}$$

from which, since s commute with **P** and **E**, we get

$$\mathbf{P} \cdot \mathbf{E} \times \mathbf{s} = \mathbf{s} \cdot \mathbf{P} \times \mathbf{E} = -\mathbf{s} \cdot \mathbf{E} \times \mathbf{P} = \mathbf{E} \times \mathbf{s} \cdot \mathbf{P},$$

Using this relation and taking the square of (4.9) we obtain

$$\frac{1}{2}m|\dot{\mathbf{Q}}|^2 = \frac{1}{2m}\left[|\mathbf{P} - e\mathbf{A}|^2 - 2a\mathbf{E} \cdot (\mathbf{P} - e\mathbf{A}) \times \mathbf{s} + a^2|\mathbf{E} \times \mathbf{s}|^2\right],$$

which to the order of our approximation yields

$$\frac{1}{2}m|\dot{\mathbf{Q}}|^2 = \frac{1}{2m}|\mathbf{P} - e\mathbf{A}|^2 - a\mathbf{E} \cdot \dot{\mathbf{Q}} \times \mathbf{s}.$$

and the hamiltonian (4.8) becomes

$$H = \frac{1}{2m}|\mathbf{P} - e\mathbf{A}(\mathbf{Q})|^2 + eV(\mathbf{Q}) - a\mathbf{E}(\mathbf{Q}) \cdot \dot{\mathbf{Q}} \times \mathbf{s} - b\mathbf{B}(\mathbf{Q}) \cdot \mathbf{s}. \tag{4.10}$$

Experience shows that the constants a and b are fixed for a particle with given mass, spin and charge. One may therefore take the intrinsic parameters of a particle with spin to be the mass m, the charge e, and the "value" s of the spin.

For a central scalar potential $V(|\mathbf{Q}|)$ one finds that, in position space, the electric field is the multiplicative operator $|\boldsymbol{\xi}|^{-1}V'(|\boldsymbol{\xi}|)\boldsymbol{\xi}$, that is, $\mathbf{E}(\mathbf{Q}) = |\mathbf{Q}|^{-1}V'(|\mathbf{Q}|)\mathbf{Q}$. To the level of approximation that we are using, we can replace $\dot{\mathbf{Q}}$ by $\mathbf{P}/m$ in the third term of the expression (4.10) for H. Since $\mathbf{Q} \times \mathbf{P}$ equals the orbital angular momentum, this term becomes

$$(a/m)|\mathbf{Q}|^{-1}V'(|\mathbf{Q}|)\mathbf{L} \cdot \mathbf{s}.$$

If the magnetic field is constant, one can take the vector potential to be $\mathbf{A}(\mathbf{Q}) = \frac{1}{2}\mathbf{B} \times \mathbf{Q}$. Inserting this value into (4.10) and expanding the square we obtain

$$H = \frac{|\mathbf{P}|^2}{2m} + eV(Q) + \mathbf{B} \cdot \left(\frac{e}{2m}\mathbf{L} + b\mathbf{s}\right) + \frac{a}{m}\frac{V'(Q)}{Q}\mathbf{L} \cdot \mathbf{s} + \frac{e^2}{4}|\mathbf{B} \times \mathbf{Q}|^2,$$

where we have used the easily verifiable relations $\mathbf{P} \times \mathbf{Q} = -\mathbf{Q} \times \mathbf{P}$ and $\mathbf{P} \cdot \mathbf{B} \times \mathbf{Q} = \mathbf{B} \cdot \mathbf{Q} \times \mathbf{P} = \mathbf{B} \cdot \mathbf{L}$, and put $Q = |\mathbf{Q}|$.

If the particle is an electron, the spin terms are of the form predicted by Dirac's relativistic theory of the electron in the low energy limit. In conventional S.I. units, this theory gives the following values of the constants: $a = e/2mc^2, b = e/m$, where c is the speed of light.

Thus, for an electron in a central scalar potential and a constant magnetic field, the hamiltonian (4.10) becomes

$$H = \frac{|\mathbf{P}|^2}{2m} + eV(Q) + \frac{e}{2m}\left[\mathbf{B}\cdot(\mathbf{L}+2\mathbf{s}) + \frac{V'(Q)}{mc^2 Q}\mathbf{L}\cdot\mathbf{s}\right] + \frac{e^2}{4}|\mathbf{B}\times\mathbf{Q}|^2.$$

The reader will recognize the third term as the Zeeman term, the fourth one as the spin-orbit coupling and the last one as the paramagnetic term. The low energy approximation of Dirac's theory predicts two more terms in the hamiltonian of an electron in a central field which, for an atomic electron, are comparable with the spin-orbit term. One of them, $-|\mathbf{P}|^2/8m^3c^2$ arises also in the expansion of the classical relativistic hamiltonian. The other $\nabla^2 V/8m^2c^2$, the so called *Darwin term*, indicates that in quantum mechanics the field-electron interaction is not strictly local.

5. Example: a neutral particle in a magnetic field

As an example we consider a neutral particle with spin 1/2 in an external inhomogeneous magnetic field **B**. This example not only has intrinsic interest in the analysis of a Stern-Gerlach experiment but also illustrates in detail how to work with operators that have a continuous spectrum on a Hilbert space regarded as the tensor product of an infinite- and a finite-dimensional one.

According to equation (4.10), the hamiltonian has the form

$$\frac{|\mathbf{P}|^2}{2m} - b\mathbf{B}(\mathbf{Q})\cdot\mathbf{s}.$$

We shall assume the magnetic field to be directed along the 3-axis and to have a constant gradient in the same direction: $\mathbf{B} = B_0(1+\gamma\mathbf{Q}\cdot\mathbf{e}_3)\mathbf{e}_3$. The hamiltonian becomes then

$$\frac{P_1^2+P_2^2}{2m} + \frac{P_3^2}{2m} - bB_0(1+\gamma Q_3)s_3.$$

We shall ignore the uninteresting first term that corresponds to a free particle in two dimensions and study a system with hamiltonian

$$H = \frac{P^2}{2m} - bB_0(1+\gamma Q)s_3,$$

where, for brevity, the subscript has been dropped from the components P_3 and Q_3.

The Hilbert space is the tensor product of an infinite-dimensional space $\mathbb{H}^0$ with irreducible set of observables $\{P, Q\}$, and a two-dimensional spin space $\mathbb{H}^I$

with spin operator s. In the P-s_3 representation a vector $|f\rangle \in \mathbb{H} = \mathbb{H}^0 \otimes \mathbb{H}^I$ is a function $\langle\mu, \eta|f\rangle = f_\mu(\eta)$ of the continuous variable $\eta \in (-\infty, \infty)$ and of the discrete variable $\mu = \pm\frac{1}{2}$. We write it as the column matrix

$$\mathbf{f}(\eta) = (f_\mu)(\eta) = \begin{pmatrix} f_{\frac{1}{2}}(\eta) \\ f_{-\frac{1}{2}}(\eta) \end{pmatrix},$$

where the boldface letter denotes a column matrix. An operator A becomes a distribution $A_{\mu\mu'}(\eta, \eta') = \langle\mu, \eta|A|\mu', \eta'\rangle$ of the discrete variables μ, $\mu' = \pm\frac{1}{2}$ and the continuous variables $\eta, \eta' \in (-\infty, \infty)$, which can be written as a square matrix

$$\mathbf{A}(\eta, \eta') = (A_{\mu\mu'})(\eta, \eta') = \begin{pmatrix} A_{\frac{1}{2},\frac{1}{2}}(\eta, \eta') & A_{\frac{1}{2},-\frac{1}{2}}(\eta, \eta') \\ A_{-\frac{1}{2},\frac{1}{2}}(\eta, \eta') & A_{-\frac{1}{2},-\frac{1}{2}}(\eta, \eta') \end{pmatrix}.$$

Again the boldface letter denotes a 2×2 matrix. The vector $|g\rangle = A|f\rangle$ becomes then the column matrix with elements

$$g_\mu(\eta) = \sum_{\mu'} \int_{-\infty}^{\infty} d\eta' \, A_{\mu\mu'}(\eta, \eta') f_{\mu'}(\eta').$$

In the case being considered the hamiltonian has the off-diagonal terms equal zero because it commutes with s_3. We can therefore write

$$\mathbf{H}(\eta, \eta') = \begin{pmatrix} H_{\frac{1}{2}}(\eta, \eta') & 0 \\ 0 & H_{-\frac{1}{2}}(\eta, \eta') \end{pmatrix}.$$

Instead of thinking of the operators as a matrix of distributions we can think of them as a matrix of operators on the function space and write

$$(Af)_\mu(\eta) = \sum_{\mu'} \hat{A}_{\mu\mu'} f_{\mu'}(x)$$

which can also be written as a matrix equation

$$((Af)_\mu)(\eta) = \hat{\mathbf{A}}\mathbf{f}(\eta).$$

For the hamiltonian we have

$$\hat{\mathbf{H}} = (\hat{H}_{\mu\mu'}) = \begin{pmatrix} \hat{H}_{\frac{1}{2}} & 0 \\ 0 & \hat{H}_{-\frac{1}{2}} \end{pmatrix},$$

where

$$\hat{H}_{\pm\frac{1}{2}} = \frac{\eta^2}{2m} \mp \tfrac{1}{2} b B_0 \left(1 + i\gamma \frac{d}{d\eta}\right).$$

The generalized eigenvalue equation for the hamiltonian becomes in this notation

$$\frac{\eta^2}{2m}\,\psi^\epsilon_\mu(\eta) - \omega_\mu\left(1 + i\gamma\frac{d}{d\eta}\right)\psi^\epsilon_\mu(\eta) = \epsilon\psi^\epsilon_\mu(\eta).$$

Here, $\omega_\mu = \frac{1}{2}\mathrm{sgn}(\mu) b B_0$, and we have denoted by $|\psi^\epsilon\rangle$ an eigenvector of $\mathbb{H}$ belonging to the eigenvalue ϵ. With the replacements $-\omega_\mu\gamma \to m\gamma, \epsilon + \omega_\mu \to \epsilon$, this equation becomes identical with the eigenvalue equation for a particle in a constant field of force discussed in Section VI.5. Transcribing the solution there obtained we have

$$\psi^\epsilon_\mu(\eta) = \kappa^\epsilon_\mu \exp\left[\frac{i}{\omega_\mu\gamma}\left((\epsilon + \omega_\mu) - \frac{\eta^3}{6m}\right)\right].$$

The integration constants κ^ϵ_μ are constrained by the orthonormality relation,

$$\sum_\mu \int_{-\infty}^{\infty} d\eta\ \psi^{\epsilon\,*}_\mu(\eta)\psi^{\epsilon'}_\mu(\eta) = \delta(\epsilon - \epsilon').$$

Substituting in the integrand the two solutions $\psi^\epsilon_{\frac{1}{2}}$ and $\psi^\epsilon_{-\frac{1}{2}}$ one obtains after performing the integration

$$\left(|\kappa^\epsilon_{\frac{1}{2}}|^2 + |\kappa^\epsilon_{-\frac{1}{2}}|^2\right) 2\pi|\omega_\mu\gamma|\delta(\epsilon - \epsilon') = \delta(\epsilon - \epsilon').$$

The indeterminacy of the $|\kappa^\epsilon_\mu|$ shows that the spectrum is degenerate. It is however easy to select two orthogonal eigenvectors $|\psi^\epsilon_1\rangle, |\psi^\epsilon_2\rangle$ belonging to the eigenvalue ϵ. It suffices to take

$$\psi^\epsilon_{(1)}(\eta) = \begin{pmatrix} \psi^\epsilon_{\frac{1}{2}}(\eta) \\ 0 \end{pmatrix}, \qquad \psi^\epsilon_{(2)}(\eta) = \begin{pmatrix} 0 \\ \psi^\epsilon_{-\frac{1}{2}}(\eta) \end{pmatrix},$$

with

$$\psi^\epsilon_\mu(\eta) = \frac{1}{\sqrt{2\pi|\omega_\mu\gamma|}} \exp\left[\frac{i}{\omega_\mu\gamma}\left((\epsilon + \omega_\mu) - \frac{\eta^3}{6m}\right)\right].$$

The evolution operator

$$U = e^{-iHt} = \int_{-\infty}^{\infty} d\epsilon\ e^{-i\epsilon t}\left(|\psi^\epsilon_{(1)}\rangle\langle\psi^\epsilon_{(1)}| + |\psi^\epsilon_{(2)}\rangle\langle\psi^\epsilon_{(2)}|\right)$$

can be explicitly written immediately by referring to equation (VI.5.3):

$$\mathbf{U}(\eta, \eta') = \begin{pmatrix} U_{\frac{1}{2}}(\eta, \eta') & 0 \\ 0 & U_{-\frac{1}{2}}(\eta, \eta') \end{pmatrix},$$

where

$$U_\mu(\eta,\eta') = \exp\left(-i\,\frac{\eta^3 - \eta'^3}{6\omega_\mu m\gamma}\right)\delta(\eta - \eta' - \omega_\mu\gamma t). \tag{5.1}$$

If the preparation state of the system is given by the matrix with elements

$$\rho_{0,\mu\mu'}(\eta,\eta') = \langle\mu,\eta|\rho_0|\mu',\eta'\rangle,$$

then the state at time t can easily be found from equation (VI.5.4) to be given by the matrix

$$\boldsymbol{\rho}_t(\eta,\eta') = \begin{pmatrix} \rho_{t,\frac{1}{2},\frac{1}{2}}(\eta,\eta') & \rho_{t,\frac{1}{2},-\frac{1}{2}}(\eta,\eta') \\ \rho_{t,-\frac{1}{2},\frac{1}{2}}(\eta,\eta') & \rho_{t,-\frac{1}{2},-\frac{1}{2}}(\eta,\eta') \end{pmatrix},$$

with elements

$$\rho_{t,\mu\mu'}(\eta,\eta') = \exp\left[-\frac{it}{2m}(\eta + \eta' - \omega_\mu\gamma t)(\eta - \eta')\right]\rho_{0,\mu\mu'}(\eta - \omega_\mu\gamma t,\, \eta' - \omega_\mu\gamma t),$$

which completely solves the dynamical problem.

In particular if the system is prepared in a pure state given by the vector

$$\boldsymbol{\psi}_0(\eta) = \begin{pmatrix} \psi_{0,\frac{1}{2}}(\eta) \\ \psi_{0,-\frac{1}{2}}(\eta) \end{pmatrix},$$

then the state at time t is pure and given by the vector

$$\boldsymbol{\psi}_t(\eta) = \begin{pmatrix} \psi_{t,\frac{1}{2}}(\eta) \\ \psi_{t,-\frac{1}{2}}(\eta) \end{pmatrix}, \qquad \psi_{t,\mu}(\eta) = \int_{-\infty}^{\infty} d\eta'\, U_\mu(\eta,\eta')\psi_{0,\mu}(\eta'),$$

which, by use of the expression (5.1) for U_μ, yields

$$\psi_{t,\mu}(\eta) = \exp\left(\frac{i\gamma}{2m}\omega_\mu\eta t^2\right)\psi_{0,\mu}(\eta - \omega_\mu\gamma t), \tag{5.2}$$

where we have dropped the unimportant phase factor

$$\exp\left\{-\frac{it}{2m}\left[\eta^2 + \tfrac{1}{3}(\omega_\mu\gamma t)^2\right]\right\}$$

common to both elements of the column matrix.

Some interesting properties of this solution will be discussed in Chapter X where we shall regard the system just considered as a composite system.

X SYSTEMS COMPOSED OF DIFFERENT SUBSYSTEMS

Roughly speaking, we think of a system as being composed of subsystems if its physical properties consist of all possible combinations of the physical properties of the subsystems. In other words, the set of system observables is constituted by the functions, in some appropriate sense, of the subsystem observables. For example, in a system of two particles, the energy is an observable which is a function of the momenta and the coordinates of the constituent particles. The observables of the system include the subsystem observables. In the above example the kinetic energies of each particle are such observables.

Recall from Section IV.5 that two systems are identical if they share a maximal set and the same intrinsic parameters; they also share, then, fundamental sets of observables. Systems that are not identical we call different.

There is a fundamental difference between systems formed by different subsystems and systems formed by identical ones. In the former case, it is in principle possible to perform measurements of subsystem observables that distinguish between the subsystems. In the latter, such possibility does not exist because all the subsystems are characterized by the same maximal set of observables and the same intrinsic parameters. For example, if in a system of two spinless neutral particles we perform a joint measurement of the momentum and the kinetic energy of a particle we can infer from it the mass. If the two particles have different masses, then we can assign the measured values unequivocally to one of the particles. Obviously this is impossible if the particles have the same mass.

It is a fact, amply verified by experience, that the two kinds of composite systems require essentially different mathematical descriptions. We discuss in this chapter composite systems formed by different subsystems. Systems formed by identical subsystems will be studied in Chapter XI.

1. The description of systems formed by different subsystems

Suppose that two different systems $\mathcal{S}^{(1)}$ and $\mathcal{S}^{(2)}$ are characterized by the fundamental sets of observables $X^{(1)} = \{X_i^{(1)}\}$, $i = 1, \ldots n_1$, on $\mathbb{H}^{(1)}$ and $X^{(2)} = \{X_j^{(2)}\}$, $j = 1, \ldots n_2$, on $\mathbb{H}^{(2)}$, and by the sets of intrinsic parameters $\mu^{(1)} = \{\mu_k^{(1)}\}$, $k = 1, \ldots, m_1$, and $\mu^{(2)} = \{\mu_l^{(2)}\}$, $l = 1, \ldots, m_2$, respectively. The set of hermitean operators $X = \{X_i^{(1)} \otimes I^{(2)}, I^{(1)} \otimes X_j^{(2)}\}$ on $\mathbb{H} = \mathbb{H}^{(1)} \otimes \mathbb{H}^{(2)}$ is then fundamental. To show that this is the case we have to show that (i) no member of X is a function of the others, and that (ii) X is irreducible.

Property (i) follows because, since $X^{(\nu)}$, $\nu = 1, 2$, is fundamental, none of its members is a function of the others. To see that (ii) holds, note first that, according to C.3.4.3, one can decompose any observable in terms of the operators

$$|b_i^{(1)}\rangle\langle b_{i'}^{(1)}| \otimes |c_j^{(2)}\rangle\langle c_{j'}^{(2)}| = |b_i^{(1)}\rangle|c_j^{(2)}\rangle\langle c_{j'}^{(2)}|\langle b_{i'}^{(1)}|$$

formed with the vectors of a product basis in $\mathbb{H}$. Thus, every operator can be written as a linear combination of product operators. All we have to do then is to show that the only product operators on $\mathbb{H}$ that commute with every operator of the set $\{X_i^{(1)} \otimes I^{(2)}, I^{(1)} \otimes X_j^{(2)}\}$ are the multiples of the identity. Let $A^{(1)} \otimes B^{(2)}$ be an operator that commutes with all the operators of the set. In case either $A^{(1)}$ or $B^{(2)}$ equal zero then $A^{(1)} \otimes B^{(2)}$ is the 0 multiple of the identity and the assertion is proved. We can next assume that neither of the factor operators is zero. We then have

$$\left[A^{(1)} \otimes B^{(2)}, X_i^{(1)} \otimes I^{(2)}\right] = \left[A^{(1)}, X_i^{(1)}\right] \otimes B^{(2)} = 0.$$

It follows from A.4.4.1 that $[A^{(1)}, X_i^{(1)}] = 0$; since $\{X_i^{(1)}\}$ is an irreducible set on $\mathbb{H}^{(1)}$, $A^{(1)} = a_1 I^{(1)}$. Similarly by considering the commutator $[A^{(1)} \otimes B^{(2)}, I^{(1)} \otimes X_j^{(2)}]$ one shows that $B^{(2)} = a_2 I^{(2)}$ and therefore $A^{(1)} \otimes B^{(2)} = a_1 a_2 I$ as was asserted.

If a system $\mathcal{S}$ has X as a fundamental set of observables and intrinsic parameters $\mu = \{\mu_k^{(1)}, \mu_l^{(2)}\}$, we say that $\mathcal{S}$ is composed of the subsystems $\mathcal{S}^{(1)}$ and $\mathcal{S}^{(2)}$. For the time being it is convenient to use the notation $A_1 = A^{(1)} \otimes I^{(2)}$, $A_2 = I^{(1)} \otimes A^{(2)}$. An observable of the form A_ν, $\nu = 1, 2$, is the observable of the system $\mathcal{S}$ corresponding to the observable $A^{(\nu)}$ of subsystem $\mathcal{S}^{(\nu)}$. Experimentally one can think of A_ν and $A^{(\nu)}$ as being measured by identical instruments. The first observable is measured in an experiment in which the composite system $\mathcal{S}$ is prepared, while the second is measured in an experiment that prepares the system $\mathcal{S}^{(\nu)}$.

From now on, when the meaning is clear from the context, we follow a common convention and denote the operators on the single-system spaces by the same symbol as the corresponding operator on the composite system space. Thus A_1 stands for both the operators $A^{(1)} \otimes I^{(2)}$ on $\mathbb{H}$ and $A^{(1)}$ on $\mathbb{H}^{(1)}$. We shall denote the hamiltonian of the system by $H_t = H(X, t)$.

According to the preceding definition, the particle with spin can be regarded as a composite system. In the notation of Section IX.3, one subsystem is a spinless particle with fundamental set of observables $\{\mathbf{Q}^0, \mathbf{P}^0\}$ on $\mathbb{H}^0$, mass m and charge e, the other a spin system of spin s and fundamental set formed by components of the spin observable $\mathbf{s}^I$ on $\mathbb{H}^I$. We shall make use of this point of view in the example of Section 4, but first we discuss at some length the important case of a system formed by two different particles.

2. A system of two different particles

Imagine an experiment in which the measuring instrument is a set of counters similar to the one introduced in Chapter VI. If in every experimental run exactly two counters are activated, we say that the experimental setup prepares a system of two particles. The set of counters defines a (composite) observable $\{\mathbf{Q}_1, \mathbf{Q}_2\}$. We assume that the particles constitute different subsystems and call $\mathbf{Q}_\nu$, $\nu = 1, 2$, the position of particle ν. If $\{\mathbf{Q}_1, \mathbf{Q}_2\}$ is maximal, then the system is constituted by two particles with no internal degrees of freedom. In the following we shall be considering both this case and the case in which the particles have spin. In the latter case, an idealized measuring instrument will consist of a set of Stern-Gerlach devices of vanishingly small dimensions as described in Section IX.3. In each experimental run exactly two devices are activated resulting in a measurement of the maximal observable $\{\mathbf{Q}_1, \mathbf{s}_1, \mathbf{Q}_2, \mathbf{s}_2\}$.

In the study of the single particle, we found the relations between the position, the velocity and the spin observables of two experiments linked by a galilean transformation. We obtained those relations by comparing the readings of the measuring instruments in both experiments. Since we are using now the same measuring instruments, those relations still apply. We shall base our discussion on them as well as on the general relations obeyed by the generators of the Galilei group.

Consider, then, the more general case of a system of two particles with spin. According to the previous section, one can take the fundamental sets of each particle to consist of the position, momentum, and spin observable: $X_\nu = \{\mathbf{P}_\nu, \mathbf{Q}_\nu, \mathbf{s}_\nu\}$, and the intrinsic parameters to be given by the mass, spin and charge: $\mu_\nu = \{m_\nu, s_\nu, e_\nu\}$, since the particles are assumed to be different, $\mu_1 \neq \mu_2$. The corresponding fundamental set of the composite system of two particles is then $X = \{\mathbf{P}_1, \mathbf{Q}_1, \mathbf{s}_1, \mathbf{P}_2, \mathbf{Q}_2, \mathbf{s}_2\}$, and its intrinsic parameters are $\mu = \{m_1, s_1, e_1, m_2, s_2, e_2\}$.

Next we examine what does galilean relativity tell us about a system of two different particles. First we shall consider a system of two spinless particles and later an isolated system formed by two particles with spin.

A system of two spinless particles. – In studying this system we allow for the presence of field sources in the experimental setup and apply, therefore, the discussion of Sections V.5 and V.6 as well as the results of Chapter VII pertaining to the transformation of the position observable.

For convenience we repeat here the commutation relations for the generators of the kinematical Galilei group. As before we call **P** and -**G** the generators of displacements and homogeneous galilean transformations, and, as in Chapter VII, we denote by **L** the generator of rotations. For brevity, and to avoid confusion between the subscripts denoting components and those denoting the subsystems, we use vector notation and write the commutation relations as

$$[\mathbf{P},\mathbf{P}] = 0, \qquad [\mathbf{G},\mathbf{G}] = 0, \tag{2.1}$$

$$[\mathbf{L},\mathbf{L}] = i\epsilon\mathbf{L}, \qquad [\mathbf{L},\mathbf{P}] = i\epsilon\mathbf{P}, \qquad [\mathbf{L},\mathbf{G}] = i\epsilon\mathbf{G}, \tag{2.2}$$

$$[\mathbf{G},\mathbf{P}] = iM\mathbf{1}. \tag{2.3}$$

where ϵ is the fully antisymmetric third rank tensor with components ϵ_{ijk}.

We need also the commutation relations among the single-particle fundamental observables, $\mathbf{P}_\nu, \mathbf{Q}_\nu,\ \nu = 1,2$,

$$[\mathbf{Q}_\nu,\mathbf{Q}_{\nu'}] = 0, \qquad [\mathbf{P}_\nu,\mathbf{P}_{\nu'}] = 0, \qquad [\mathbf{Q}_\nu,\mathbf{P}_{\nu'}] = i\delta_{\nu\nu'}\mathbf{1}. \tag{2.4}$$

Because, as mentioned at the beginning of this section, the position and velocity observables have the same behaviour under a galilean transformation as they do in the one-particle case, the following commutation relations between them and the generators apply:

$$[\mathbf{Q}_\nu,\mathbf{P}] = i\mathbf{1}, \qquad [\mathbf{L},\mathbf{Q}_\nu] = i\epsilon\mathbf{Q}_\nu, \qquad [\mathbf{Q}_\nu,\mathbf{G}] = 0, \tag{2.5}$$

$$[\dot{\mathbf{Q}}_\nu,\mathbf{G}] = -i\mathbf{1}, \tag{2.6}$$

with the velocities defined by

$$\dot{\mathbf{Q}}_\nu = i[H_t,\mathbf{Q}_\nu]. \tag{2.7}$$

We now examine what the commutation relations (2.1)-(2.6) tell us about the generators. In so doing it is often convenient to replace the maximal pair $\{\mathbf{Q}_1,\mathbf{Q}_2\}$

by the pair $\{\mathbf{Q}, \mathbf{q}\}$, also maximal, defined by

$$\mathbf{Q} = \frac{m_1\mathbf{Q}_1 + m_2\mathbf{Q}_2}{m_1 + m_2}, \qquad \mathbf{q} = \mathbf{Q}_2 - \mathbf{Q}_1. \tag{2.8}$$

We call $\mathbf{q}$ the relative position, while $\mathbf{Q}$ will shortly be identified as the centre of mass coordinate.

In search of an observable that we could identify with the total momentum, we put

$$\mathbf{P} = \mathbf{P}_1 + \mathbf{P}_2 + \mathbf{P}' \tag{2.9}$$

and substitute this expression into the first equation (2.5). We find that $\mathbf{P}'$ commutes with $\mathbf{Q}_1$ and $\mathbf{Q}_2$. Since the last pair of operators is maximal, we conclude that $\mathbf{P}'$ is a function of $\mathbf{Q}_1$ and $\mathbf{Q}_2$ only. Furthermore the first relation (2.1) requires that

$$[\mathbf{P}_1 + \mathbf{P}_2, \mathbf{P}'] + [\mathbf{P}', \mathbf{P}_1 + \mathbf{P}_2] = 0. \tag{2.10}$$

We show next that this condition permits us to remove the term $\mathbf{P}'$ from (2.9) by means of a unitary transformation $S_1(\mathbf{Q}_1, \mathbf{Q}_2)$ that depends only on $\mathbf{Q}_1$ and $\mathbf{Q}_2$ and does not change, therefore, the position observables.

Let us introduce the position space of functions of the two vector variables $\boldsymbol{\xi}_\nu$, $\nu = 1, 2$, and let $f(\boldsymbol{\xi}_1, \boldsymbol{\xi}_2) = \langle \boldsymbol{\xi}_1, \boldsymbol{\xi}_2 | f \rangle$. In this space we have in obvious notation

$$(\mathbf{P}_\nu f)(\boldsymbol{\xi}_1, \boldsymbol{\xi}_2) = -i\nabla_\nu f(\boldsymbol{\xi}_1, \boldsymbol{\xi}_2).$$

It is convenient to make the change of variables

$$\boldsymbol{\xi} = \frac{m_1\boldsymbol{\xi}_1 + m_2\boldsymbol{\xi}_2}{m_1 + m_2}, \qquad \mathbf{x} = \boldsymbol{\xi}_2 - \boldsymbol{\xi}_1$$

appropriate to the function space representation induced by the operators $\mathbf{Q}$ and $\mathbf{q}$ of (2.8), we obtain

$$((\mathbf{P}_1 + \mathbf{P}_2)f)(\boldsymbol{\xi}, \mathbf{x}) = -i\nabla_\xi f(\boldsymbol{\xi}, \mathbf{x}),$$

and the condition (2.10) becomes

$$\nabla_\xi \times \mathbf{P}'(\boldsymbol{\xi}, \mathbf{x}) = 0,$$

which means that $\mathbf{P}'(\boldsymbol{\xi}, \mathbf{x})$ is of the form

$$\mathbf{P}'(\boldsymbol{\xi}, \mathbf{x}) = \nabla_\xi F_1(\boldsymbol{\xi}, \mathbf{x}).$$

It is now an easy matter to verify (cf. the calculation at the end of Section VII.1) that the unitary transformation sought is

$$S_1(\mathbf{Q}, \mathbf{q}) = e^{iF_1(\mathbf{Q}, \mathbf{q})}.$$

We shall assume, then, that $\mathbf{P}'$ has been removed and write

$$\mathbf{P} = \mathbf{P}_1 + \mathbf{P}_2 \tag{2.11}$$

Quite naturally $\mathbf{P}$ is called the *total momentum* of the two-particle system.

Looking for something that could be called the total angular momentum, we put

$$\mathbf{L} = \mathbf{L}_1 + \mathbf{L}_2 + \mathbf{L}', \qquad \mathbf{L}_\nu = \mathbf{Q}_\nu \times \mathbf{P}_\nu, \tag{2.12}$$

one finds from the commutation relations of $\mathbf{L}$ with $\mathbf{Q}_\nu$ and $\mathbf{P}$, that $\mathbf{L}'$ is a function of $\mathbf{q}$ only. Furthermore, one can show from the first equation (2.2) that, in position space, $\mathbf{L}'$ is of the form

$$\mathbf{L}'(\mathbf{x}) = \mathbf{x} \times \nabla_x F_2(\mathbf{x}).$$

As a result there exists an unitary transformation S_2 depending on $\mathbf{q}$ only that removes $\mathbf{L}'$. It is given by

$$S_2(\mathbf{q}) = e^{iF_2(\mathbf{q})}.$$

On can therefore write the generator of rotations in the form

$$\mathbf{L} = \mathbf{L}_1 + \mathbf{L}_2 \tag{2.13}$$

and call it the *total angular momentum* of the system.

We now turn our attention to the generator of homogeneous galilean transformations.

According to equation (2.5), $\mathbf{G}$ commutes with both $\mathbf{Q}_1$ and $\mathbf{Q}_2$ and is therefore a function of them only. Let us write

$$\mathbf{G} = m_1\mathbf{Q}_1 + m_2\mathbf{Q}_2 + \mathbf{G}'(\mathbf{Q}_1, \mathbf{Q}_2). \tag{2.14}$$

In this case no further conditions are imposed on $\mathbf{G}'$ by the second commutation relation (2.1), which means that $\mathbf{G}'$ cannot be removed by a unitary transformation that commutes with the positions. Inserting this expression into (2.3) one gets

$$[\mathbf{G}', \mathbf{P}] = i(m_1 + m_2 - M)\mathbf{1}. \tag{2.15}$$

Equations (2.11) and (2.13)-(2.15) are all the conditions that we can find on the generators of the kinematical Galilei group on the basis of our experimental definition of a system of two spinless particles.

To find restrictions on the hamiltonian, one can replace the above expression for $\mathbf{G}$ into (2.7) and use the definition (2.6); the resulting restrictions are not very illuminating, for they allow the hamiltonian to be any scalar function of all the observables of the fundamental set.

One can make more headway with the further reasonable assumption that the single-particle momenta $\mathbf{P}_\nu$ transform under the operations of the Galilei group in the same way as the momentum of a one-particle system does. That is, we assume that

$$[\mathbf{P}_\nu, \mathbf{P}] = 0, \qquad [\mathbf{L}, \mathbf{P}_\nu] = i\boldsymbol{\epsilon}\mathbf{P}_\nu, \qquad [\mathbf{G}, \mathbf{P}_\nu] = im_\nu \mathbf{1}. \tag{2.16}$$

We stress that these relations are *additional* assumptions that, although borne out by experience, do not stem from the consideration of the idealized experimental setup used to define the system.

The first consequence of these assumptions is that the terms denoted by a prime in equations (2.9), (2.12) and (2.14) commute with all members of the irreducible set $\{\mathbf{P}_1, \mathbf{Q}_1, \mathbf{P}_2, \mathbf{Q}_2\}$ and are therefore zero (or unimportant multiples of the identity). As a consequence, equations (2.11) and (2.13) are obeyed without the need to perform unitary transformations. Furthermore, with $\mathbf{G}' = 0$, equation (2.15) yields the pleasing result

$$M = m_1 + m_2.$$

which, interpreting M as the mass of the system, expresses the additivity of mass.

To proceed further, it is convenient to work with the irreducible set of observables $\{\mathbf{Q}, \mathbf{P}, \mathbf{q}, \mathbf{p}\}$. The first and the third ones were defined in equation (2.8). In view of the additivity of mass we can rewrite $\mathbf{Q}$ as

$$\mathbf{Q} = \frac{1}{M}(m_1\mathbf{Q}_1 + m_2\mathbf{Q}_2)$$

and feel justified in calling it the *centre of mass* coordinate. The total momentum $\mathbf{P}$ was defined in (2.11). The last observable of the set is the *reduced momentum* $\mathbf{p}$ defined by

$$\frac{\mathbf{p}}{m} = \frac{\mathbf{P}_2}{m_2} - \frac{\mathbf{P}_1}{m_1}, \qquad m = \frac{m_1 m_2}{M},$$

where m is the *reduced mass*. The only commutation relations among the members of the set that are different from zero are

$$[\mathbf{Q}, \mathbf{P}] = i\mathbf{1}, \qquad [\mathbf{q}, \mathbf{p}] = i\mathbf{1}. \tag{2.17}$$

Furthermore $\mathbf{q}$ and $\mathbf{p}$ commute with $\mathbf{G}$ and their commutation relations with $\mathbf{L}$ are the characteristic ones of vector operators.

In terms of the new observables, the generator of homogeneous galilean transformations becomes now

$$\mathbf{G} = M\mathbf{Q}, \tag{2.18}$$

and the equations (2.6) and (2.7) for the velocities and the hamiltonian

$$[\dot{\mathbf{Q}}, \mathbf{G}] = -i\mathbf{1}, \qquad [\dot{\mathbf{q}}, \mathbf{G}] = 0 \tag{2.19}$$

and

$$\dot{\mathbf{Q}} = i[H_t, \mathbf{Q}], \qquad \dot{\mathbf{q}} = i[H_t, \mathbf{q}]. \tag{2.20}$$

Let us put

$$M\dot{\mathbf{Q}} = \mathbf{P} + \mathbf{A}, \qquad m\dot{\mathbf{q}} = \mathbf{p} + \mathbf{a} \tag{2.21}$$

and substitute these together with (2.18) into (2.19). One finds that both $\mathbf{A}$ and $\mathbf{a}$ commute with $\mathbf{Q}$. It follows from (2.17) that they cannot be functions of $\mathbf{P}$. Thus $\mathbf{A} = \mathbf{A}(\mathbf{Q}, \mathbf{q}, \mathbf{p})$, $\mathbf{a} = \mathbf{a}(\mathbf{Q}, \mathbf{q}, \mathbf{p})$.

Write the hamiltonian as

$$H_t = \tfrac{1}{2}M|\dot{\mathbf{Q}}|^2 + \tfrac{1}{2}m|\dot{\mathbf{q}}|^2 + H'. \tag{2.22}$$

Because $\mathbf{A}$ commutes with $\mathbf{Q}$, we find from (2.17) and (2.21) that

$$i\left[\tfrac{1}{2}M|\dot{\mathbf{Q}}|^2, \mathbf{Q}\right] = \dot{\mathbf{Q}},$$

and, inserting (2.22) into the first equation (2.20), that

$$[H', \mathbf{Q}] = 0.$$

Because $\mathbf{a}$ does not commute with $\mathbf{q}$ no further restrictions on the hamiltonian are found if one inserts (2.22) into the second equation (2.20).

Thus, with the additional assumption on the single-particle momenta, we find that the most general form of the hamiltonian of a system formed by two spinless particles is

$$H_t = \frac{|\mathbf{P} + \mathbf{A}_t|^2}{2M} + \frac{|\mathbf{p} + \mathbf{a}_t|^2}{2m} + V_t$$

where $\mathbf{A}_t$ and $\mathbf{a}_t$ are vector functions and V_t a scalar function of $\mathbf{Q}$, $\mathbf{q}$, and $\mathbf{p}$. We have added the subscript t because all the above functions can depend explicitly on the time that enters into the calculation only as a parameter. Also, the commutation relations of the positions and momenta with $\mathbf{L}$ have been implicitly taken into account in determining the vectorial character of the functions.

In terms of the single-particle positions and momenta the above hamiltonian can be written

$$H_t = \frac{|\mathbf{P}_1 + \mathbf{A}_1|^2}{2m_1} + \frac{|\mathbf{P}_2 + \mathbf{A}_2|^2}{2m_2} + V_t \tag{2.23}$$

where $\mathbf{A}_1 = (m_1/M)\mathbf{A}_t - \mathbf{a}_t$, $\mathbf{A}_2 = (m_2/M)\mathbf{A}_t + \mathbf{a}_t$. If there is no interaction between the particles, the hamiltonian is just the sum of single-particle hamiltonians of the form (VII.1.7) and $\mathbf{A}_1 = \mathbf{A}_1(\mathbf{Q}_1)$, $\mathbf{A}_2 = \mathbf{A}_2(\mathbf{Q}_2)$, $V_t = V_1(\mathbf{Q}_1) + V_2(\mathbf{Q}_2)$. It can be seen that galilean invariance allows for an interaction that depends not only on the relative positions of the particles but also on the reduced momentum. This is not surprising for in the case of two charged particles the magnetic field at one particle would depend on the relative velocity of the other.

It is of interest to note that the system of two particles being discussed satisfies the definition of a particle with internal degrees of freedom in Section IX.1. Indeed, in the notation of that section, the Hilbert space $\mathbb{H}$ can be regarded as a tensor product $\mathbb{H}^0 \otimes \mathbb{H}^I$ such that the centre of mass position $\mathbf{Q}$ and total momentum $\mathbf{P}$ form an irreducible set on $\mathbb{H}^0$, while the relative position $\mathbf{q}$ and reduced momentum $\mathbf{p}$ form an irreducible set on the "internal" space $\mathbb{H}^I$, and are preserved by a galilean transformation.

We also note that, with a redefinition of V_t, the hamiltonian (2.23) can also be written in the simpler form

$$H_t = \frac{|\mathbf{P} + \mathbf{A}_t(\mathbf{Q}, \mathbf{q}, \mathbf{p})|^2}{2M} + \frac{|\mathbf{p}|^2}{2m} + V_t(\mathbf{Q}, \mathbf{q}, \mathbf{p}). \tag{2.24}$$

If the system is isolated, then, as we know from Section VIII.5, the hamiltonian is the time independent generator of time translations and must obey the relations

$$[\mathbf{P}, H] = 0, \qquad [\mathbf{L}, H] = 0, \qquad [\mathbf{G}, H] = i\mathbf{P}.$$

In this case the functions $\mathbf{A}$, and V must be time independent. Moreover, the first of the above commutation relations implies that they can only be functions of $\mathbf{q}$ and $\mathbf{p}$ while the third one requires $\mathbf{A}$ to be zero. The second one simply says that the hamiltonian H is a scalar and is already obeyed by the hamiltonian (2.24). Thus, for an isolated system of two spinless particles we have

$$H = \frac{|\mathbf{P}|^2}{2M} + \frac{|\mathbf{p}|^2}{2m} + W(\mathbf{q}, \mathbf{p}).$$

The scalar function W represents the interaction between the particles. As mentioned in Section IX.4, space inversion and time reversal are dynamical symmetries of all known systems that can be treated non-relativistically. For such systems, we will see in the next subsection that the interaction can only depend on $|\mathbf{q}|$ and that, therefore, their hamiltonian is of the form

$$H = \frac{|\mathbf{P}|^2}{2M} + \frac{|\mathbf{p}|^2}{2m} + W(|\mathbf{q}|). \tag{2.25}$$

A system of two particles with spin. – Consider now an *isolated* system formed by two particles with spin. In this case, the generator of rotations is the total angular momentum

$$\mathbf{J} = \mathbf{L} + \mathbf{s}, \qquad \mathbf{s} = \mathbf{s}_1 + \mathbf{s}_2$$

which should replace $\mathbf{L}$ in the commutation relations (2.2) and (2.16). However, because $\mathbf{Q}_1, \mathbf{Q}_2, \mathbf{P}_1, \mathbf{P}_2$, and $\mathbf{G}$ commute with both $\mathbf{s}_1$ and $\mathbf{s}_2$ those commutation relations also hold as they stand. Galilean relativity and the assumption on the transformation of the single-particle momenta embodied in the commutation relations (2.16) lead to a hamiltonian of the form (2.25), except that the interaction W can now depend also on the spins $\mathbf{s}_1$ and $\mathbf{s}_2$ of the particles:

$$H = \frac{|\mathbf{P}|^2}{2M} + \frac{|\mathbf{p}|^2}{2m} + W(\mathbf{q}, \mathbf{p}, \mathbf{s}_1, \mathbf{s}_2).$$

We now determine the form of the interaction potential W subject to the additional conditions that both space inversion and time reversal be also dynamical symmetries. They require that

$$W(-\mathbf{q}, -\mathbf{p}, \mathbf{s}_1, \mathbf{s}_2) = W(\mathbf{q}, -\mathbf{p}, -\mathbf{s}_1, -\mathbf{s}_2) = W(\mathbf{q}, \mathbf{p}, \mathbf{s}_1, \mathbf{s}_2)$$

We shall work only to first order in the reduced momentum and the spins. This is appropriate from a non-relativistic point of view, for the momentum and spin dependence of the interactions found in nature are of electromagnetic origin and one does not expect higher order terms to be given correctly.

Besides the identity, we have at our disposal the following operators to construct an interaction potential with the required properties.

With the vector observable $\mathbf{q}$ we can form the scalars $f_1(q)$, and the vectors $f_2(q)\mathbf{q}$, where $q = |\mathbf{q}|$, and f_1 and f_2 are arbitrary functions. The scalars are invariant under both space inversion and time reversal while the vectors change sign under the first and are invariant under the second.

To first order, we can form no scalars with $\mathbf{p}$ and only the vector $\mathbf{p}$ itself which changes sign under both operations.

With the spins, we can construct to first order the following operators: the scalar $\mathbf{s}_1 \cdot \mathbf{s}_2$ ($|\mathbf{s}_1|^2$ and $|\mathbf{s}_2|^2$ are multiples of the identity), invariant under both operations; the vectors $\mathbf{s}_1$ and $\mathbf{s}_2$ that are invariant under space inversion and change sign under time reversal, the vector $\mathbf{s}_1 \times \mathbf{s}_2$, invariant under them; and the tensor $\mathbf{s}_1\mathbf{s}_2$, also invariant under both. However, not all components of this tensor are independent, for the distinct elements of its antisymmetric part are the components of $\mathbf{s}_1 \times \mathbf{s}_2$ and its trace is just $\mathbf{s}_1 \cdot \mathbf{s}_2$. Only its symmetric, traceless part

$$\mathbf{t} = \tfrac{1}{2}(\mathbf{s}_1\mathbf{s}_2 + \mathbf{s}_2\mathbf{s}_1) - \tfrac{1}{3}(\mathbf{s}_1 \cdot \mathbf{s}_2)\mathbf{1}$$

is independent from the other spin quantities.

Combining the quantities in the three last paragraphs we find that, up to an arbitrary multiplicative function of q, the following independent expressions are the only scalars that are invariant under both space inversion and time reversal: I, $\mathbf{s}_1 \cdot \mathbf{s}_2$, the contraction of $\mathbf{qq}$ with $\mathbf{t}$,

$$\sum_{ij} q_i q_j t_{ij} = (\mathbf{q} \cdot \mathbf{s}_1)(\mathbf{q} \cdot \mathbf{s}_2) - \tfrac{1}{3}q^2(\mathbf{s}_1 \cdot \mathbf{s}_2),$$

and the triple products $\mathbf{q} \cdot (\mathbf{s}_1 \times \mathbf{p})$ and $\mathbf{q} \cdot (\mathbf{s}_2 \times \mathbf{p})$.

Note that $\mathbf{s}_1$ and $\mathbf{s}_2$ commute with both $\mathbf{q}$ and $\mathbf{p}$. Because of this and because only different components of $\mathbf{q}$ and $\mathbf{p}$ are multiplied together in the triple products, the latter obey the rules of ordinary vector multiplication. From this property it is easy to verify that all the above operators are hermitean and commute with q.

We conclude that for an isolated system of two particles for which space inversion and time reversal are dynamical symmetries, the most general form of the interaction potential, to first order in the reduced momentum and the spins, is a sum

$$W = W_0 + W_1 + W_2 + W_3 + W_4 \tag{2.26}$$

with

$$W_0 = f_0(q), \qquad W_1 = f_1(q)(\mathbf{s}_1 \cdot \mathbf{s}_2),$$

$$W_2 = f_2(q)\left[(\mathbf{q} \cdot \mathbf{s}_1)(\mathbf{q} \cdot \mathbf{s}_2) - \tfrac{1}{3}q^2(\mathbf{s}_1 \cdot \mathbf{s}_2)\right]$$

$$W_3 = f_3(q)(\mathbf{s}_1 + \mathbf{s}_2) \cdot \mathbf{q} \times \mathbf{p}, \qquad W_4 = f_4(q)(\mathbf{s}_1 - \mathbf{s}_2) \cdot \mathbf{q} \times \mathbf{p}$$

where the functions $f_k(q)$ are arbitrary, and, for reasons that will become apparent in the next chapter, we find it convenient to write W_3 and W_4 in terms of the sum and difference of the spins.

3. Some useful representations

In this section we construct explicitly some frequently used representations of the two-particle Hilbert space $\mathbb{H}$. They will be found useful in the study of systems of identical particles to which the following chapter is devoted. We shall consider first the general case in which there are external fields present. Later we shall present the simpler case of an isolated system, where the dynamical problem can be reduced to the problem of a single particle in a central field.

In case the particles have no spin one often takes the function space representation induced by the maximal composite operator $\{\mathbf{Q}_1, \mathbf{Q}_2\}$ on $\mathbb{H}$. With

$$\mathbf{Q}_\nu = \int d^3\xi_\nu \, \boldsymbol{\xi}_\nu |\boldsymbol{\xi}_\nu\rangle\langle\boldsymbol{\xi}_\nu|,$$

we have

$$\{\mathbf{Q}_1, \mathbf{Q}_2\} = \iint d^3\xi_1 d^3\xi_2 \, \{\boldsymbol{\xi}_1, \boldsymbol{\xi}_2\}|\boldsymbol{\xi}_1, \boldsymbol{\xi}_2\rangle\langle\boldsymbol{\xi}_1, \boldsymbol{\xi}_2|.$$

A vector $|f\rangle \in \mathbb{H}$ is represented by a function of two vector variables $f(\boldsymbol{\xi}_1, \boldsymbol{\xi}_2\} = \langle\boldsymbol{\xi}_1, \boldsymbol{\xi}_2|f\rangle$, while an operator A is represented by a distribution $\langle\boldsymbol{\xi}_1, \boldsymbol{\xi}_2|A|\boldsymbol{\xi}'_1, \boldsymbol{\xi}'_2\rangle$, so that

$$(Af)(\boldsymbol{\xi}_1, \boldsymbol{\xi}_2) = \iint d^3\xi'_1 d^3\xi'_2 \, \langle\boldsymbol{\xi}_1, \boldsymbol{\xi}_2|A|\boldsymbol{\xi}'_1, \boldsymbol{\xi}'_2\rangle f(\boldsymbol{\xi}'_1, \boldsymbol{\xi}'_2\}. \tag{3.1}$$

In particular, the centre of mass position $\mathbf{Q}$ and the total momentum $\mathbf{P}$ are represented by the distributions

$$\langle\boldsymbol{\xi}_1, \boldsymbol{\xi}_2|\mathbf{Q}|\boldsymbol{\xi}'_1, \boldsymbol{\xi}'_2\rangle = \frac{m_1\boldsymbol{\xi}_1 + m_2\boldsymbol{\xi}_2}{m_1 + m_2} \delta(\boldsymbol{\xi}_1 - \boldsymbol{\xi}'_1)\delta(\boldsymbol{\xi}_2 - \boldsymbol{\xi}'_2),$$

$$\langle\boldsymbol{\xi}_1, \boldsymbol{\xi}_2|\mathbf{P}|\boldsymbol{\xi}'_1, \boldsymbol{\xi}'_2\rangle = -i\left[\delta(\boldsymbol{\xi}_2 - \boldsymbol{\xi}'_2)\nabla_1\delta(\boldsymbol{\xi}_1 - \boldsymbol{\xi}'_1) + \delta(\boldsymbol{\xi}_1 - \boldsymbol{\xi}'_1)\nabla_2\delta(\boldsymbol{\xi}_2 - \boldsymbol{\xi}'_2)\right],$$

where ∇_ν is the gradient operator on the space of functions of $\boldsymbol{\xi}_\nu$. Equivalently, we have

$$(\mathbf{Q}f)(\boldsymbol{\xi}_1, \boldsymbol{\xi}_2) = \frac{m_1\boldsymbol{\xi}_1 + m_2\boldsymbol{\xi}_2}{m_1 + m_2} f(\boldsymbol{\xi}_1, \boldsymbol{\xi}_2),$$

$$(\mathbf{P}f)(\boldsymbol{\xi}_1, \boldsymbol{\xi}_2) = -i\left[\nabla_1 f(\boldsymbol{\xi}_1, \boldsymbol{\xi}_2) + \nabla_2 f(\boldsymbol{\xi}_1, \boldsymbol{\xi}_2)\right].$$

If the particles have spin, a function-space representation for the composite system is defined by the maximal observable $\{\mathbf{Q}_1, s_{13}, \mathbf{Q}_2, s_{23}\}$, where $s_{\nu 3}$ is the component of $\mathbf{s}_\nu$ along the 3-axis. Let $|\mu_\nu\rangle$, $\mu_\nu = s_\nu, s_\nu - 1, \ldots, -s_\nu$, be the eigenvectors and eigenvalues of $s_{\nu 3}$. We have, then,

$$\{\mathbf{Q}_1, s_{13}, \mathbf{Q}_2, s_{23}\} = \sum_{\mu_1}\sum_{\mu_2}\iint d^3\xi_1 d^3\xi_2 \, \{\boldsymbol{\xi}_1, \mu_1, \boldsymbol{\xi}_2, \mu_2\}|\boldsymbol{\xi}_1, \mu_1, \boldsymbol{\xi}_2, \mu_2\rangle\langle\boldsymbol{\xi}_1, \mu_1, \boldsymbol{\xi}_2, \mu_2|.$$

A vector $|f\rangle$ is represented by a function $f_{\mu_1,\mu_2}(\boldsymbol{\xi}_1, \boldsymbol{\xi}_2) = \langle \boldsymbol{\xi}_1, \mu_1, \boldsymbol{\xi}_2, \mu_2|f\rangle$ of the two continuous vector variables $\boldsymbol{\xi}_\nu$ and of the two discrete variables μ_ν. This function is often thought of as the column matrix

$$\begin{pmatrix} f_{s_1,s_2}(\xi_1, \xi_2) \\ f_{s_1-1,s_2-1}(\xi_1, \xi_2) \\ \cdots \\ f_{-s_1,-s_2}(\xi_1, \xi_2) \end{pmatrix}$$

with $N = (2s_1 + 1)(2s_2 + 2)$ elements. In this representation an operator A is represented by a distribution $\langle \boldsymbol{\xi}_1, \mu_1, \boldsymbol{\xi}_2, \mu_2|A|\boldsymbol{\xi}'_1, \mu'_1, \boldsymbol{\xi}'_2, \mu'_2\rangle$ which can be regarded as an $N \times N$ square matrix whose elements are distributions of the four continuous vector variables $\boldsymbol{\xi}_\nu, \boldsymbol{\xi}'_\nu$. The generalization of the expression (3.1) to this case is

$$(Af)_{\mu_1,\mu_2}(\boldsymbol{\xi}_1, \boldsymbol{\xi}_2) = \sum_{\mu'_1}\sum_{\mu'_2} \iint d^3\xi'_1 d^3\xi'_2 \, \langle \boldsymbol{\xi}_1, \mu_1, \boldsymbol{\xi}_2, \mu_2|A|\boldsymbol{\xi}'_1, \mu'_1, \boldsymbol{\xi}'_2, \mu'_2\rangle f_{\mu'_1,\mu'_2}(\boldsymbol{\xi}'_1, \boldsymbol{\xi}'_2).$$

If an operator B is not a function of the spins, i.e., $B = B(\mathbf{P}, \mathbf{Q})$ alone, then the matrix is diagonal. As a result we have

$$(Bf)_{\mu_1,\mu_2}(\boldsymbol{\xi}_1, \boldsymbol{\xi}_2) = \iint d^3\xi'_1 d^3\xi'_2 \langle \boldsymbol{\xi}_1, \boldsymbol{\xi}_2|B|\boldsymbol{\xi}'_1, \boldsymbol{\xi}'_2\rangle f_{\mu_1,\mu_2}(\boldsymbol{\xi}'_1, \boldsymbol{\xi}'_2). \tag{3.2}$$

where $\langle \boldsymbol{\xi}_1, \boldsymbol{\xi}_2|\mathrm{B}|\boldsymbol{\xi}'_1, \boldsymbol{\xi}'_2\rangle$ is a distribution identical to the distribution that represents the operator $B(\mathbf{P}, \mathbf{Q})$ on the space of a system composed of spinless particles. Similar remarks apply to an operator $C = C(\mathbf{s}_1, \mathbf{s}_2)$ that depends on the spins alone. In this case we have instead of (3.2)

$$(Cf)_{\mu_1,\mu_2}(\boldsymbol{\xi}_1, \boldsymbol{\xi}_2) = \sum_{\mu'_1}\sum_{\mu'_2} \langle \mu_1, \mu_2|C|\mu'_1, \mu'_2\rangle f_{\mu'_1,\mu'_2}(\boldsymbol{\xi}_1, \boldsymbol{\xi}_2),$$

where $\langle \mu_1, \mu_2|C|\mu'_1, \mu'_2\rangle$ are the matrix elements that represent the operator $C(\mathbf{s})$ defined on the N-dimensional space of a system composed of two spin systems.

These remarks become obvious when one regards the Hilbert space of a system composed of two particles with spin as the tensor product of the space of a system of two spinless particles times the space of a system composed of two spin systems. We shall find this point of view useful when in Chapter XI we study a system composed of two identical particles.

In case the system is isolated and the particles have no internal degrees of freedom, it is advantageous to write the hamiltonian in the form (2.25). The dynamical problem can then be reduced to the problem of a single particle in a central field. As remarked following equation (2.23) the Hilbert space $\mathbb{H}$ can be written as the tensor product $\mathbb{H}^0 \otimes \mathbb{H}^I$ where $\mathbb{H}^0$ is a space on which the centre of mass coordinate $\mathbf{Q}$ and the total momentum $\mathbf{P}$ form an irreducible set, while the relative coordinate $\mathbf{q}$ and momentum $\mathbf{p}$ constitute an irreducible set on $\mathbb{H}^I$. Accordingly, we write

$$H = H^0 \otimes I^I + I^0 \otimes H^I \tag{3.3}$$

with

$$H^0 = \frac{|\mathbf{P}|^2}{2M}, \qquad H^I = \frac{|\mathbf{p}|^2}{2m} + W_0(|\mathbf{q}|).$$

The solutions of the eigenvalue problem $H|\phi\rangle = E|\phi\rangle$ are of the form $E = E^0 + E^I$, $|\phi\rangle = |\gamma\rangle|\psi\rangle$ where E^0, $|\gamma\rangle$ and E^I, $|\psi\rangle$ are solutions of the eigenvalue equations $H^0|\gamma\rangle = E^0|\gamma\rangle$ in $\mathbb{H}^0$ and $H^I|\psi\rangle = E^I|\psi\rangle$ in $\mathbb{H}^I$, respectively. The first of these equations is just a free particle equation so that it has as eigenvectors the eigenkets $|\boldsymbol{\eta}\rangle$ of the total momentum $\mathbf{P}$ with corresponding eigenvalues $E^0 = |\boldsymbol{\eta}|^2/2M$. The second equation is identical with the eigenvalue equation for the hamiltonian of a single particle of mass m, momentum $\mathbf{p}$ and position $\mathbf{q}$ in a central potential $W_0(q)$. We shall refer to this hamiltonian as the "centre of mass hamiltonian". As in Section VII.8 we decompose the space $\mathbb{H}^I$ into the product $\mathbb{H}_r \otimes \mathbb{H}_\Omega$ of a radial and an angular space. Using the polar coordinates introduced in that section, we find the solutions

$$E^I = E_{nl}, \qquad |\psi\rangle = |\psi_n, l, m_3\rangle = |\psi_{nl}\rangle|l, m_3\rangle.$$

Here, $|l, m_3\rangle \in \mathbb{H}_\Omega$ are eigenvectors of l^2 and the component l_3 of the relative angular momentum $\mathbf{l} = \mathbf{q} \times \mathbf{p}$, and E_{nl}, $|\psi_{nl}\rangle \in \mathbb{H}_r$ are the solutions of the eigenvalue problem

$$\left(\frac{\Pi_r^2}{2m} + \frac{l(l+1)}{2m} q^{-2} + W_0(q)\right)|\psi_{nl}\rangle = E_{nl}|\psi_{nl}\rangle, \tag{3.4}$$

where $l = 0, 1, \ldots$ and Π_r is the momentum conjugate to q that obeys $[q, \Pi_r] = i$. This equation is identical to the equation (VII.9.2) that describes a single particle in a central field. Since E_{nl} is independent of the eigenvalues m_3 of l_3 it has, in general, a degeneracy of $2l+1$ (larger degeneracies may occur for special forms of W_0).

As an example consider a system of two particles interacting through a potential

$$W_0(q) = \tfrac{1}{2} m\omega^2 (q - a)^2,$$

where a is a constant. This system, of interest in molecular spectroscopy, is used as an approximation to describe the rotational and vibrational levels of a diatomic molecule near the minimum of the interatomic potential which occurs when the nuclei are a distance a apart. The polar function space representation (VII.9.3) of the eigenvalue equation (3.4) becomes in this case

$$\left[-\frac{1}{2m}\frac{d^2}{dx^2}+\frac{l(l+1)}{2m(a+x)^2}+\tfrac{1}{2}m\omega^2x^2\right]\chi_{nl}(x)=E_{nl}\chi_{nl}(x)$$

where we have put $x=r-a$. The eigenfunctions $\chi_{nl}(x)$ must be square integrable on $(-a,\infty)$ and vanish at $x=-a$. In most cases of interest $|\chi_{nl}(x)|$ is negligibly small except for a range of values of the variable in which $x\ll a$. To first approximation we can then neglect x in the denominator of the second term and obtain

$$E_{nl}=E_n+\frac{l(l+1)}{2ma^2},\qquad \chi_{nl}(x)=\chi_n(x)$$

where E_n and $\chi_n(x)$ are the solutions of

$$\left(-\frac{d^2}{dx^2}+\tfrac{1}{2}m\omega^2x^2\right)\chi_n(x)=E_n\chi_n(x).$$

This equation is identical to the eigenvalue equation for a one-dimensional harmonic oscillator, but the two eigenvalue problems differ in the conditions that the eigenfunctions must satisfy. However, within the approximation in which we are working, the interval $(-a,\infty)$ of $\chi_n(x)$ can obviously be extended to $(-\infty,\infty)$, and the two eigenvalue problems become identical. The energies E_n take then the values $\omega(n+\frac{1}{2})$, $n=0,1,\ldots$, and we have

$$E_{nl}=\omega(n+\tfrac{1}{2})+\frac{l(l+1)}{2ma^2},\qquad n,l=0,1,2,\ldots$$

If the the particles have spin, it is no longer possible to reduce the dynamical problem to a single particle problem even if the system is isolated . It is still convenient, however, to write the Hilbert space $\mathbb{H}$ as the product $\mathbb{H}^0\otimes\mathbb{H}^C$, where $\mathbb{H}^0$ is as above and $\mathbb{H}^C$ is itself the tensor product of the space $\mathbb{H}^I$ considered before and the $(2s_1+1)(2s_2+1)$-dimensional spin space $\mathbb{H}^s$.

If the interaction is spin independent and given by the potential $W_0(Q)$, the energy levels will still be given by (3.4) but each will have an additional degeneracy of $(2s_1+1)(2s_2+1)$. In general, this degeneracy will be in part removed if there is a spin dependent interaction.

Imagine, for example, that, besides $W_0(q)$, there is also a spin dependent interaction of the form

$$W_1=f(q)(\mathbf{s}_1\cdot\mathbf{s}_2).$$

Putting $\mathbf{s} = \mathbf{s}_1 + \mathbf{s}_2$, we can write by (IX.3.3) $\mathbf{s}_1 \cdot \mathbf{s}_2 = \frac{1}{2}(s^2 - \vartheta)$, $\vartheta = s_1(s_1 + 1) + s_2(s_2 + 1)$, so that, in polar coordinates, we have the following centre of mass hamiltonian

$$H^C = \frac{\Pi_r^2}{2m} + \frac{l^2}{2mq^2} + W_0(q) + \tfrac{1}{2}f(q)(s^2 - \vartheta)$$

on $\mathbb{H}^C = \mathbb{H}^I \otimes \mathbb{H}^s$. This hamiltonian commutes with the (relative) orbital angular momentum $\mathbf{l} = \mathbf{q} \times \mathbf{p}$ and the total spin $\mathbf{s}$. Let $|s, \mu\rangle \in \mathbb{H}^s$ be eigenvectors of s^2 and s_3, and $|l, m_3\rangle \in \mathbb{H}_\Omega$ of l^2 and l_3. With $|\psi_{nls}\rangle \in \mathbb{H}_r$, we write an eigenvector of H^C as $|\psi_{nls}\rangle|l, m_3\rangle|s, \mu\rangle$ and find that $|\psi_{nls}\rangle$ obeys the eigenvalue equation

$$\left[\frac{\Pi_r^2}{2m} + \frac{l(l+1)q^{-2}}{2m} + W_0(q) + \tfrac{1}{2}f(q)[s(s+1) - \vartheta]\right]|\psi_{nls}\rangle = E_{nls}|\psi_{nls}\rangle.$$

The energies depend on l and s but not on m_3 and μ. Therefore, barring "accidental" degeneracies due to special forms of W_0 and f the eigenvalue E_{nls} will have degeneracy $(2l + 1)(2s + 1)$.

4. The state of a subsystem

In the previous section we have been concerned mostly with the observables of a system $\mathcal{S}$ composed of two different subsystems $\mathcal{S}^{(1)}$ and $\mathcal{S}^{(2)}$. An observable of $\mathcal{S}$ is a hermitean operator A on $\mathbb{H} = \mathbb{H}^{(1)} \otimes \mathbb{H}^{(2)}$ which is not in general the tensor product of an operator on $H^{(1)}$ times an operator on $H^{(2)}$. Examples of such observables in a two-particle system are the total momentum and the hamiltonian. There are, however, many observables of physical interest which are of the form $A^{(1)} \otimes I^{(2)}$ or $I^{(1)} \otimes A^{(2)}$ as, for example, the single-particle momentum or position in a two-particle system. In a general composite system, observables of the latter type will be referred to as subsystem observables. It will now be necessary to return to our original explicit notation that distinguishes between the operators $A_1 = A^{(1)} \otimes I^{(2)}$ and $A_2 = I^{(1)} \otimes A^{(2)}$ on $\mathbb{H}$, and the operators $A^{(1)}$ on $\mathbb{H}^{(1)}$ and $A^{(2)}$ on $\mathbb{H}^{(2)}$, respectively.

Let the state of $\mathcal{S}$ be ρ and let $\rho^{(1)}$ be a state of $\mathcal{S}^{(1)}$ that is, a positive hermitean operator with unit trace on $\mathbb{H}^{(1)}$. We say that $\rho^{(1)}$ is a state of $\mathcal{S}^{(1)}$ corresponding to the state ρ of $\mathcal{S}$ if for every subsystem observable A_1 we have

$$\operatorname{Tr}(\rho^{(1)}A^{(1)}) = \operatorname{Tr}(\rho A_1). \tag{4.1}$$

A similar definition holds, of course, for $\mathcal{S}^{(2)}$:

$$\operatorname{Tr}(\rho^{(2)}A^{(2)}) = \operatorname{Tr}(\rho A_2). \tag{4.2}$$

The following considerations provide a physical motivation for the definition. Although an experimental setup in which $\mathcal{S}$ is prepared is obviously different from another in which $\mathcal{S}^{(1)}$ alone is prepared, one can conceive of experiments on, say, a two-particle system in which the positions of the two particles are consistently measured to be so far away from each other that it would appear reasonable to think of the two particles as being unaffected by each other at the measurement time. It seems meaningful then to speak of the state of one of the particles at the measurement time. Of course the predictions of the outcome of the measurement must be identical whether one regards the particle as a system by itself or as a subsystem of the overall system. The identity of these predictions is precisely what the definition embodies.

First we remark that it follows from equation (4.1) that, for a given system state ρ, the corresponding subsystem state $\rho^{(1)}$ is unique because two such states $\rho^{(1)}$ and $\sigma^{(1)}$ must obey $\mathrm{Tr}\,(\rho^{(1)}A^{(1)}) = \mathrm{Tr}\,(\sigma^{(1)}A^{(1)})$ for all observables $A^{(1)}$.

Suppose next that $\rho^{(1)}$ and $\rho^{(2)}$ are given; what can we say about ρ? It is easy to verify that the operator $\rho^{(1)} \otimes \rho^{(2)}$ is a solution of equations (4.1) and (4.2). But this solution is not unique if neither of the subsystem states is pure. To see this, decompose each of them into two states,

$$\rho^{(1)} = a_1\sigma^{(1)} + (1-a_1)\tau^{(1)}, \qquad 0 \leq a_1 \leq 1,$$

$$\rho^{(2)} = a_2\sigma^{(2)} + (1-a_2)\tau^{(2)}, \qquad 0 \leq a_2 \leq 1.$$

It can be easily verified that

$$\rho = b\sigma^{(1)} \otimes \sigma^{(2)} + c\sigma^{(1)} \otimes \tau^{(2)} + d\tau^{(1)} \otimes \sigma^{(2)} + e\tau^{(1)} \otimes \tau^{(2)} \tag{4.3}$$

satisfies (4.1) and (4.2) provided that the positive constant $b, c, d, e,$ obey the conditions

$$c = a_1 - b, \qquad d = a_2 - b, \qquad e = 1 - a_1 - a_2 + b. \tag{4.4}$$

Since this equation is satisfied by a whole range of positive values of b that ensure the positiveness of the other constants, we have shown that there are infinitely many states of the system of the form (4.3) for given states $\rho^{(1)}, \rho^{(2)}$ of the subsystems.

It remains to see what happens if one of the subsystem states, say $\rho^{(1)}$ is pure and cannot, therefore, be decomposed. In that case, the above argument still holds, except that we must put $a_1 = 1$ and $\sigma^{(1)} = \rho^{(1)}$. We can also accept that some of the constants be zero but not negative. Putting $a_1 = 1$ into the equations (4.4) we find $e = -a_2 + b = -d$ which, by the non-negativeness of the constants, implies that $e = d = 0$, and $b = a_2$, $c = 1 - a_2$. Substituting these values in (4.3) we obtain the unique solution

$$\rho = a_2\rho^{(1)} \otimes \sigma^{(2)} + (1 - a_2)\rho^{(1)} \otimes \tau^{(2)} = \rho^{(1)} \otimes \rho^{(2)}.$$

This completes the proof of the following answer to the question.

The state ρ of a composite system is uniquely determined by the states $\rho^{(1)}$ and $\rho^{(2)}$ of the subsystems if and only if at least one of the subsystem states is pure. In the latter case, $\rho = \rho^{(1)} \otimes \rho^{(2)}$.

In physical language one says that the interaction between subsystems introduces correlations that cannot be inferred from measurements performed on the subsystems alone. The result should not be surprising, for it describes a situation which occurs also in classical statistical mechanics where the knowledge of the probability densities for two complementary subsets of the coordinates and momenta of a system does not, in general, permit the determination of the probability density for the whole set of coordinates and momenta. The last remarks suggest that if the state ρ of the overall system is known, it is then possible to determine uniquely the states $\rho^{(1)}$ and $\rho^{(2)}$ of the subsystems. This is indeed the case as we shall see next.

If ρ is not pure, it can be decomposed into a sum $\rho = \alpha\rho_a + \beta\rho_b$. If $\rho_a^{(1)}$ and $\rho_b^{(1)}$ are the states of $\mathcal{S}^{(1)}$ corresponding to the states ρ_a and ρ_b of $\mathcal{S}$, the linearity of the trace tells us that the state of $\mathcal{S}^{(1)}$ that corresponds to $\mathcal{S}$ is $\rho^{(1)} = \alpha\rho_a^{(1)} + \beta\rho_b^{(2)}$. A similar relation holds for the state of $\mathcal{S}^{(2)}$. Since every ρ can be decomposed into a linear sum of pure states, it follows that we need only to consider the case in which ρ is a pure state. As in A.4.7, we shall here be regarding the vectors of $\mathbb{H} = \mathbb{H}^{(1)} \otimes \mathbb{H}^{(2)}$ as transformations from $\mathbb{H}^{(1)}$ into $\mathbb{H}^{(2)}$ and shall use the notation of A.4.8.2.

Let us then assume that $\rho = |\mathbf{\Psi}\rangle\langle\mathbf{\Psi}|$, $|\mathbf{\Psi}\rangle \in \mathbb{H}$. Then, $\mathrm{Tr}\,(\rho A_1) = \langle\mathbf{\Psi}|A_1|\mathbf{\Psi}\rangle$, and $\rho^{(1)}$ is determined by requiring that

$$\mathrm{Tr}\,(\rho^{(1)}A^{(1)}) = \langle\mathbf{\Psi}|A_1|\mathbf{\Psi}\rangle$$

for all $A^{(1)}$. Since $A_1 = A^{(1)}\otimes I^{(2)}$, and $A^{(1)}$ is hermitean, the relation $\langle\mathbf{\Psi}|B\otimes C|\mathbf{\Psi}\rangle = \mathrm{Tr}\,\Big(B(\mathbf{\Psi}^{\ddagger}C^{\dagger}\mathbf{\Psi})\Big)$ gives

$$\mathrm{Tr}\,(\rho^{(1)}A^{(1)}) = \mathrm{Tr}\,(\mathbf{\Psi}^{\ddagger}\mathbf{\Psi}A^{(1)}).$$

From this and a similar calculation we conclude that

$$\rho^{(1)} = \mathbf{\Psi}^{\ddagger}\mathbf{\Psi}, \qquad \rho^{(2)} = \mathbf{\Psi}\mathbf{\Psi}^{\ddagger}.$$

If $|\mathbf{\Psi}\rangle$ is a product vector $|\mathbf{\Psi}\rangle = |\psi^{(1)}\rangle|\psi^{(2)}\rangle$, then, according to A.4.8.1, the above result gives $\rho^{(1)} = |\psi^{(1)}\rangle\langle\psi^{(1)}|$, $\rho^{(2)} = |\psi^{(2)}\rangle\langle\psi^{(2)}|$. This result, together with the one given above in italics proves that the state of the system is the pure state $|\psi^{(1)}\rangle|\psi^{(2)}\rangle$ if and only if the subsystem states are the pure states $|\psi^{(1)}\rangle$ and $|\psi^{(2)}\rangle$.

Return now to the case in which the pure state $|\mathbf{\Psi}\rangle$ of the system is not of the product form. We introduce an orthonormal basis in $\mathbb{H}^{(1)}$ formed by eigenvectors of $\rho^{(1)} = \mathbf{\Psi}^\ddagger\mathbf{\Psi}$. We shall denote by $|\chi_r^{(1)}\rangle$ the eigenvectors, if any, belonging to the eigenvalue zero; the others will be denoted by $|\phi_m^{(1)}\rangle$. We can then write

$$\mathbf{\Psi}^\ddagger\mathbf{\Psi} = \sum_m \lambda_m |\phi_m^{(1)}\rangle\langle\phi_m^{(1)}|, \tag{4.5}$$

where some of the λ_m may be equal. We then have by A.4.7.5

$$\langle\mathbf{\Psi}\chi_r^{(1)}|\mathbf{\Psi}\chi_r^{(1)}\rangle = \langle\chi_r^{(1)}|\mathbf{\Psi}^\ddagger\mathbf{\Psi}|\chi_r^{(1)}\rangle = 0,$$

that is, $|\mathbf{\Psi}\chi_r^{(1)}\rangle$ is the zero vector in $\mathbb{H}^{(2)}$. Also,

$$\langle\mathbf{\Psi}\phi_m^{(1)}|\mathbf{\Psi}\phi_n^{(1)}\rangle = \langle\phi_n^{(1)}|\mathbf{\Psi}^\ddagger\mathbf{\Psi}|\phi_m^{(1)}\rangle = \lambda_m\delta_{mn}.$$

Therefore, the vectors

$$|\phi_m^{(2)}\rangle = \frac{1}{\sqrt{\lambda_m}}|\mathbf{\Psi}\phi_m^{(1)}\rangle \tag{4.6}$$

form an orthonormal set (in general not a basis) on $\mathbb{H}^2$. In addition, from

$$\mathbf{\Psi}\mathbf{\Psi}^\ddagger|\phi_m^{(2)}\rangle = \frac{1}{\sqrt{\lambda_m}}\mathbf{\Psi}\mathbf{\Psi}^\ddagger|\mathbf{\Psi}\phi_m^{(1)}\rangle = \lambda_m|\phi_m^{(2)}\rangle,$$

we conclude that $|\phi_m^{(2)}\rangle$ is an eigenvector of $\rho^{(2)} = \mathbf{\Psi}\mathbf{\Psi}^\ddagger$ corresponding to the eigenvalue λ_m. We have thus shown that $\sqrt{\lambda_m}\mathbf{\Psi}$ transforms each eigenvector of $\rho^{(1)}$ belonging to the non-zero eigenvalue λ_m into an eigenvector of $\rho^{(2)}$ belonging to the same eigenvalue. Similarly one can show that if $|\phi_n^{(2)}\rangle$ is an eigenvector of $\rho^{(2)}$ belonging to the non-zero eigenvalue μ_n then

$$|\phi_n^{(1)}\rangle = \frac{1}{\sqrt{\lambda_m}}|\mathbf{\Psi}^\ddagger\phi_n^{(2)}\rangle$$

is an eigenvector of $\rho^{(1)}$ belonging to the same eigenvalue. By applying $|\mathbf{\Psi}\rangle$ to the last equation we find that $|\phi_n^{(2)}\rangle$ is of the form (4.6). Thus, every eigenvector of $\rho^{(2)}$ belonging to a non-zero eigenvalue is of that form. We conclude that the positive eigenvalues of $\rho^{(1)}$ and $\rho^{(2)}$ coincide and have the same multiplicity. Therefore,

$$\mathbf{\Psi}\mathbf{\Psi}^\ddagger = \sum_m \lambda_m|\phi_m^{(2)}\rangle\langle\phi_m^{(2)}|. \tag{4.7}$$

If zero is not an eigenvalue of $\rho^{(2)} = \boldsymbol{\Psi}\boldsymbol{\Psi}^\ddagger$, then the set $\{|\phi_m^{(2)}\rangle\}$ is a basis in $\mathbb{H}^{(2)}$. Otherwise one can extend the set to a basis by adding an orthonormal set of eigenvectors $|\chi_s^{(2)}\rangle$ belonging to the eigenvalue zero.

Construct now a basis in $\mathbb{H}$ which is the product of the bases we have introduced in $\mathbb{H}^{(1)}$ and $\mathbb{H}^{(2)}$, and expand $|\boldsymbol{\Psi}\rangle$ in terms of it. From the expression (see A.4.7.2)

$$\left(\langle\phi^{(2)}|\langle\phi^{(1)}|\right)|\boldsymbol{\Psi}\rangle = \langle\phi^{(2)}|\boldsymbol{\Psi}\phi^{(1)}\rangle = \langle\phi^{(1)}|\boldsymbol{\Psi}^\ddagger\phi^{(2)}\rangle$$

we find that the only non-zero components of $|\boldsymbol{\Psi}\rangle$ are of the form

$$\left(\langle\phi_m^{(2)}|\langle\phi_n^{(1)}|\right)|\boldsymbol{\Psi}\rangle = \langle\phi_m^{(2)}|\boldsymbol{\Psi}\phi_n^{(1)}\rangle = \sqrt{\lambda_m}\,\delta_{mn}.$$

Therefore, the expansion of $|\boldsymbol{\Psi}\rangle$ in terms of the product basis is

$$|\boldsymbol{\Psi}\rangle = \sum_m \sqrt{\lambda_m}\,|\phi_m^{(1)}\rangle|\phi_m^{(2)}\rangle. \tag{4.8}$$

Collecting together the terms that belong to equal eigenvalues in the expressions (4.5) and (4.7) for the subsystem states, we can rewrite them as

$$\rho^{(1)} = \sum_m \lambda_m R_m^{(1)}, \qquad \rho^{(2)} = \sum_m \lambda_m R_m^{(2)}, \tag{4.9}$$

where the eigenprojections $R_m^{(1)}$ on $\mathbb{H}^{(1)}$ and $R_m^{(2)}$ on $\mathbb{H}^{(2)}$ both belonging to λ_m have the same dimension. Evidently we have

$$\langle\phi_m^{(1)}|\rho^{(1)}|\phi_m^{(1)}\rangle = \langle\phi_m^{(2)}|\rho^{(2)}|\phi_m^{(2)}\rangle = \lambda_m.$$

Suppose that $\mathcal{S}$ is prepared in the state $|\boldsymbol{\Psi}\rangle$ given by (4.8), and that, using the basis vectors as eigenvectors, one constructs the two subsystem observables

$$A^{(1)} = \sum_m \alpha_m^{(1)}|\phi_m^{(1)}\rangle\langle\phi_m^{(1)}|, \qquad A^{(2)} = \sum_m \alpha_m^{(2)}|\phi_m^{(2)}\rangle\langle\phi_m^{(2)}|,$$

where all the non-zero eigenvalues $\alpha_m^{(1)}, \alpha_m^{(2)}$ are non-degenerate (the orthonormal sets $\{|\chi_r^{(1)}\rangle\}$ and $\{|\chi_s^{(2)}\rangle\}$ are bases in the eigensubspaces belonging to the eigenvalue zero of $A^{(1)}$ and $A^{(2)}$, respectively). The corresponding observables A_1 and A_2 on $\mathbb{H}$ commute and are therefore commensurable. They can be regarded as a single observable $\{A_1, A_2\}$ with eigenvectors $|\phi_m^{(1)}\rangle|\phi_n^{(2)}\rangle$, $|\phi_m^{(1)}\rangle|\chi_s^{(2)}\rangle$, $|\chi_r^{(1)}\rangle|\phi_n^{(2)}\rangle$, and $|\chi_r^{(1)}\rangle|\chi_s^{(2)}\rangle$ belonging, respectively, to the composite eigenvalues $\{\alpha_m^{(1)}, \alpha_n^{(2)}\}$, $\{\alpha_m^{(1)}, 0\}$, $\{0, \alpha_n^{(2)}\}$ and $\{0, 0\}$. The probability $\mathcal{P}(\alpha_m^{(1)}\alpha_n^{(2)}|\boldsymbol{\Psi})$ of obtaining the values

$\alpha_m^{(1)}, \alpha_n^{(2)}$ upon a joint measurement of A_1 and A_2 is, then

$$\mathcal{P}(\alpha_m^{(1)}\alpha_n^{(2)}|\Psi) = \left|\left(\langle\phi_m^{(2)}|\langle\phi_n^{(1)}|\right)|\Psi\rangle\right|^2 = \lambda_m\delta_{mn},$$

while,

$$\mathcal{P}(\alpha_m^{(1)},0|\Psi) = \mathcal{P}(0,\alpha_n^{(2)}|\Psi) = \mathcal{P}(0,0|\Psi) = 0.$$

It can be seen that there is a perfect statistical correlation between the values that would be obtained in a joint measurement of both observables.

In the next section we illustrate these results by applying them to a simple example that has often been used as a model in dicussions of the measurement problem.

5. Example: states of a particle in a magnetic field

We return to the example of a neutral particle in an inhomogeneous magnetic field for which the dynamical problem was solved in Section IX.5, that is, a system with hamiltonian

$$H = \frac{P^2}{2m} - bB_0(1+\gamma Q)s_3,$$

where Q is the component of the position along the 3-axis and $B_0(1+\gamma Q)$ is the magnetic field directed along it. We found there that if the system is prepared in a pure state given in momentum-spin representation by the vector

$$\Psi_0(\eta) = \begin{pmatrix} \psi_{0,\frac{1}{2}}(\eta) \\ \psi_{0,-\frac{1}{2}}(\eta) \end{pmatrix},$$

then, the state at time t is

$$\Psi_t(\eta) = \begin{pmatrix} \psi_{t,\frac{1}{2}}(\eta) \\ \psi_{t,-\frac{1}{2}}(\eta) \end{pmatrix}, \tag{5.1}$$

where according to equation (IX.5.2)

$$\psi_{t,\mu}(\eta) = \exp\left(i\frac{\gamma\omega_\mu}{2m}\eta t^2\right)\psi_{0,\mu}(\eta - \gamma\omega_\mu t)$$

and

$$\omega_{\frac{1}{2}} = \omega, \qquad \omega_{-\frac{1}{2}} = -\omega, \qquad \omega = \tfrac{1}{2}bB_0.$$

We shall regard the system as a composite system $\mathcal{S}$ formed by a subsystem $\mathcal{S}^{(1)}$ consisting of a neutral particle in one dimension, and a subsystem $\mathcal{S}^{(2)}$ which is a spin 1/2 system. $\mathcal{S}^{(1)}$ has a fundamental set of observables formed by the momentum $P^{(1)}$ and the position $Q^{(1)}$, while any two components of the spin observable $\mathbf{s}^{(1)}$ constitute a fundamental set of $\mathcal{S}^{(2)}$. In our notation, $P = P^{(1)} \otimes I^{(2)}$, $Q = Q^{(1)} \otimes I^{(2)}$ and $\mathbf{s} = I^{(1)} \otimes \mathbf{s}^{(2)}$.

Suppose the system is prepared at $t = 0$ in a pure state $|\mathbf{\Psi}_0\rangle$ which is of the product form: $|\mathbf{\Psi}_0\rangle = |\psi_0^{(1)}\rangle|\psi_0^{(2)}\rangle$, where the spin state $|\psi_0^{(2)}\rangle$ corresponds to a component $+1/2$ along the direction given by the polar angles θ, and φ. In the s_3-representation we have by (II.3.7)

$$|\psi_0^{(2)}\rangle = \begin{pmatrix} e^{-\frac{1}{2}i\varphi}\cos\frac{1}{2}\theta \\ e^{\frac{1}{2}i\varphi}\sin\frac{1}{2}\theta \end{pmatrix}.$$

If we put $\langle\eta|\psi_0^{(1)}\rangle = \psi_0(\eta)$, the state of $\mathcal{S}$ in momentum-spin space is

$$\mathbf{\Psi}_0(\eta) = \begin{pmatrix} \psi_0(\eta)e^{-\frac{1}{2}i\varphi}\cos\frac{1}{2}\theta \\ \psi_0(\eta)e^{\frac{1}{2}i\varphi}\sin\frac{1}{2}\theta \end{pmatrix}.$$

The state $|\mathbf{\Psi}_t\rangle$ at time t is given by (5.1) as

$$\mathbf{\Psi}_t(\eta) = \begin{pmatrix} \psi_0(\eta - \gamma\omega t)e^{-\frac{1}{2}i(\varphi-\gamma\omega\eta t^2/m)}\cos\frac{1}{2}\theta \\ \psi_0(\eta + \gamma\omega t)e^{\frac{1}{2}i(\varphi-\gamma\omega\eta t^2/m)}\sin\frac{1}{2}\theta \end{pmatrix}, \tag{5.2}$$

which is patently not of the product form because it cannot be written as a function of η multiplied by a column matrix independent of η.

Next we want to find the spin state corresponding to $|\mathbf{\Psi}_t\rangle$. We shall first describe a general procedure that was not included in the previous section not to complicate the already lengthy exposition.

In general, let $|\mathbf{\Psi}\rangle$ represent the pure state in the space $\mathbb{H}$ of an overall system $\mathcal{S}$ consisting of the subsystems $\mathcal{S}^{(1)}$ and $\mathcal{S}^{(2)}$ with spaces $\mathbb{H}^{(1)}$ and $\mathbb{H}^{(2)}$. The density operator on $\mathbb{H}$ is then $\rho = |\mathbf{\Psi}\rangle\langle\mathbf{\Psi}|$, while the density operators on $\mathbb{H}^{(1)}$ and $\mathbb{H}^{(2)}$are respectively $\rho^{(1)} = \mathbf{\Psi}^{\ddagger}\mathbf{\Psi}$ and $\rho^{(2)} = \mathbf{\Psi}\mathbf{\Psi}^{\ddagger}$. Let $\{|b_i\rangle\}, \{|c_j\rangle\}$ be arbitrary orthonormal bases in the subsystem spaces. The matrix elements of $\rho^{(1)}$ and $\rho^{(2)}$ in their respective spaces are

$$\rho_{ii'}^{(1)} = \langle b_i|\mathbf{\Psi}^{\ddagger}\mathbf{\Psi}|b_{i'}\rangle = \langle\mathbf{\Psi}b_{i'}|\mathbf{\Psi}b_i\rangle \tag{5.3}$$

$$\rho_{jj'}^{(2)} = \langle c_j|\mathbf{\Psi}\mathbf{\Psi}^{\ddagger}|c_{j'}\rangle = \langle\mathbf{\Psi}^{\ddagger}c_{j'}|\mathbf{\Psi}^{\ddagger}c_j\rangle. \tag{5.4}$$

The matrix elements of the state of $\mathcal{S}$ with respect to the product basis $\{|b_i, c_j\rangle\}$ are

$$\rho_{ij,i'j'} = \langle b_i, c_j|\mathbf{\Psi}\rangle\langle\mathbf{\Psi}|b_{i'}, c_{j'}\rangle.$$

Using the expression for the scalar product of two vectors in the product space given in A.4.7.2, we obtain

$$\langle b_i, c_j|\Psi\rangle = \sum_k \langle(\langle c_j|\langle b_i|)\, b_k|\Psi b_k\rangle = \sum_k \langle b_i|b_k\rangle\langle c_j|\Psi b_k\rangle = \langle c_j|\Psi b_i\rangle$$

so that

$$\rho_{ij,i'j'} = \langle c_j|\Psi b_i\rangle\langle \Psi b_{i'}|c_{j'}\rangle.$$

Summing over terms with $j' = j$, we get

$$\sum_j \rho_{ij,i'j} = \langle \Psi b_{i'}|\Psi b_i\rangle = \langle b_i|\Psi^\dagger\Psi|b_{i'}\rangle$$

Therefore (5.3) can be written

$$\rho^{(1)}_{ii'} = \sum_j \rho_{ij,i'j}. \tag{5.5}$$

A similar calculation yields

$$\rho^{(2)}_{jj'} = \sum_i \rho_{ij,ij'}. \tag{5.6}$$

One often expresses the relations (5.5) and (5.6) in words by saying that the subsystems states are the "partial traces" of the system state. Since an arbitrary state of $\mathcal{S}$ can always be written as a linear sum of pure states, it is easy to see that formulae (5.5) and (5.6) apply even in the general case in which the state ρ is not pure.

Returning to our example, let us abbreviate equation (5.2) by putting

$$\psi_{\frac{1}{2}}(\eta) = \psi_0(\eta - \gamma\omega t)e^{\frac{1}{2}i\gamma\omega\eta t^2/m} \tag{5.7}$$

$$\psi_{-\frac{1}{2}}(\eta) = \psi_0(\eta + \gamma\omega t)e^{-\frac{1}{2}i\gamma\omega\eta t^2/m} \tag{5.8}$$

so that

$$\Psi_t(\eta) = \begin{pmatrix} \psi_{\frac{1}{2}}(\eta)e^{-\frac{1}{2}i\varphi}\cos\frac{1}{2}\theta \\ \psi_{-\frac{1}{2}}(\eta)e^{\frac{1}{2}i\varphi}\sin\frac{1}{2}\theta \end{pmatrix}$$

The corresponding abstract vector $|\Psi_t\rangle$ can be expressed as a linear combination of product vectors

$$|\Psi_t\rangle = e^{-\frac{1}{2}i\varphi}\cos(\tfrac{1}{2}\theta)|\psi_{\frac{1}{2}}\rangle|\tfrac{1}{2}\rangle + e^{\frac{1}{2}i\varphi}\sin(\tfrac{1}{2}\theta)|\psi_{-\frac{1}{2}}\rangle|-\tfrac{1}{2}\rangle, \tag{5.9}$$

where $|\psi_{\frac{1}{2}}\rangle$ and $|\psi_{-\frac{1}{2}}\rangle$ are vectors in $\mathbb{H}^{(1)}$ with momentum space representation $\langle\eta|\psi_{\frac{1}{2}}\rangle = \psi_{\frac{1}{2}}(\eta)$, and $\langle\eta|\psi_{-\frac{1}{2}}\rangle = \psi_{-\frac{1}{2}}(\eta)$, while $|\frac{1}{2}\rangle$ and $|-\frac{1}{2}\rangle$ are the eigenvectors of s_3 in $\mathbb{H}^{(2)}$,

$$|\tfrac{1}{2}\rangle = \begin{pmatrix} 1 \\ 0 \end{pmatrix}, \qquad |-\tfrac{1}{2}\rangle = \begin{pmatrix} 0 \\ 1 \end{pmatrix}.$$

Using the decomposition (5.9) we obtain the vectors of $\mathbb{H}^{(1)}$,

$$|\Psi_t^\dagger\ \tfrac{1}{2}\rangle = e^{-\frac{1}{2}i\varphi}\cos(\tfrac{1}{2}\theta)|\psi_{\frac{1}{2}}\rangle, \qquad |\Psi_t^\dagger\ -\tfrac{1}{2}\rangle = e^{\frac{1}{2}i\varphi}\sin(\tfrac{1}{2}\theta)|\psi_{-\frac{1}{2}}\rangle$$

from which the matrix representing the spin state $\rho^{(2)}$ can be calculated with the help of (5.4)

$$\rho^{(2)} = \begin{pmatrix} \cos^2\frac{1}{2}\theta & \frac{1}{2}\kappa e^{-i\varphi}\sin\theta \\ \frac{1}{2}\kappa^* e^{i\varphi}\sin\theta & \sin^2\frac{1}{2}\theta \end{pmatrix}, \tag{5.10}$$

where, by the definitions (5.7) and (5.8),

$$\begin{aligned} k = \langle\psi_{-\frac{1}{2}}|\psi_{\frac{1}{2}}\rangle &= \int_{-\infty}^{\infty} d\eta\ \psi^*_{-\frac{1}{2}}(\eta)\psi_{\frac{1}{2}}(\eta) \\ &= \int_{-\infty}^{\infty} d\eta\ e^{i\gamma\omega\eta t^2/m}\psi_0(\eta+\gamma\omega t)\psi_0(\eta-\gamma\omega t), \end{aligned} \tag{5.11}$$

and we have taken into account that, by normalization,

$$\langle\psi_{\frac{1}{2}}|\psi_{\frac{1}{2}}\rangle = \int_{-\infty}^{\infty} d\eta\ |\psi_0(\eta-\gamma\omega t)|^2 = \int_{-\infty}^{\infty} d\eta\ |\psi_0(\eta)|^2 = 1$$

and, similarly, $\langle\psi_{-\frac{1}{2}}|\psi_{-\frac{1}{2}}\rangle = 1$.

To calculate the state of $\mathcal{S}^{(1)}$ we use the generalization of formula (5.3) to the space of generalized vectors. The generalization is obtained by replacing formally the $|b_i\rangle$ by the generalized eigenvectors $|\eta\rangle$ of the momentum. The formula then reads

$$\rho^{(1)}(\eta,\eta') = \langle\eta|\Psi_t^\dagger\Psi_t|\eta'\rangle = \langle\Psi_t\eta'|\Psi_t\eta\rangle.$$

From the expression (5.9), the vector $|\Psi_t\eta\rangle \in \mathbb{H}^{(2)}$ is computed to be

$$|\Psi_t\eta\rangle = e^{-\frac{1}{2}i\varphi}\cos(\tfrac{1}{2}\theta)\psi_{\frac{1}{2}}(\eta)|\tfrac{1}{2}\rangle + e^{\frac{1}{2}i\varphi}\sin(\tfrac{1}{2}\theta)\psi_{-\frac{1}{2}}(\eta)|-\tfrac{1}{2}\rangle,$$

from which we get the following expression for the state of $\mathcal{S}^{(1)}$ in momentum space

$$\rho^{(1)}(\eta,\eta') = \cos^2(\tfrac{1}{2}\theta)\psi_{\frac{1}{2}}(\eta)\psi^*_{\frac{1}{2}}(\eta') + \sin^2(\tfrac{1}{2}\theta)\psi_{-\frac{1}{2}}(\eta)\psi^*_{-\frac{1}{2}}(\eta'). \qquad (5.12)$$

The reader may wish to verify this result by direct use of the expressions (5.5) and (5.6).

According to the discussion surrounding equations (4.8) and (4.9) of the last section, $\rho^{(1)}$ and $\rho^{(2)}$ must have the same non-zero eigenvalues. We verify next that this is the case in the present example.

From the secular equation of the matrix (5.10), the eigenvalues of $\rho^{(2)}$ are easily found to be

$$\lambda_{1,2} = \tfrac{1}{2}\left(1 \pm \sqrt{\cos^2\theta + |\kappa|^2\sin^2\theta}\right) \qquad (5.13)$$

with κ given by (5.11).

The eigenvalue equation $\rho^{(1)}|\phi\rangle = \lambda|\phi\rangle$ reads in momentum space

$$\int_{-\infty}^{\infty} d\eta'\ \rho^{(1)}(\eta,\eta')\phi(\eta') = \lambda\phi(\eta').$$

Substituting the expression (5.12) for $\rho^{(1)}$ into this equation, we get

$$a_{\frac{1}{2}}\cos^2(\tfrac{1}{2}\theta)\psi_{\frac{1}{2}}(\eta) + a_{-\frac{1}{2}}\sin^2(\tfrac{1}{2}\theta)\psi_{-\frac{1}{2}}(\eta) = \lambda\phi(\eta) \qquad (5.14)$$

where

$$a_{\pm\frac{1}{2}} = \int_{-\infty}^{\infty} d\eta'\ \psi^*_{\pm\frac{1}{2}}(\eta')\phi(\eta').$$

Inserting the expression (5.14) for $\phi(\eta)$ into the last integral and taking into account the definition of κ, we obtain, if $\lambda \neq 0$, the system of homogeneous equations for $a_{\pm\frac{1}{2}}$

$$\left[\cos^2(\tfrac{1}{2}\theta) - \lambda\right] a_{\frac{1}{2}} + \kappa^*\sin^2(\tfrac{1}{2}\theta)a_{-\frac{1}{2}} = 0,$$

$$\kappa\cos^2(\tfrac{1}{2}\theta)a_{\frac{1}{2}} + \left[\sin^2(\tfrac{1}{2}\theta) - \lambda\right] a_{-\frac{1}{2}} = 0,$$

whose compatibility condition is seen to give precisely the eigenvalues (5.13). One can now compute the eigenfunctions of both $\rho^{(1)}$ and $\rho^{(2)}$ and verify the formulae (4.8) and (4.9). We shall do this in a particular case in which is both interesting and simple.

Suppose that the initial state of $\mathcal{S}^{(1)}$ is represented by a narrow momentum wave packet $\psi_0(\eta)$ concentrated around $\eta = 0$. Then, for large enough t, the quantity κ defined in equation (5.11) will be equal to zero because the functions in the integrand will have no overlap, that is, $|\psi_{\frac{1}{2}}\rangle$ and $|\psi_{-\frac{1}{2}}\rangle$ will be orthogonal. In this case, the eigenvalues (5.13) become

$$\lambda_1 = \cos^2 \tfrac{1}{2}\theta, \qquad \lambda_2 = \sin^2 \tfrac{1}{2}\theta.$$

The spin state (5.10) reduces to

$$\rho^{(2)} = \begin{pmatrix} \cos^2 \frac{1}{2}\theta & 0 \\ 0 & \sin^2 \frac{1}{2}\theta \end{pmatrix},$$

that is,

$$\rho^{(2)} = \cos^2(\tfrac{1}{2}\theta)|\tfrac{1}{2}\rangle\langle\tfrac{1}{2}| + \sin^2(\tfrac{1}{2}\theta)|-\tfrac{1}{2}\rangle\langle-\tfrac{1}{2}|, \tag{5.15}$$

a formula which is a particular case of (4.9). Since $|\psi_{\frac{1}{2}}\rangle$ and $|\psi_{-\frac{1}{2}}\rangle$ are orthogonal, so are the one-dimensional projections $|\psi_{\frac{1}{2}}\rangle\langle\psi_{\frac{1}{2}}|$ and $|\psi_{-\frac{1}{2}}\rangle\langle\psi_{-\frac{1}{2}}|$, and the formula (5.12), which can be written

$$\rho^{(1)} = \cos^2(\tfrac{1}{2}\theta)|\psi_{\frac{1}{2}}\rangle\langle\psi_{\frac{1}{2}}| + \sin^2(\tfrac{1}{2}\theta)|\psi_{-\frac{1}{2}}\rangle\langle\psi_{-\frac{1}{2}}|,$$

is also of the form (4.9). Moreover, $e^{-\frac{1}{2}i\varphi}|\frac{1}{2}\rangle$ and $e^{\frac{1}{2}i\varphi}|-\frac{1}{2}\rangle$ are also normalized eigenvectors of $\rho^{(2)}$, so that the expression (5.9) for $|\Psi_t\rangle$ is already of the form (4.8).

We conclude this section with some remarks that illustrate the discussion at the end of Section 2 on the correlation between the measurements on the constituent subsystems.

The probability that a measurement of the spin at time t will yield the value $\frac{1}{2}$ is $\mathrm{Tr}\left(\rho^{(2)}|\frac{1}{2}\rangle\langle\frac{1}{2}|\right) = \cos^2 \frac{1}{2}\theta$, by (5.15). This probability is equal to the probability of obtaining a positive value of the 3-component of the momentum. Indeed, by (5.12), the latter probability is

$$\int_0^\infty d\eta\ \rho(\eta, \eta) = \cos^2 \tfrac{1}{2}\theta$$

because $\psi_{\frac{1}{2}}(\eta)$ and $\psi_{-\frac{1}{2}}(\eta)$ are respectively concentrated in $(0, \infty)$ and $(-\infty, 0)$.

The correlation between measurements of the 3-component of the spin and the sign of the 3-component of the momentum is perfect in the following sense: let $\mathcal{P}(\Delta, \mu)$ be the probability for obtaining the 3-component of the momentum in the interval Δ and the value μ of s_3 in a joint measurement of both observables. Let $\Delta_+ = (0,\infty)$, $\Delta_- = (-\infty,0)$, then, $\mathcal{P}(\Delta_+,\frac{1}{2}) = \mathcal{P}(\Delta_-,-\frac{1}{2}) = 1$, $\mathcal{P}(\Delta_+,-\frac{1}{2}) = \mathcal{P}(\Delta_-,\frac{1}{2}) = 0$, These values can be verified immediately from the formula

$$\mathcal{P}(\Delta,\mu) = \int_{\Delta} d\eta \; |\langle \eta, \mu | \Psi_t \rangle|^2$$

and the expression (5.9) for $|\Psi_t\rangle$. Thus, in a Stern-Gerlach experiment on can measure the component of the spin of a particle along the magnetic field gradient by measuring the sign of the corresponding component of the momentum, that is, by determinig if the particle is deflected towards or away from the field gradient.

XI SYSTEMS OF IDENTICAL PARTICLES

As already remarked, it is an amply verified experimental fact that the properties of composite systems are strongly dependent on whether the constituent subsystems are identical or different. At the root of this dependence lies the fact that there is no way of experimentally deciding whether a certain measurement of a property of a subsystem should be attributed to one subsystem and not to another.

In the classical description of a system of identical particles (i.e, of the same mass) it is always possible, in principle, to label each particle with its initial position and velocity and, by following their well defined trajectories during the motion, identify it at any time. Such a labelling is not possible in the quantum mechanical description because a particle does not have a well defined trajectory and, in principle, a measurement of its position could yield any value (fortunately, in practice, particles prepared in an experiment can be ignored by an experimenter in a distant laboratory because they have a negligible probability of being found in it).

The basic rules of quantum mechanics are strong enough that, by simply taking seriously the fact that identical systems are experimentally indistinguishable, lead to the explanation of the observed differences between the properties of systems formed by identical or different subsystems. To study how this comes about is the object of the present chapter.

1. The description of systems formed by two identical subsystems

As was already implied in the discussion of Section X.1, the mathematical description of composite systems there introduced fails when the component subsystems are identical. To understand the reason for this failure, let us consider a composite system $\mathcal{C}$ formed by two identical subsystems $\mathcal{S}$. Each of the subsystems will be characterized by the same fundamental set $X = \{X_\kappa\}$ of observables on a Hilbert space $\mathbb{H}$. Assume for the moment that, as was the case with a system composed of different subsystems, the composite system is characterized by the fundamental set of observables $\mathbf{X} = \{X_\kappa \otimes I, I \otimes X_{\kappa'}\}$ on the Hilbert space $\mathbb{T} = \mathbb{H} \otimes \mathbb{H}$. Because the tensor product is not symmetric in its factors, the mathematical description distinguishes between the operators $X_\kappa \otimes I$ and $I \otimes X_\kappa$ which represent the single-subsystem observable X_κ when measured in an experiment on the composite system. This distinction is however not warranted by the physical situation, for, since both subsystems are characterized by identical sets of observables and intrinsic parameters, there is no measurement that can distinguish between them.

It turns out that starting with the space $\mathbb{T}$ and the irreducible set $\mathbf{X}$ it is possible to construct spaces and irreducible sets of operators on them that appropriately take into account the physical indistinguishability of the subsystems. To construct those spaces we start by imagining that is possible to describe the states of the composite system on the tensor product space $\mathbb{T}$ in a way similar to the one discussed in Chapter X.

The factor spaces in the product $\mathbb{H} \otimes \mathbb{H}$ are identical copies of each other and mathematically they are distinguished exclusively by their order. We shall denote corresponding operators and vectors in each copy by the same letter. When they appear in a product the order will show to which copy they belong. For example, in the product $|f\rangle|g\rangle \in \mathbb{T}$, $|f\rangle$ is to be regarded as a vector of the first factor and $|g\rangle$ as one of the second (however, following the usual notation for product bras, we write $\langle f, g| = \langle g|\langle f|$, where $\langle f|$ is a bra of the first factor space and $\langle g|$ a bra of the second). Occasionally, when referring to $\mathbb{H}$ we may want to specify whether we are regarding it as the first or the second factor in the product $\mathbb{H} \otimes \mathbb{H}$; we shall do so by writing $\mathbb{H}^{(1)}$ or $\mathbb{H}^{(2)}$, respectively. In those occasions we shall distinguish (unphysically) between the corresponding subsystems by referring to them as $\mathcal{S}^{(1)}$ and $\mathcal{S}^{(2)}$. General vectors and operators of the product space $\mathbb{T}$ will be denoted by boldface letters.

Imagine that the state of the overall system $\mathcal{C}$ is a pure state represented by the product vector $|\phi\rangle|\psi\rangle$, it follows that the state of one of the subsystems is $|\phi\rangle$ and that of the other $|\psi\rangle$. Such would also be the case if the state of the overall system were represented by the vector $|\psi\rangle|\phi\rangle$.

We would like to consider next the more general case in which the system $\mathcal{C}$ is in an arbitrary pure state. To this end let us introduce the linear *transposition operator* $\mathbf{\Pi}$ on $\mathbb{T}$ which is defined for a product vector $|\phi\rangle|\psi\rangle$ by

$$\mathbf{\Pi}|\phi\rangle|\psi\rangle = |\psi\rangle|\phi\rangle.$$

The action of $\mathbf{\Pi}$ on an arbitrary vector $|\mathbf{\Phi}\rangle \in \mathbb{T}$ can be determined by linearity: let

$\{|b_i\rangle|b_j\rangle\}$ be a product basis and

$$|\mathbf{\Phi}\rangle = \sum_{ij} \Phi_{ij}|b_i\rangle|b_j\rangle, \tag{1.1}$$

then

$$\mathbf{\Pi}|\mathbf{\Phi}\rangle = \sum_{ij} \Phi_{ij}\mathbf{\Pi}|b_i\rangle|b_j\rangle = \sum_{ij} \Phi_{ij}|b_j\rangle|b_i\rangle.$$

Referring to the definition of conjugate vectors in A.4.6 we see that the last term in this expression is the conjugate of the sum in (1.1). Therefore we have

$$\mathbf{\Pi}|\mathbf{\Phi}\rangle = |\mathbf{\Phi}^{\ddagger}\rangle.$$

An expression which is meaningful because from $\mathbb{T} = \mathbb{H} \otimes \mathbb{H}$ it follows that $\mathbb{T}^{\ddagger} = \mathbb{T}$ and therefore both $|\mathbf{\Phi}\rangle$ and $|\mathbf{\Phi}^{\ddagger}\rangle$ can be regarded as vectors of $\mathbb{T}$.

Suppose now that the overall system is in the pure state $|\mathbf{\Phi}\rangle$. According to the results of Section X.4, the states of the subsystems $\mathcal{S}^{(1)}$ and $\mathcal{S}^{(2)}$are then the mixtures $\rho^{(1)} = \mathbf{\Phi}^{\ddagger}\mathbf{\Phi}$ on $\mathbb{H}^{(1)}$ and $\rho^{(2)} = \mathbf{\Phi}\mathbf{\Phi}^{\ddagger}$ on $\mathbb{H}^{(2)}$, respectively. If on the other hand the system were in the state $|\mathbf{\Phi}^{\ddagger}\rangle = \mathbf{\Pi}|\mathbf{\Phi}\rangle$ then the states of the subsystems $\mathcal{S}^{(1)}$ and $\mathcal{S}^{(2)}$would be $\rho^{(1)} = \mathbf{\Phi}\mathbf{\Phi}^{\ddagger}$ and $\rho^{(2)} = \mathbf{\Phi}^{\ddagger}\mathbf{\Phi}$. Thus, application of the transposition operator to the system states merely results in the interchange of the states of the subsystems. Since these are not distinguishable, $|\mathbf{\Phi}\rangle$ and $|\mathbf{\Phi}^{\ddagger}\rangle = \mathbf{\Pi}|\mathbf{\Phi}\rangle$ must represent the same system state. Since a general state $\boldsymbol{\rho}$ of the system can always be written as a linear function of pure states, we conclude that $\boldsymbol{\rho}$ and $\mathbf{\Pi}\boldsymbol{\rho}\mathbf{\Pi}$ must also represent the same system state.

Before proceeding further let us study the operator $\mathbf{\Pi}$. It is clear that $\mathbf{\Pi}^2 = \mathbf{I}$, the identity on $\mathbb{T}$. It follows from the remarks made in Chapter VIII after equation (VIII.3.4) that $\mathbf{\Pi}$ is of the form $\mathbf{\Pi} = 2\mathbf{E}_S - \mathbf{I}$, where

$$\mathbf{E}_S = \frac{1}{2}\left(\mathbf{I} + \mathbf{\Pi}\right) \tag{1.2}$$

is a projection on $\mathbb{T}$. Its action on the general vector $|\mathbf{\Phi}\rangle$ of equation (1.1) is

$$\mathbf{E}_S|\mathbf{\Phi}\rangle = \sum_{ij} \frac{1}{2}\Phi_{ij}\left(|b_i\rangle|b_j\rangle + |b_j\rangle|b_i\rangle\right) = \textstyle\sum_{ij} \frac{1}{2}(\Phi_{ij} + \Phi_{ji})|b_i\rangle|b_j\rangle, \tag{1.3}$$

that is,

$$\mathbf{E}_S|\mathbf{\Phi}\rangle = \frac{1}{2}\left(|\mathbf{\Phi}\rangle + |\mathbf{\Phi}^{\ddagger}\rangle\right) \tag{1.4}$$

and in particular

$$\mathbf{E}_S|\phi\rangle|\psi\rangle = \frac{1}{2}\left(|\phi\rangle|\psi\rangle + |\psi\rangle|\phi\rangle\right). \tag{1.5}$$

The projection $\mathbf{E}_S$ projects on a subspace $\mathbb{S} \subset \mathbb{T}$, called the *symmetric* subspace, which is the eigensubspace of $\mathbf{\Pi}$ belonging to the eigenvalue 1.

The vectors in $\mathbb{S}$ are called *symmetric*. A symmetric vector satisfies $\mathbf{E}_S|\mathbf{\Psi}\rangle = |\mathbf{\Psi}\rangle$ and, by (1.3), its components Ψ_{ij} with respect to an arbitrary product basis must obey the relation $\Psi_{ij} = \Psi_{ji}$.

It is convenient to introduce a notation for a normalized symmetric vector in the ray of $\mathbf{E}_S|\phi\rangle|\psi\rangle$. We shall denote it by $|\phi, \psi; \mathbb{S}\rangle$. Thus, if $|\phi\rangle$ and $|\psi\rangle$ are orthonormal we write

$$|\phi, \phi; \mathbb{S}\rangle = |\phi\rangle|\phi\rangle, \qquad |\phi, \psi; \mathbb{S}\rangle = \frac{1}{\sqrt{2}}\left(|\phi\rangle|\psi\rangle + |\phi\rangle|\psi\rangle\right). \tag{1.6}$$

The orthonormal set formed by the vectors $|b_i, b_j; \mathbb{S}\rangle$, $j \geq i$ constitutes an orthonormal basis on $\mathbb{S}$ since, as shown by (1.3), every symmetric vector can be written as a linear combination of the vectors of the set.

The orthogonal complement $\mathbb{A} = \mathbb{S}^{\perp}$ of the symmetric subspace is the eigensubspace of $\mathbf{\Pi}$ belonging to the eigenvalue -1. It is called the *antisymmetric* subspace. To it there corresponds the projection $\mathbf{E}_A = \mathbf{I} - \mathbf{E}_S$, or

$$\mathbf{E}_A = \tfrac{1}{2}(\mathbf{I} - \mathbf{\Pi})$$

whose action is given by

$$\mathbf{E}_A|\mathbf{\Phi}\rangle = \sum_{ij} \tfrac{1}{2}\Phi_{ij}\left(|b_i\rangle|b_j\rangle - |b_j\rangle|b_i\rangle\right) = \sum_{ij} \tfrac{1}{2}(\Phi_{ij} - \Phi_{ji})|b_i\rangle|b_j\rangle$$

or

$$\mathbf{E}_A|\mathbf{\Phi}\rangle = \tfrac{1}{2}\left(|\mathbf{\Phi}\rangle - |\mathbf{\Phi}^{\ddagger}\rangle\right)$$

which, for a product vector, simplifies to

$$\mathbf{E}_A|\phi\rangle|\psi\rangle = \tfrac{1}{2}\left(|\phi\rangle|\psi\rangle - |\psi\rangle|\phi\rangle\right).$$

The vectors in $\mathbb{A}$ are called *antisymmetric*. The components Ω_{ij} of an antisymmetric vector $|\mathbf{\Omega}\rangle = \mathbf{E}_A|\mathbf{\Omega}\rangle$ obey $\Omega_{ij} = -\Omega_{ji}$. An orthonormal basis on $\mathbb{A}$ is formed by the set of vectors $|b_i, b_j; \mathbb{A}\rangle$, $j > i$, where in a notation similar to the one introduced above we have

$$|\phi, \psi; \mathbb{A}\rangle = \frac{1}{\sqrt{2}}\left(|\phi\rangle|\psi\rangle - |\phi\rangle|\psi\rangle\right) \tag{1.7}$$

if $|\phi\rangle$ and $|\psi\rangle$ are orthonormal.

We now examine which operators on $\mathbb{T}$ can represent observables of the composite system $\mathcal{C}$. An "acceptable" observable $\mathbf{A}$ must be an operator on $\mathbb{T}$ that gives the same result when measured with the system prepared in the state $\boldsymbol{\rho}$ or in the state $\mathbf{\Pi}\boldsymbol{\rho}\mathbf{\Pi}$ because both these operators represent the same state. We must therefore have

$$\langle \mathbf{A} \rangle = \mathrm{Tr}\,(\boldsymbol{\rho}\mathbf{A}) = \mathrm{Tr}\,(\mathbf{\Pi}\boldsymbol{\rho}\mathbf{\Pi}\mathbf{A}) = \mathrm{Tr}\,(\boldsymbol{\rho}\mathbf{\Pi}\mathbf{A}\mathbf{\Pi})$$

for every state $\boldsymbol{\rho}$. It follows that

$$\mathbf{A} = \mathbf{\Pi}\mathbf{A}\mathbf{\Pi},$$

that is, $\mathbf{A}$ must commute with $\mathbf{\Pi}$.

The operators that commute with $\mathbf{\Pi}$ are those that commute with its eigenprojections $\mathbf{E}_A$ and $\mathbf{E}_S$. An arbitrary operator $\mathbf{B}$ on $\mathbb{T}$ can always be written

$$\mathbf{B} = \mathbf{E}_S\mathbf{B}\mathbf{E}_S + \mathbf{E}_A\mathbf{B}\mathbf{E}_A + \mathbf{E}_S\mathbf{B}\mathbf{E}_A + \mathbf{E}_A\mathbf{B}\mathbf{E}_S.$$

Clearly only the first two operators of the right member commute with $\mathbf{\Pi}$. We define the operators $\mathbf{B}^S = \mathbf{E}_S\mathbf{B}\mathbf{E}_S$, $\mathbf{B}^A = \mathbf{E}_A\mathbf{B}\mathbf{E}_A$. The first of these operators can be regarded as an operator on the symmetric subspace $\mathbb{S} \subset \mathbb{H}$, the second as one on the antisymmetric subspace $\mathbb{A} \subset \mathbb{H}$. We call them the symmetric and the antisymmetric restrictions of $\mathbf{B}$. If $\mathbf{A} = \mathbf{A}^S$ ($\mathbf{A} = \mathbf{A}^A$), $\mathbf{A}$ is a symmetric (antisymmetric) operator.

Consider a product operator $A \otimes B$ on $\mathbb{T}$. By using the relations (1.2) and (1.4) for the projection $\mathbf{E}_S$ one can easily find the action of the symmetric restriction $(A \otimes B)_S = \mathbf{E}_S A \otimes B \mathbf{E}_S$ on a product vector

$$(A \otimes B)_S|\phi\rangle|\psi\rangle = (A \otimes B + B \otimes A)\,(|\phi\rangle|\psi\rangle + |\psi\rangle|\phi\rangle)\,.$$

This relation can be extended by linearity to an arbitrary vector $|\mathbf{\Phi}\rangle$ of $\mathbb{T}$ to yield

$$(A \otimes B)_S|\mathbf{\Phi}\rangle = (A \otimes B + B \otimes A)\left(|\mathbf{\Phi}\rangle + |\mathbf{\Phi}^{\ddagger}\rangle\right)$$

which, if $|\mathbf{\Psi}\rangle \in \mathbb{S}$ reduces to

$$(A \otimes B)_S|\mathbf{\Psi}\rangle = \frac{1}{2}\,(A \otimes B + B \otimes A)\,|\mathbf{\Psi}\rangle.$$

A similar procedure gives the action of the antisymmetric restriction,

$$(A \otimes B)_A|\mathbf{\Phi}\rangle = (A \otimes B + B \otimes A)\left(|\mathbf{\Phi}\rangle - |\mathbf{\Phi}^{\ddagger}\rangle\right)$$

which for $|\mathbf{\Omega}\rangle$ antisymmetric gives

$$(A \otimes B)_A|\mathbf{\Omega}\rangle = \frac{1}{2}\,(A \otimes B + B \otimes A)\,|\mathbf{\Omega}\rangle$$

It is clear from these formulae that operators of the form $A \otimes B + B \otimes A$ are acceptable.

At the beginning of this section we introduced the fundamental set $X = \{X_\kappa\}$ of observables of a subsystem S and the corresponding irreducible set of observables $\mathbf{X} = \{X_\kappa \otimes I, I \otimes X_{\kappa'}\}$ on $\mathbb{T}$. The latter generates the irreducible algebra $\mathcal{B}$ of operators on $\mathbb{T}$. We call $\mathcal{B}^S$ and $\mathcal{B}^A$ the sets formed, respectively, by the symmetric and antisymmetric restrictions of all the operators of $\mathcal{B}$, and write

$$\mathcal{B}^S = \mathbf{E}_S \mathcal{B} \mathbf{E}_S, \qquad \mathcal{B}^A = \mathbf{E}_A \mathcal{B} \mathbf{E}_A.$$

These sets contain all the admissible operators. It is clear that neither set is irreducible on $\mathbb{T}$, for they are both reduced by the subspaces $\mathbb{S}$ and $\mathbb{A}$. They cannot, therefore, be used to define a system with Hilbert space $\mathbb{T}$. However, $\mathcal{B}^S$ regarded as a set of operators on $\mathbb{S}$ is an irreducible algebra and, similarly, $\mathcal{B}^A$ is an irreducible algebra on $\mathbb{A}$.

To see this, remember that we know from Section IV.3 that an algebra is irreducible if and only if any given vector is transformed into any other given vector by some operator of the algebra. From this criterion and the irreducibility of $\mathcal{B}$ it follows that one of the vectors of any given pair in $\mathbb{S}$ can be transformed into the other by an operator of $\mathcal{B}^S$, and that, therefore, this set is an irreducible algebra on $\mathbb{S}$. A similar reasoning can be applied to $\mathcal{B}^A$.

Let us now focus our attention on the symmetric subspace $\mathbb{S}$. Among the operators on $\mathbb{T}$ we can select in many ways irreducible sets that generate the algebra $\mathcal{B}$. It is not in general true that the restriction to $\mathbb{S}$ of an arbitrary such set is irreducible on $\mathbb{S}$. However, if the single-system fundamental set X is chosen as to be formed by two maximal non-commuting observables X and Y (possibly composite in the sense of A.3.6), then we can construct an irreducible set on $\mathbb{S}$ that can be taken as the "natural" fundamental set of observables for a composite system with Hilbert space $\mathbb{S}$.

Let the two single-system observables be

$$X = \sum_i \lambda_i |\psi_i\rangle\langle\psi_i|, \qquad Y = \sum_i \mu_i |\phi_i\rangle\langle\phi_i|,$$

we construct the following two composite maximal operators on $\mathbb{S}$

$$\mathbf{X}^S = \sum_i \sum_{j \geq i} \{\lambda_i, \lambda_j\} |\psi_i, \psi_j; \mathbb{S}\rangle\langle\psi_i, \psi_j; \mathbb{S}|,$$

$$\mathbf{Y}^S = \sum_i \sum_{j \geq i} \{\mu_i, \mu_j\} |\phi_i, \phi_j; \mathbb{S}\rangle\langle\phi_i, \phi_j; \mathbb{S}|,$$

where the notation (1.6) has been used. The operator $\mathbf{X}^S$ ($\mathbf{Y}^S$) corresponds to an observable of $\mathcal{C}$ measured by a composite instrument consisting of two parts, each of which determines the value of the single-particle observable X (Y).

Using the expression (1.6) for the basis vectors and manipulating the dummy indices, one can easily verify that the above pair of operators are nothing more than the symmetric restrictions of the operators on $\mathbb{T}$ defined by

$$\mathbf{X} = \sum_{ij} |\lambda_i, \lambda_j| |\psi_i, \psi_j\rangle\langle\psi_i, \psi_j|, \tag{1.8}$$

$$\mathbf{Y} = \sum_{ij} |\mu_i, \mu_j| |\phi_i, \phi_j\rangle\langle\phi_i, \phi_j|. \tag{1.9}$$

where $|a, b|$ stands for a non-ordered pair: $|a, b| = \{a, b\}$ for $b \geq a$ and $|a, b| = \{b, a\}$ for $b < a$.

The pair of operators $\{\mathbf{X}^S, \mathbf{Y}^S\}$ is irreducible on $\mathbb{S}$ because they have different bases of eigenvectors with no common ones. It can therefore be taken to be a fundamental set of observables that define a system with Hilbert space $\mathbb{S}$.

Turning our attention to the antisymmetric subspace $\mathbb{A}$ we can similarly construct on it the following irreducible set of maximal operators, in the notation of (1.7),

$$\mathbf{X}^A = \sum_i \sum_{j \geq i} \{\lambda_i, \lambda_j\} |\psi_i, \psi_j; \mathbb{A}\rangle\langle\psi_i, \psi_j; \mathbb{A}|,$$

$$\mathbf{Y}^A = \sum_i \sum_{j \geq i} \{\mu_i, \mu_j\} |\phi_i, \phi_j; \mathbb{A}\rangle\langle\phi_i, \phi_j; \mathbb{A}|,$$

which can be taken as the fundamental set of observables that defines a system on the Hilbert space $\mathbb{A}$. The operators of this set are simply the antisymmetric restrictions of the operators on $\mathbb{T}$ defined by (1.8) and (1.9).

The above discussion shows that there are two alternative mathematical descriptions of a system composed of two identical subsystems. As is well known there are, in fact, in nature two classes of systems composed of identical particles that correspond to the two alternatives. The composite systems with a symmetric Hilbert space are formed by particles with integral spin. They are said to obey *Bose-Einstein statistics* and the component particles are called *bosons*. The composite systems whose Hilbert space is antisymmetric obey *Fermi-Dirac statistics* and its constituent particles are called *fermions*. Non-relativistic quantum mechanics gives no clue on the correlation between spin and statistics. In relativistic quantum field theory it can be derived from some natural symmetry assumptions.

Many of the systems that can be treated as particles in experiments that involve a certain range of energies, are found to be constituted by more elementary particles when the energies are increased. It is not difficult to show that such "composite particles" should be treated as bosons if their constituent particles include an even number of fermions, and as fermions if the number of fermions among their constituents is odd. In the first case the total internal angular momentum, which is defined as the spin of the (composite) particle, is an integer while in the second it is a half integer. Therefore the spin so defined obeys the spin-statistics correlation of the previous paragraph.

After the examples to be presented in the next section, the remainder of the chapter will be devoted to the generalization of the results obtained here to systems composed of more than two identical subsystems.

2. Example: a system of two identical particles

The results of the preceding section will now be applied to two examples of systems formed by two identical particles. It is convenient to consider first the case in which the particles have no spin. The results that will be obtained are not only of interest in their own right but will also be useful for the study of the second example to be considered: a system composed of two particles with spin.

A system composed of two identical spinless particles. – In this case we shall take the fundamental set of single-particle observables to consist of the position $\mathbf{Q}$ and the momentum $\mathbf{P}$ which form an irreducible set of maximal operators on the single-particle space $\mathbb{H}$. Because the particles are bosons, the Hilbert space of the composite system C is the symmetric subspace $\mathbb{S} \in \mathbb{T} = \mathbb{H} \otimes \mathbb{H}$.

For $\mathbb{T}$ we use the position representation discussed in Section X.3. A vector $|\mathbf{\Phi}\rangle \in \mathbb{T}$ is represented by the function of two vector variables $\mathbf{\Phi}(\boldsymbol{\xi}_1, \boldsymbol{\xi}_2) = (\langle\boldsymbol{\xi}_2|\langle\boldsymbol{\xi}_1|)\,|\mathbf{\Phi}\rangle = \langle\boldsymbol{\xi}_1, \boldsymbol{\xi}_2|\mathbf{\Phi}\rangle$.

Applying the projection $\mathbf{E}_S$ to the product ket $|\boldsymbol{\xi}_1\rangle|\boldsymbol{\xi}_2\rangle$ we obtain according to (1.5)

$$\mathbf{E}_S|\boldsymbol{\xi}_1\rangle|\boldsymbol{\xi}_2\rangle = \frac{1}{2}\left(|\boldsymbol{\xi}_1\rangle|\boldsymbol{\xi}_2\rangle + |\boldsymbol{\xi}_2\rangle|\boldsymbol{\xi}_1\rangle\right)$$

so that the projection onto $\mathbb{S}$ of the vector represented by the function $\mathbf{\Phi}(\boldsymbol{\xi}_1, \boldsymbol{\xi}_2)$ is the symmetric function

$$(\mathbf{E}_S\mathbf{\Phi})(\boldsymbol{\xi}_1, \boldsymbol{\xi}_2) = \langle\boldsymbol{\xi}_1, \boldsymbol{\xi}_2|\mathbf{E}_S|\mathbf{\Phi}\rangle = \frac{1}{2}\left[\mathbf{\Phi}(\boldsymbol{\xi}_1, \boldsymbol{\xi}_2) + \mathbf{\Phi}(\boldsymbol{\xi}_2, \boldsymbol{\xi}_1)\right].$$

Thus, in this representation the space $\mathbb{S}$ is the Hilbert space of square integrable symmetric functions of two vector variables. If $\{\psi_i(\boldsymbol{\xi}_1)\}$ is the position space representation of a basis in $\mathbb{H}$, the corresponding symmetric basis in $\mathbb{S}$ consists of the functions

$$\boldsymbol{\Psi}_{ii}(\boldsymbol{\xi}_1, \boldsymbol{\xi}_2) = \psi_i(\boldsymbol{\xi}_1)\psi_i(\boldsymbol{\xi}_2),$$

$$\boldsymbol{\Psi}_{ij}(\boldsymbol{\xi}_1, \boldsymbol{\xi}_2) = \frac{1}{\sqrt{2}}\left[\psi_i(\boldsymbol{\xi}_1)\psi_j(\boldsymbol{\xi}_2) + \psi_j(\boldsymbol{\xi}_1)\psi_i(\boldsymbol{\xi}_2)\right], \qquad j > i.$$

A generalization of (1.8), (1.9) yields for a fundamental set of composite system observables the symmetric restrictions $\mathbf{X}^S$ and $\mathbf{Y}^S$ of the operators

$$\mathbf{X} = \iint d^3\xi_1 d^3\xi_2 \; |\boldsymbol{\xi}_1, \boldsymbol{\xi}_2||\boldsymbol{\xi}_1, \boldsymbol{\xi}_2\rangle\langle\boldsymbol{\xi}_1, \boldsymbol{\xi}_2|,$$

$$\mathbf{Y} = \iint d^3\eta_1 d^3\eta_2 \; |\boldsymbol{\eta}_1, \boldsymbol{\eta}_2||\boldsymbol{\eta}_1, \boldsymbol{\eta}_2\rangle\langle\boldsymbol{\eta}_1, \boldsymbol{\eta}_2|.$$

They correspond to instruments that measure respectively the positions and the momenta of the two particles. One can write $\mathbf{X}^S$ and $\mathbf{Y}^S$ as operators on $\mathbb{S}$ in the form

$$\mathbf{X}^S = \iint_{\boldsymbol{\xi}_1 < \boldsymbol{\xi}_2} d^3\xi_1 d^3\xi_2 \; \{\boldsymbol{\xi}_1, \boldsymbol{\xi}_2\}|\boldsymbol{\xi}_1, \boldsymbol{\xi}_2; \mathbb{S}\rangle\langle\boldsymbol{\xi}_1, \boldsymbol{\xi}_2; \mathbb{S}|, \tag{2.1}$$

$$\mathbf{Y}^S = \iint_{\boldsymbol{\eta}_1 < \boldsymbol{\eta}_2} d^3\eta_1 d^3\eta_2 \; \{\boldsymbol{\eta}_1, \boldsymbol{\eta}_2\}|\boldsymbol{\eta}_1, \boldsymbol{\eta}_2; \mathbb{S}\rangle\langle\boldsymbol{\eta}_1, \boldsymbol{\eta}_2; \mathbb{S}| \tag{2.2}$$

by generalizing the notation (1.6):

$$|\boldsymbol{\xi}_1, \boldsymbol{\xi}_2; \mathbb{S}\rangle = \frac{1}{\sqrt{2}}\left(|\boldsymbol{\xi}_1\rangle|\boldsymbol{\xi}_2\rangle + |\boldsymbol{\xi}_1\rangle|\boldsymbol{\xi}_2\rangle\right), \qquad \boldsymbol{\xi}_2 > \boldsymbol{\xi}_1,$$

where $\boldsymbol{\xi}_2 > \boldsymbol{\xi}_1$ represents the octant of euclidean space where every component of $\boldsymbol{\xi}_2$ is greater than the corresponding component of $\boldsymbol{\xi}_1$.

If the positions of the two particles are measured to be $\boldsymbol{\xi}_1$ and $\boldsymbol{\xi}_2$ one assigns naturally the coordinate $\frac{1}{2}(\boldsymbol{\xi}_1 + \boldsymbol{\xi}_2)$ to the centre of mass. This means that the centre of mass observable is the symmetric restriction $\mathbf{Q}_0^S$ of the operator

$$\mathbf{Q}_0 = \iint d^3\xi_1 d^3\xi_2 \; \frac{\boldsymbol{\xi}_1 + \boldsymbol{\xi}_2}{2} \; |\boldsymbol{\xi}_1, \boldsymbol{\xi}_2\rangle\langle\boldsymbol{\xi}_1, \boldsymbol{\xi}_2|$$

which equals $\frac{1}{2}(\mathbf{Q} \otimes I + I \otimes \mathbf{Q})$. The operator $\mathbf{Q}_0$ on $\mathbb{T}$ has, of course the form that would be expected if the particles were different. As an operator on $\mathbb{S}$ the expression of the centre of mass is

$$\mathbf{Q}_0^S = \iint_{\boldsymbol{\xi}_1 < \boldsymbol{\xi}_2} d^3\xi_1 d^3\xi_2 \; \frac{\boldsymbol{\xi}_1 + \boldsymbol{\xi}_2}{2} \; |\boldsymbol{\xi}_1, \boldsymbol{\xi}_2; \mathbb{S}\rangle\langle\boldsymbol{\xi}_1, \boldsymbol{\xi}_2; \mathbb{S}| \tag{2.3}$$

which is patently a function of $\mathbf{X}^S$.

The total momentum of the system is defined by an expression similar to (2.3). However, the identity of the particles makes it impossible to define operators on $\mathbb{S}$ that represent the relative position or reduced momentum, for the corresponding operators on $\mathbb{T}$ do not commute with the transposition operator $\mathbf{\Pi}$. Their magnitudes, however do commute with $\mathbf{\Pi}$ and their restrictions can therefore be defined without difficulty.

We have exhibited the operators (2.1)-(2.3) to stress how it is possible to regard them as operator on the Hilbert space $\mathbb{S}$ without reference to the product space $\mathbb{T}$. However in many problems involving systems of a few particles, it is often convenient to work with the corresponding operators on $\mathbb{T}$ and then study their restrictions to $\mathbb{S}$.

We shall now apply this procedure to the study of an isolated system composed of two identical particles. As in the discussion of Section X.3 we regard the space $\mathbb{T}$ as the direct product $\mathbb{H}^0 \otimes \mathbb{H}^I$ of the space $\mathbb{H}^0$ where $\mathbf{Q}_0$ and $\mathbf{P}_0 = \mathbf{P} \otimes I + I \otimes \mathbf{P}$ form an irreducible set, and the space $\mathbb{H}^I$ where one such set consists of the operators $\mathbf{q} = \mathbf{Q} \otimes I - I \otimes \mathbf{Q}$ and $\mathbf{p} = \mathbf{P} \otimes I - I \otimes \mathbf{P}$.

It is clear that the transposition operator $\mathbf{\Pi}$ on $\mathbb{T}$ commutes with both $\mathbf{Q}_0$ and $\mathbf{P}_0$ and that $\mathbf{\Pi q \Pi} = -\mathbf{q}, \mathbf{\Pi p \Pi} = -\mathbf{p}$. It follows that $\mathbf{\Pi} = I^0 \otimes \Pi^I$, where I^0 is the identity on $\mathbb{H}^0$, and (from Section VIII.3 and the property $\mathbf{\Pi}^2 = \mathbf{I}$) that Π^I is the parity operator on $\mathbb{H}^I$. Substituting this form of $\mathbf{\Pi}$ into the definition (1.2) of the projection $\mathbf{E}_S$ on the symmetric subspace we find

$$\mathbf{E}_S = I^0 \otimes \tfrac{1}{2}(I^I + \Pi^I). \tag{2.4}$$

According to the remarks in Section VIII.3 following equation (VIII.3.4), the second factor in the last product is the projection on the positive parity subspace $\mathbb{H}^I_+ \subset \mathbb{H}^I$. It follows that the Hilbert space $\mathbb{S}$ of the system can be written as the product

$$\mathbb{S} = \mathbb{H}^0 \otimes \mathbb{H}^I_+. \tag{2.5}$$

It is now easy to discuss the dynamical problem for the system under consideration by making use of the results obtained in the study of the similar problem (with different particles) in Section X.3. In the present case the hamiltonian is the symmetric restriction of the hamiltonian (X.3.3). In view of the result (2.5), this symmetric restriction is

$$H^S = H^0 \otimes I^I_+ + I^0 \otimes H^I_+$$

where, if we call m the mass of each particle,

$$H^0 = \frac{|\mathbf{P}|^2}{4m}$$

and H^I_+ is the positive parity restriction of

$$H^I = \frac{|\mathbf{p}|^2}{m} + W(|\mathbf{q}|). \tag{2.6}$$

In the notation of Section X.3, to the eigenvector $|\psi_n, l, m\rangle = |\psi_{nl}\rangle|l,m\rangle$ there corresponds the eigenvalue E_{nl} solution of

$$\left(\frac{\Pi_r^2}{m} + \frac{l(l+1)}{m} q^{-2} + W(q)\right)|\psi_{nl}\rangle = E_{nl}|\psi_{nl}\rangle.$$

Since $|\psi_{nl}\rangle$ depends only on $q = |\mathbf{q}|$, the parity of $|\psi_n, l, m\rangle$ is the parity of the eigenvector $|l, m\rangle$ of the reduced angular momentum which, as we know from Section VIII.3, equals $(-1)^l$. The operator H^I_+ is therefore of the form

$$H^I_+ = \sum_n \sum_{\lambda=0}^{\infty} E_{n,2\lambda} \sum_{m=-2\lambda}^{2\lambda} |\psi_n, 2\lambda, m\rangle\langle\psi_n, 2\lambda, m|.$$

In this case, the effect of the identity of the particles on the spectrum of the hamiltonian is to efface all the energy levels corresponding to odd values of l.

It was mentioned in Section X.3 that a hamiltonian like (2.6) is used as a first approximation to model the rotational and vibrational levels of a diatomic molecule. It would be wrong to conclude that in a molecule with two identical nuclei it is the odd l levels that are suppressed because such a molecule is not a two-particle system, it also contains electrons; as a result equation (2.4) no longer holds. It is still true that (in the absence of nuclear spin) alternating rotational levels are suppressed but whether these levels correspond to even or odd values of l depends on the electronic state.

A system of two identical particles with spin. – When each of the two particles that constitute the system have spin s, we can take as the Hilbert space $\mathbb{T}$ the tensor product $\mathbb{H}^0 \otimes \mathbb{H}^C$, where $\mathbb{H}^0$ is as above and $\mathbb{H}^C$ is itself the tensor product of the space $\mathbb{H}^I$ of the previous example and the $(2s+1)^2$-dimensional spin space $\mathbb{H}^S$. On $\mathbb{H}^0$ one can again take $\mathbf{Q}_0$ and $\mathbf{P}_0$ to be the fundamental set of observables. On $\mathbb{H}^C$, the fundamental set of observables can be taken to be $\{\mathbf{q}, \mathbf{p}, \mathbf{s}_1, \mathbf{s}_2\}$ or, equivalently, $\{\mathbf{q}, \mathbf{p}, S^2, S_3\}$, where S^2 is the square of the total spin $\mathbf{S} = \mathbf{s}_1 + \mathbf{s}_2$ and S_3 its component along the 3-axis. If the spin of the particles is an integer (half integer), the Hilbert space of the system is the symmetric subspace $\mathbb{S} \subset \mathbb{T}$ (antisymmetric subspace $\mathbb{A} \subset \mathbb{T}$).

The transposition operator on $\mathbb{T}$ can be decomposed as the product $\mathbf{\Pi} = I^0 \otimes \Pi^I \otimes \mathbf{\Pi}^S$, where Π^I is, as before, the parity operator and $\mathbb{H}^I$, and $\mathbf{\Pi}^S$ the transposition operator on $\mathbb{H}^S$. We denote the eigenvectors of S^2 and S_3 by $|S, M\rangle \in \mathbb{H}^S$. The corresponding eigenvalues are $S(S+1)$, $S = 0, 1, \ldots, 2s$, and $M = -S, -S+1, \ldots, S$. The action of $\mathbf{\Pi}^S$ on $|S, M\rangle$ can be found by using the Clebsch-Gordan coefficients to write this vector in terms of the basis in $\mathbb{H}^S$ formed by the products of the eigenvectors $|s, \mu\rangle$ of the single particle spins:

$$|S, M\rangle = \sum_{\mu+\mu'=M} \langle s, s, \mu, \mu' | S, M\rangle |s, \mu\rangle |s, \mu'\rangle.$$

Upon application of $\mathbf{\Pi}^S$ we obtain after interchanging the dummy indices μ and μ'

$$\mathbf{\Pi}^S |S, M\rangle = \sum_{\mu+\mu'=M} \langle s, s, \mu', \mu | S, M\rangle |s, \mu\rangle |s, \mu'\rangle$$

and by use of the relation (IX.3.4) among the Clebsch-Gordan coefficients we find

$$\mathbf{\Pi}^S |S, M\rangle = (-1)^{2s-S} |S, M\rangle.$$

Thus, if the particles have half-integral spin, the eigenvectors belonging to odd and even S lie in the symmetric and antisymmetric subspaces of $\mathbb{H}^S$, respectively. For particles with integral spin the situation is reversed.

The operators on $\mathbb{T}$, with symmetric (antisymmetric) restrictions that are observables of the system, must commute with the transposition operator $\mathbf{\Pi}$. When an isolated system of two different particles with spin was studied in Chapter X, it was found that the interaction potential (X.2.26) was the sum of terms of five types. In the present case, only the type denoted there by W_4 (the one that contains the difference of the spins) does not commute with $\mathbf{\Pi}$. The term W_4 cannot, therefore, appear in the hamiltonian of two identical particles.

3. The description of systems of many identical particles

In this section the results of Section 1 are generalized to a composite system C consisting of many (more than two) identical subsystems. We shall think of the subsystems as particles and call $\mathbb{H}$ the single-particle Hilbert space and n the number of particles that constitute the system. As before we start by considering the tensor product space $\mathbb{T}$ formed with n identical copies of $\mathbb{H}$.

$$\mathbb{T} = \mathbb{H} \otimes \mathbb{H} \otimes \ldots \otimes \mathbb{H}.$$

The different copies of $\mathbb{H}$ are distinguished only by their position in the above product. If convenient, the copy in the νth position will be denoted by $\mathbb{H}^{(\nu)}$. With this convention we can write the above product as

$$\mathbb{T} = \bigotimes_{\nu=1}^{n} \mathbb{H}^{(\nu)}$$

In general the copy to which a single-particle vector or operator belongs will be apparent but, if necessary for clarity, they will be labelled with the appropriate superscript; e.g., $|\phi\rangle^{(\nu)}, A^{(\nu)}$. General vectors and operators of $\mathbb{T}$ will be denoted by boldface letters.

As in the two-particle case we start by imagining that it is possible to describe the system $\mathcal{C}$ in the product space $\mathbb{T}$. Suppose that $\mathcal{C}$ is in the pure product state

$$|\phi_{i_1}\phi_{i_2}\ldots\phi_{i_n}\rangle = |\phi_{i_1}\rangle|\phi_{i_2}\rangle\ldots|\phi_{i_n}\rangle \tag{3.1}$$

(in this case the full notation

$$|\phi_{i_1}^{(1)}\phi_{i2}^{(2)}\ldots\phi_{i_n}^{(n)}\rangle = |\phi_{i_1}\rangle^{(1)}|\phi_{i_2}\rangle^{(2)}\ldots|\phi_{i_n}\rangle^{(n)}$$

is unnecessary). This means that one of the subsystems will be in state $|\phi_{i_1}\rangle$, another in state $|\phi_{i_2}\rangle, \ldots$, another in state $|\phi_{i_n}\rangle$. Because of the identity of the particles, such would also be the case for any arbitrary choice of the order of the factors $|\phi_{i1}\rangle, |\phi_{i_2}\rangle, \ldots, |\phi_{i_n}\rangle$ in the product (3.1).

To elucidate the structure that the identity of particles imposes on the space $\mathbb{T}$ we are led to consider the permutations of a set of objects.

A permutation of n ordered objects which puts the object in position α_1 into position 1, the one in α_2 into $2, \ldots$ the one in α_n into n, is usually denoted by

$$\begin{pmatrix} 1 & 2 & \ldots & n \\ \alpha_1 & \alpha_2 & \ldots & \alpha_n \end{pmatrix}.$$

Clearly the order of the columns in this symbol is irrelevant. Columns consisting of two identical symbols are often omitted. The product of two permutations is defined in the obvious manner

$$\begin{pmatrix} 1 & 2 & \ldots & n \\ \alpha_1 & \alpha_2 & \ldots & \alpha_n \end{pmatrix}\begin{pmatrix} \alpha_1 & \alpha_2 & \ldots & \alpha_n \\ \beta_1 & \beta_2 & \ldots & \beta_n \end{pmatrix} = \begin{pmatrix} 1 & 2 & \ldots & n \\ \beta_1 & \beta_2 & \ldots & \beta_n \end{pmatrix}$$

and is non-commutative. With this definition of product, the permutations of n objects are easily seen to form a group, the symmetric group $\mathfrak{S}_n$, that has $n!$ elements. In the above notation the unit element has two identical rows and the inverse of a permutation is obtained by interchanging the rows. A general permutation will be denoted by $\mathbf{\Pi}$ (with primes if necessary) and its inverse by $\mathbf{\Pi}^{-1}$.

A permutation that can be written in the form

$$\begin{pmatrix} \beta_1 & \beta_2 & \dots & \beta_{m-1} & \beta_m \\ \beta_2 & \beta_3 & \dots & \beta_m & \beta_1 \end{pmatrix}$$

is called a *cycle* of order m; it is written in abbreviated form as $(\beta_1\beta_2\beta_3\dots\beta_{m-1}\beta_m)$.

Any permutation can be decomposed in a unique way into a product of cycles that contain no common element; two such cycles commute. For example,

$$\begin{pmatrix} 123456789 \\ 921564837 \end{pmatrix} = (19783)(645)(2). \tag{3.2}$$

A cycle of length 2 is a *transposition*. Every cycle of length l can be decomposed into a product of $l-1$ transpositions, e.g.,

$$(19783) = (19)(17)(18)(13).$$

It follows that every permutation can be written as a product of transpositions. If the number of transpositions in this decomposition is even (odd), the permutation is said to be of even (odd) parity. For example, the permutation (3.2) is even. A permutation and its inverse have the same parity; the product of two permutations of the same (opposite) parity is even (odd). We shall denote the parity of $\mathbf{\Pi}$ by a quantity δ_Π, which has the value $+1$ (-1) if $\mathbf{\Pi}$ is even (odd). Half the permutations in $\mathfrak{S}_n$ are even and half are odd. A general even (odd) permutation will be written $\mathbf{\Pi}_+$ $(\mathbf{\Pi}_-)$.

A permutation of the form $\mathbf{\Pi}'\mathbf{\Pi}\mathbf{\Pi}'^{-1}$ is said to be *conjugate* to $\mathbf{\Pi}$ by $\mathbf{\Pi}'$. The set of all permutations conjugate to a given one is called a *class*. Each permutation belongs to a single class. All the permutations in a class have the same parity which we shall call the parity of the class.

Consider a partition of the integer n into a sum of k integers

$$l_1 + l_2 + \dots + l_k = n, \qquad l_1 \geq l_2 \geq \dots \geq l_k \geq 1.$$

If $\mathbf{\Pi}$ can be decomposed into k cycles of lengths $l_1, \dots, l_k$, so can each of its conjugates. A class is therefore characterized by a partition of the integer n, and the number K of classes equals the number of such partitions. The unit permutation forms a class by itself.

To each permutation $\mathbf{\Pi}$ we make correspond a linear permutation operator on $\mathbb{T}$ (which we denote by the same symbol as the permutation) whose action on a product vector is defined by

$$\begin{pmatrix} 1 & 2 & \dots & n \\ \alpha_1 & \alpha_2 & \dots & \alpha_n \end{pmatrix} |\phi_{i_1}\phi_{i_2}\dots\phi_{i_n}\rangle = |\phi_{i_{\alpha_1}}\phi_{i_{\alpha_2}}\dots\phi_{i_{\alpha_n}}\rangle.$$

Its action on any vector $|\mathbf{\Phi}\rangle \in \mathbb{T}$ can then be found by linearity once $|\mathbf{\Phi}\rangle$ is written as a linear combination of vectors of a product basis. With this definition the product of two permutation operators corresponds to the product of the corresponding permutations, and the identity $\mathbf{I}$ on $\mathbb{T}$ to the unit permutation. We will often fail to distinguish between a permutation operator and the permutation it represents.

It is clear from the definition that a permutation operator $\mathbf{\Pi}$ changes an orthonormal basis into another and, being linear, is therefore unitary: $\mathbf{\Pi}^\dagger = \mathbf{\Pi}^{-1}$.

An extension of the argument used in the two-particle case shows that an acceptable observable $\mathbf{A}$ on $\mathbb{T}$ must commute with every permutation operator, i.e., $\mathbf{\Pi A \Pi}^{-1} = \mathbf{A}$, for every $\mathbf{\Pi}$. A general permutation is not an acceptable observable but we can construct functions of the permutations that are. One example is the hermitean operator $\mathbf{\Gamma}$, called a *class sum*, defined by

$$\mathbf{\Gamma} = \sum_{\Pi\in\Gamma} \mathbf{\Pi},$$

where the summation runs over the k_Γ permutations in the class $\mathbf{\Gamma}$ (we use the same symbol for the class and the class sum). The class sums, commute with every $\mathbf{\Pi}$, that is, $\mathbf{\Pi\Gamma\Pi}^{-1} = \mathbf{\Gamma}$. A relation that becomes obvious when one realizes that the operation $\mathbf{\Pi}(\cdot)\mathbf{\Pi}^{-1}$ merely rearranges the terms in the sum. It follows that the class sums commute among themselves. The class sum of the unit permutation is $\mathbf{\Gamma}_1 = \mathbf{I}$.

An operator $\mathbf{A}$ that is a function of the $\mathbf{\Pi}$s alone is a linear combination $\sum c_\Pi \mathbf{\Pi}$ of them (because every product of permutations is a permutation). If furthermore $\mathbf{\Pi A \Pi}^{-1} = \mathbf{A}$ for every $\mathbf{\Pi}$, then all conjugates of $\mathbf{\Pi}$ in the sum must have the same coefficient. This means that $\mathbf{A}$ can be written as a linear combination of class sums

$$\mathbf{A} = \sum_\Gamma c_\Gamma \mathbf{\Gamma}. \tag{3.3}$$

We shall therefore restrict our attention to the K class sums in our study of functions of the $\mathbf{\Pi}s$ that are acceptable observables.

Among the operators of the type of (3.3), the sums of all permutations of the same parity play an important role in the theory,

$$\mathbf{S}_+ = \sum_{\Pi_+} \mathbf{\Pi}_+ = \sum_{\Gamma_+} \mathbf{\Gamma}_+, \qquad \mathbf{S}_- = \sum_{\Pi_-} \mathbf{\Pi}_- = \sum_{\Gamma_-} \mathbf{\Gamma}_-$$

where even and odd classes have been denoted by $\mathbf{\Gamma}_+$ and $\mathbf{\Gamma}_-$, respectively. We also write

$$\mathbf{S} = \mathbf{S}_+ + \mathbf{S}_- = \sum_{\Pi} \mathbf{\Pi} = \sum_{\Gamma} \mathbf{\Gamma}.$$

The following relations can be easily verified

$$\mathbf{\Pi}_\pm \mathbf{S}_\pm = \mathbf{S}_+, \qquad \mathbf{\Pi}_\pm \mathbf{S}_\mp = \mathbf{S}_-,$$

$$\mathbf{\Pi S} = \mathbf{S}, \qquad \mathbf{\Pi}(\mathbf{S}_+ - \mathbf{S}_-) = \delta_\Pi (\mathbf{S}_+ - \mathbf{S}_-) \tag{3.4}$$

from which one gets

$$\mathbf{S}_+^2 = \mathbf{S}_-^2 = \frac{n!}{2}\mathbf{S}_+, \qquad \mathbf{S}_+\mathbf{S}_- = \frac{n!}{2}\mathbf{S}_-, \qquad \mathbf{S}^2 = n!\,\mathbf{S}. \tag{3.5}$$

All the operators that appear in these equations commute. Moreover, $\mathbf{S}$, $\mathbf{S}_+$ and $\mathbf{S}_-$ are hermitean because the class sums are.

A product of class sums is an operator that satisfies the conditions leading to (3.3). Thus, every product, and therefore every function, of class sums is a linear combination like (3.3). In other words, the functions $\mathbf{F}$ of the class sums are the elements of a K-dimensional vector space $\mathfrak{C}$ in which the K $\mathbf{\Gamma}$s form a basis.[1]

Since all the $\mathbf{\Gamma}$s commute, there exists a complete family of orthogonal projections $\mathbf{E}_\mu$ on $\mathbb{T}$ such that every $\mathbf{\Gamma}$ is of the form

$$\mathbf{\Gamma} = \sum_{\mu} \lambda_\mu^{(\Gamma)} \mathbf{E}_\mu. \tag{3.6}$$

Evidently, every function $\mathbf{F} \in \mathfrak{C}$ can be written as linear combinations of the $\mathbf{E}_\mu$ which, being members of a spectral family, are linearly independent and constitute, therefore, a basis in $\mathfrak{C}$. Their number equals the dimension K of $\mathfrak{C}$, that is, the number of permutation classes, or of partitions of n. Because the $\mathbf{\Gamma}$s commute with every $\mathbf{\Pi}$ so do the the projections $\mathbf{E}_\mu$, that is, the corresponding subspaces $\mathbb{M}_\mu$ reduce all the permutations.[2]

Two of the projections $\mathbf{E}_\mu$ play a special role in the theory, they are

$$\mathbf{E}_S = \frac{1}{n!}\mathbf{S} = \frac{1}{n!}\sum_{\Pi} \mathbf{\Pi} \tag{3.7}$$

[1] Note that the elements of the vector space $\mathfrak{C}$ are operators on $\mathbb{T}$.

[2] The decomposition of $\mathbb{T}$ into a direct sum of the subspaces $\mathbb{M}_\mu$ corresponds, in the language of group theory, to the decomposition of the group of permutation operators into irreducible representations. The eigenvalues $\lambda_\mu^{(\Gamma)}$, Γ fixed, are, apart from a constant, the irreducible characters of the class Γ.

and

$$\mathbf{E}_A = \frac{1}{n!}(\mathbf{S}_+ - \mathbf{S}_-) = \frac{1}{n!}\sum_{\Pi} \delta_\Pi \mathbf{\Pi}. \tag{3.8}$$

One can see immediately from (3.5) that these operators are projections. To verify that they are members of the family $\{\mathbf{E}_\mu\}$ we have to show that they obey the eigenvalue equation $\mathbf{\Gamma E}_\mu = \lambda_\mu^{(\Gamma)} \mathbf{E}_\mu$. Remembering that k_Γ is the number of permutations in the class $\mathbf{\Gamma}$ of parity δ_Γ, one has by (3.4)

$$\mathbf{\Gamma E}_S = \frac{1}{n!}\sum_{\Pi\in\Gamma} \mathbf{\Pi S} = \frac{1}{n!} k_\Gamma \mathbf{S} = k_\Gamma \mathbf{E}_S,$$

$$\mathbf{\Gamma E}_A = \frac{1}{n!}\sum_{\Pi\in\Gamma} \mathbf{\Pi}(\mathbf{S}_+ - \mathbf{S}_-) = \frac{1}{n!} k_\Gamma \delta_\Gamma (\mathbf{S}_+ - \mathbf{S}_-) = \delta_\Gamma k_\Gamma \mathbf{E}_A.$$

Thus, $\mathbf{E}_S$ and $\mathbf{E}_A$ are members of the family $\{\mathbf{E}_\mu\}$. The corresponding eigenvalues of $\mathbf{\Gamma}$ are $\lambda_S^{(\Gamma)} = k_\Gamma$ and $\lambda_A^{(\Gamma)} = \delta_\Gamma k_\Gamma$. The range of $\mathbf{E}_S$ is the *symmetric subspace* $\mathbb{S} \subset \mathbb{T}$. That one of $\mathbf{E}_A$ is the *antisymmetric subspace* $\mathbb{A} \subset \mathbb{T}$. These subspaces are orthogonal because, for an odd class, they belong to different eigenvalues of $\mathbf{\Gamma}$.

According to (3.4), $\mathbf{\Pi E}_S = \mathbf{E}_S\mathbf{\Pi} = \mathbf{E}_S$ and $\mathbf{\Pi E}_A = \mathbf{E}_A\mathbf{\Pi} = \delta_\Pi \mathbf{E}_A$, for all $\mathbf{\Pi}$. These relations show that the vectors in $\mathbb{S}$ and those in $\mathbb{A}$ are eigenvectors of all the permutations. We show next that no other subspace $\mathbb{M}_\mu$ shares this property. Solving the K equations (3.6) for the projections, we find the latter as linear combinations of class sums

$$\mathbf{E}_\mu = \sum_\Gamma c_\mu^{(\Gamma)} \mathbf{\Gamma}.$$

Suppose that the vectors of $\mathbb{M}_\mu$ are eigenvectors of an arbitrary permutation $\mathbf{\Pi}$, that is, that $\mathbf{\Pi E}_\mu = \mathbf{E}_\mu \mathbf{\Pi} = \alpha_\Pi \mathbf{E}_\mu$. We have then

$$\mathbf{\Pi}\sum_\Gamma c_\mu^{(\Gamma)} \mathbf{\Gamma} = \alpha_\Pi \sum_\Gamma c_\mu^{(\Gamma)} \mathbf{\Gamma}. \tag{3.9}$$

The product $\mathbf{\Pi\Gamma}$ will consist in general of a sum of permutations not all of which are in the class $\mathbf{\Gamma}$. Let $\mathbf{\Gamma}', \mathbf{\Gamma}'', \ldots$ be the classes among which these permutations are distributed. Evidently we must have $c_\mu^{(\Gamma)} = \alpha_\Pi c_\mu^{(\Gamma')} = \alpha_\Pi c_\mu^{(\Gamma'')} = \ldots$, or, $c_\mu^{(\Gamma')} = c_\mu^{(\Gamma'')} = \ldots$. Assume that $\mathbf{\Pi}$ is odd; when we consider all the products $\mathbf{\Pi\Gamma}_+$ for all even classes $\mathbf{\Gamma}_+$ we find all the odd permutations so that the $c_\mu^{(\Gamma_-)}$ for all odd classes $\mathbf{\Gamma}_-$ must be equal. If, on the other hand, $\mathbf{\Pi}$ is even, we conclude that all the coefficients corresponding to the even classes $\mathbf{\Gamma}_+$ must be equal. Thus equation

(3.9) can be written

$$\mathbf{\Pi}\left(c_\mu^+\mathbf{S}_+ + c_\mu^-\mathbf{S}_-\right) = \alpha_\Pi\left(c_\mu^+\mathbf{S}_+ + c_\mu^-\mathbf{S}_-\right)$$

and the projection $\mathbf{E}_\mu$ has the form

$$\mathbf{E}_\mu = c_\mu^+\mathbf{S}_+ + c_\mu^-\mathbf{S}_-. \tag{3.10}$$

The hermiticity of this projection implies that c_μ^+ and c_μ^- are real, while the relation $\mathbf{E}_\mu^2 = \mathbf{E}_\mu$ gives

$$\frac{n!}{2}\left(c_\mu^{+2}\mathbf{S}_+ + 2c_\mu^+c_\mu^-\mathbf{S}_- + c_\mu^{-2}\mathbf{S}_+\right) = c_\mu^+\mathbf{S}_+ + c_\mu^-\mathbf{S}$$

from which we obtain

$$c_\mu^+ = \frac{1}{n!}, \qquad c_\mu^- = \pm\frac{1}{n!}.$$

Substituting these values into (3.10) we find, as promised, that the only two $\mathbf{E}_\mu$ that satisfy the eigenvalue equation $\mathbf{\Pi}\mathbf{E}_\mu = \alpha_\Pi\mathbf{E}_\mu$ are precisely the projections $\mathbf{E}_S$ and $\mathbf{E}_A$ of equations (3.7) and (3.8). As we know already, $\alpha_\Pi = 1$ for $\mathbf{E}_S$, and $\alpha_\Pi = \delta_\Pi$ for $\mathbf{E}_A$.

It is now time to return to the consideration of a general acceptable observable of the system. As we know, it must commute with every permutation. It therefore commutes with every class sum $\mathbf{\Gamma}$ and consequently with every projection $\mathbf{E}_\mu$, that is, it is reduced by each of the subspaces $\mathbb{M}_\mu$. It follows that there is no set of acceptable observables that is irreducible on the tensor product space $\mathbb{T}$ which, as we saw in Section 1, means that $\mathbb{T}$ is not an appropriate Hilbert space for the description of a system of identical particles.

The appropriate Hilbert space must evidently be sought among the subspaces $\mathbb{M}_\mu$. All of them except the symmetric and antisymmetric ones $\mathbb{S}$ and $\mathbb{A}$ can be eliminated if we impose the requirement that the system possesses a *maximal* observable $\mathbf{A}$. Indeed, since $\mathbf{A}$ commutes with all the permutations $\mathbf{\Pi}$, the eigenvectors of $\mathbf{A}$ must be eigenvectors of all the $\mathbf{\Pi}$s and must therefore lie in either $\mathbb{S}$ or $\mathbb{A}$.

As is well known, the last conclusion is borne out by experience. In spite of the more complicated structure of the space $\mathbb{T}$, the essential consequences of the indistinguishability of the particles are the same that were already present in the case of two particles. The Hilbert space appropriate for the description of a system of n identical particles is either the symmetric subspace $\mathbb{S} \subset \mathbb{T}$ in the case of bosons, or the antisymmetric subspace $\mathbb{A} \subset \mathbb{T}$ in the case of fermions.

Although of no interest from a fundamental point of view, the study of the remaining subspaces $\mathbb{M}_\mu$ is of considerable value in the approximate study of the structure of atoms, molecules and nuclei. We shall, however, not pursue it in this book.

As we did in the two-particle case, we shall next use a product basis in $\mathbb{T}$ to construct bases in the symmetric and antisymmetric spaces corresponding to systems of n bosons and n fermions, respectively.

A system of n identical bosons. – As we have seen, the appropriate Hilbert space to describe a system of n identical bosons is the symmetric subspace $\mathbb{S}$ of the tensor product $\mathbb{T}$ of the single particle spaces.

Let $\{|\psi_i\rangle\}$ be an orthonormal basis in the single-particle space $\mathbb{H}$. In the notation of (3.1), the product vectors $|\psi_{i_1}\psi_{i_2}\dots\psi_{i_n}\rangle$ form an orthonormal basis in $\mathbb{T}$. The projections of all these vectors on $\mathbb{S}$ span this subspace. The scalar product of two vectors, $\mathbf{E}_S|\psi_{i_1}\psi_{i2}\dots\psi_{i_n}\rangle$ and $\mathbf{E}_S|\psi_{j_1}\psi_{j2}\dots\psi_{jn}\rangle$ is given by

$$\begin{aligned}\langle\psi_{j_1}\psi_{j_2}\dots\psi_{j_n}|\mathbf{E}_S|\psi_{i_1}\psi_{i_2}\dots\psi_{i_n}\rangle &= \frac{1}{n!}\sum_{\Pi}\langle\psi_{j_1}\psi_{j2}\dots\psi_{j_n}|\mathbf{\Pi}|\psi_{i_1}\psi_{i_2}\dots\psi_{i_n}\rangle \\ &= \frac{1}{n!}\sum_{\Pi}\delta_{j_1i_1}\delta_{j_2i_2}\dots\delta_{j_ni_n},\end{aligned}$$

where the sum runs over all the permutations of $\{i_1i_2\dots i_n\}$. If this set of indices is (regardless of order) the same as $\{j_1j_2\dots j_n\}$ then the two vectors are equal, otherwise they are orthogonal. In the former case let n_r be the number of appearances of r in the set $\{i_1i_2\dots i_n\}$, then the number of terms different from zero in the last sum above is $n_1!n_2!\dots n_r!\dots$, which means that the square of the norm of the vector $\mathbf{E}_S|\psi_{i_1}\psi_{i_2}\dots\psi_{i_n}\rangle$ is $n_1!n_2!\dots n_r!\dots/n!$. Evidently we can label the corresponding normalized vector with the numbers n_r and write

$$|n_1n_2\dots n_r\dots\rangle = \frac{\sqrt{n!}}{\sqrt{n_1!n_2!\dots n_r!\dots}}\mathbf{E}_S|\psi_{i_1}\psi_{i_2}\dots\psi_{i_n}\rangle, \tag{3.11}$$

where, obviously, $n_1+n_2+\dots+n_r+\dots = n$. These vectors constitute an orthonormal basis on the n-boson Hilbert space $\mathbb{S}$. One may think of (3.11) as the pure state of a system of n bosons in which there are n_r bosons in the pure single-particle state $|\psi_{i_r}\rangle$.

Although the notation $|n_1n_2\dots n_r\dots\rangle$ is pleasing, because it automatically takes into account the symmetry resulting from the identity of the particles, it is, however, incomplete for it does not make reference to the single-particle states used to construct the basis in $\mathbb{S}$, on which the basis vectors evidently depend.

A system of n identical fermions. – In this case the appropriate Hilbert space is the antisymmetric subspace $\mathbb{A} \subset \mathbb{T}$. A basis in $\mathbb{A}$ can be constructed by applying the corresponding projection $\mathbf{E}_A$ to a basis in $\mathbb{T}$ formed by the product of the orthonormal basis vectors $|\psi_i\rangle \in \mathbb{H}$. We have

$$\mathbf{E}_A|\psi_{i_1}\psi_{i_2}\dots\psi_{i_n}\rangle = \frac{1}{n!}\sum_{\Pi}\delta_{\Pi}\mathbf{\Pi}|\psi_{i_1}\psi_{i_2}\dots\psi_{i_n}\rangle \tag{3.12}$$

where the sum runs over all the permutations of the set $\{i_1 i_2 \dots i_n\}$. Note that these vectors are zero if two or more of the indices i_r are equal. This fact is intuitively expressed by *Pauli's exclusion principle: in a system of identical fermions two particles cannot be in the same single-particle state.*

We only have to consider the sets $\{i_1 i_2 \dots i_n\}$ for which all the indices i_r are different. As before two vectors of the form (3.12) with different set of indices are orthogonal. If the sets are formed by the same indices they are either equal or they differ in sign and represent, therefore, the same pure state. The square of the norm of (3.12) is

$$\langle i_1 i_2 \dots i_n|\mathbf{E}_A|i_1 i_2 \dots i_n\rangle = \frac{1}{n!}\sum_{\Pi}\delta_{\Pi}\langle i_1 i_2 \dots i_n|\mathbf{\Pi}|i_1 i_2 \dots i_n\rangle.$$

Since all the indices are different, only the term in the sum with the unit permutation is different from zero; it equals one. The norm, therefore, equals $1/\sqrt{n!}$. As in the case of the bosons, if we call n_r the number of appearances of the index r in the set of indices $\{i_1 i_2 \dots i_n\}$ we can write the normalized basis vectors in $\mathbb{A}$ as

$$|n_1 n_2 \dots n_r \dots\rangle = \frac{1}{\sqrt{n!}}\mathbf{E}_A|\psi_{i_1}\psi_{i_2}\dots\psi_{i_n}\rangle, \tag{3.13}$$

where

$$n_r = 0, 1, \qquad n_1 + n_2 + \dots + n_r + \dots = n.$$

From (3.12) and the definition of determinant, (3.13) can also be written as a *Slater determinant*:

$$|n_1 n_2 \dots n_r \dots\rangle = \frac{1}{\sqrt{n!}}\begin{vmatrix} |\psi_{i_1}\rangle^{(1)} & |\psi_{i_1}\rangle^{(2)} & \dots & |\psi_{i_1}\rangle^{(n)} \\ |\psi_{i_2}\rangle^{(1)} & |\psi_{i_2}\rangle^{(2)} & \dots & |\psi_{i_2}\rangle^{(n)} \\ \dots & \dots & \dots & \dots \\ |\psi_{i_n}\rangle^{(1)} & |\psi_{i_n}\rangle^{(2)} & \dots & |\psi_{i_n}\rangle^{(n)} \end{vmatrix},$$

where, for clarity, we have written the superscripts that denote the order of the single-particle vectors in the product vector of $\mathbb{T}$.

As in the case of bosons we note that the vectors (3.13) obviously depend on the basis selected in the single-particle space $\mathbb{H}$.

4. Number representation of systems of identical particles

When considering systems of identical particles, it is often convenient – and sometimes more realistic – not to regard the number of particles as fixed, contrary to what has been done in the previous sections. To describe such a system we need to introduce the concept of direct sum $\mathbb{R}$ of an infinite sequence of Hilbert spaces $\mathbb{K}_n$, $n = 0, 1, \ldots$, which we denote by

$$\mathbb{R} = \bigoplus_{n=0}^{\infty} \mathbb{K}_n.$$

If for all $|\psi\rangle_n \in \mathbb{K}_n$

$$\sum_{n=0}^{\infty} \| |\psi\rangle_n \|^2 < \infty,$$

then $\mathbb{R}$ is a Hilbert space whose vectors are all the sequences (finite or infinite) $\{|\psi\rangle_n\}$ such that, if $|f\rangle = \{|\psi\rangle_n\}$, its norm is given by

$$\| |f\rangle \|^2 = \sum_{n=0}^{\infty} \| |\psi\rangle_n \|^2 < \infty.$$

If $|g\rangle = \{|\phi\rangle_n\}$, the addition, multiplication by a number and scalar product are defined respectively by

$$|f\rangle + |g\rangle = \{|\psi\rangle_n + |\phi\rangle_n\}, \qquad \alpha|f\rangle = \{\alpha|\psi\rangle_n\},$$

$$\langle f|g\rangle = \sum_{n=0}^{\infty} \langle \psi|\phi\rangle_n.$$

An operator R on $\mathbb{R}$ such that $|g\rangle = R|f\rangle$ is the direct sum of the operators K_n on $\mathbb{K}_n$ for which $|\phi\rangle_n = K_n|\psi\rangle_n$.

Let $\mathbb{G}_n$ be the Hilbert space of a system of n identical particles (the symmetric subspace of $\mathbb{T}_n$ for bosons, the antisymmetric one for fermions). The Hilbert space appropriate for the description of a system of an arbitrary number of such particles is the *Fock space* defined by

$$\mathbb{F} = \bigoplus_{n=0}^{\infty} \mathbb{G}_n. \tag{4.1}$$

In the case of bosons (fermions), $\mathbb{F}$ can be thought of as the symmetric (antisymmetric) subspace of the infinite direct sum $\mathbb{T}$ of the spaces $\mathbb{T}_n$.

$$\mathbb{T} = \bigoplus_{n=0}^{\infty} \mathbb{T}_n. \tag{4.2}$$

Regarded as a subspace of $\mathbb{F}$, $\mathbb{G}_n$ is the subspace where the state of a system that has been prepared with n particles lies. An orthonormal basis in $\mathbb{F}$ is formed by the union of orthonormal bases in $\mathbb{G}_n$, e.g., by the vectors $|n_1 n_2 \ldots n_r \ldots\rangle$ of (3.11), in the case of bosons, without any restriction on the sum of the numbers n_r. The one-dimensional subspace $\mathbb{G}_0 \subset \mathbb{F}$ is included to represent a system with no particles. We choose a normalized vector in this subspace and denote it by $|0\rangle = |00\ldots 0\ldots\rangle$ (not to be confused with the zero vector denoted simply by 0). The (necessarily pure) state represented by $|0\rangle$ is called the *vacuum* state of the system.

The vectors $|n_1 n_2 \ldots n_r \ldots\rangle$ with n_i, $i \neq r$, fixed and n_r taking all possible values (any non-negative integer for bosons, 0 or 1 for fermions) span a subspace of $\mathbb{M}_r \subset \mathbb{F}$ on which they form an orthonormal basis. On this subspace, which is associated to the single-particle state r, we can define, in analogy with the case of the harmonic oscillator, a *number operator* $\mathbf{N}_r$ by the relation

$$\mathbf{N}_r |n_1 n_2 \ldots n_r \ldots\rangle = n_r |n_1 n_2 \ldots n_r \ldots\rangle. \tag{4.3}$$

Because the basis vectors are eigenvectors of the number operators, the representation of $\mathbb{F}$ with respect to this basis is called the *number representation*. Different single-particle bases give rise to different number representations but the vacuum state is common to all of them.

Pursuing further the analogy with the harmonic oscillator we shall attempt to decompose $\mathbf{N}_r$ into the product of two mutually adjoint operators that when applied to the vector $|n_1 n_2 \ldots n_r \ldots\rangle$ transform it into other basis vectors belonging to the values $n_r \pm 1$. However, in order to proceed in this way we must treat separately systems of bosons and of fermions.

5. An assembly of bosons

We consider an assembly of bosons and use the description introduced in the previous section. In this case the Fock space (4.1) is a direct sum of the symmetric spaces of Section 3

$$\mathbb{F}_b = \bigoplus_{n=0}^{\infty} \mathbb{S}_n.$$

We define the number operator $\mathbf{N}_r$ as in (4.3)

$$\mathbf{N}_r |n_1 n_2 \ldots n_r \ldots\rangle = n_r |n_1 n_2 \ldots n_r \ldots\rangle \tag{5.1}$$

and attempt to write it as a product

$$\mathbf{N}_r = b_r^\dagger b_r \tag{5.2}$$

of mutually adjoint *creation* and *annihilation* operators $b_r^\dagger$ and b_r, respectively, that satisfy

$$b_r^\dagger|n_1n_2\dots n_r\dots\rangle = \gamma'_{n,r}|n_1n_2\dots n_r+1\dots\rangle,$$
$$b_r|n_1n_2\dots n_r\dots\rangle = \gamma_{n,r}|n_1n_2\dots n_r-1\dots\rangle,$$

for some numbers $\gamma'_{n,r}$ and $\gamma_{n,r}$. From (5.1) and (5.2) we find that $\gamma_{n,r}\gamma'_{n-1,r} = n_r$, while the requirement that the two operators be mutually adjoint implies that $\gamma'_{n-1,r} = \gamma^*_{n,r}$. Therefore $|\gamma'_{n-1,r}|^2 = |\gamma_{n,r}|^2$. Taking the numbers real, we can define the creation and annihilation operators by

$$b_r^\dagger|n_1n_2\dots n_r\dots\rangle = \sqrt{n_r+1}\,|n_1n_2\dots n_r+1\dots\rangle, \tag{5.3}$$
$$b_r|n_1n_2\dots n_r\dots\rangle = \sqrt{n_r}\,|n_1n_2\dots n_r-1\dots\rangle, \tag{5.4}$$

from which the commutation relation $[b_r, b_r^\dagger] = 1$ follows. Creation and annihilation operators belonging to different single-particle states operate on different subspaces of $\mathbb{F}_b$ and therefore commute. We thus have

$$[b_r, b_s^\dagger] = \delta_{rs}, \qquad [b_r, b_s] = [b_r^\dagger, b_s^\dagger] = 0. \tag{5.5}$$

From (5.4) one sees that the application of any annihilation operator to the vacuum state yields the zero vector

$$b_r|0\rangle = 0.$$

Any basis vector can be obtained by repeated application of creation operators to the vacuum state. If the latter is normalized, we obtain appropriately normalized basis vectors from the formula

$$|n_1n_2\dots n_r\dots\rangle = \frac{(b_1^\dagger)^{n_1}(b_2^\dagger)^{n_2}\dots(b_r^\dagger)^{n_r}\dots}{\sqrt{n_1!n_2!\dots n_r!\dots}}|0\rangle.$$

One should not forget that this vector is also given by (3.11) with $n = n_1 + n_2 + \dots + n_r + \dots$ For example,

$$|0_10_2\dots 1_m\dots\rangle = b_m|0\rangle = |\psi_m\rangle, \tag{5.6}$$
$$|0_10_2\dots 1_m\dots 1_n\dots\rangle = b_mb_n|0\rangle = \frac{1}{\sqrt{2}}\left(|\psi_m\psi_n\rangle + |\psi_n\psi_m\rangle\right),$$

where the notation indicates that all the n_r are zero except those corresponding to the basis vectors $|\psi_m\rangle$ and $|\psi_n\rangle$ (only $|\psi_m\rangle$ in the first equation) which are equal to one.

The creation and annihilation operators have been defined for a particular number representation of $\mathbb{F}_b$ induced by a particular basis in the single-particle space $\mathbb{H}$. A different basis in $\mathbb{H}$ will induce a different set of creation and annihilation operators but, as has already been remarked, the vacuum vector $|0\rangle$ is common to all number representations. We now study the relation between the creation and annihilation operators corresponding to different number representations.

Let $\{b_j^\dagger, b_j\}$ and $\{\beta_i^\dagger, \beta_i\}$ be the sets of operators induced by the single-particle bases $\{|\psi_j\rangle\}$ and $\{|\phi_i\rangle\}$, respectively. By (5.6) we can write $b_j^\dagger|0\rangle = |\psi_j\rangle$ and $\beta_i^\dagger|0\rangle = |\phi_i\rangle$ so that the equation $|\phi_i\rangle = \Sigma_j \langle\psi_j|\phi_i\rangle|\psi_j\rangle$ becomes

$$\beta_i^\dagger|0\rangle = \sum_j \langle\psi_j|\phi_i\rangle b_j^\dagger|0\rangle$$

which is satisfied if we take

$$\beta_i^\dagger = \sum_j \langle\psi_j|\phi_i\rangle b_j^\dagger, \qquad \beta_i = \sum_j \langle\phi_i|\psi_j\rangle b_j. \tag{5.7}$$

It is easy to verify that commutation relations similar to (5.5) are satisfied by $\beta_i^\dagger$ and β_i, and that the products $\beta_{i_1}^\dagger \beta_{i_2}^\dagger \ldots |0\rangle$ give the appropriate combinations of products of single-particle states $|\phi_i\rangle$.

One encounters often operators on $\mathbb{F}_b$ which are the symmetrized sums of operators defined on the tensor product of a few single-particle spaces. For example, the momentum of an assembly of bosons is the sum of the single-particle momenta and, assuming pair interactions, its energy is the sum of energy operators in the space of a pair of particles. Such operators have a simple and very useful expression in terms of a set of creation and annihilation operators which we discuss next.

Let $\{|\psi_i\rangle\}$ be the basis in $\mathbb{H}$ that induces the set $\{b_i, b_i^\dagger\}$ of creation and annihilation operators on $\mathbb{F}_b$ and let B be an observable on the single-particle space $\mathbb{H}$ which, in terms of the matrix elements $B_{ij} = \langle\psi_i|B|\psi_j\rangle$, is written

$$B = \sum_{ij} B_{ij}|\psi_i\rangle\langle\psi_j|.$$

On the infinite direct sum $\mathbb{T}$ given by (4.2) we construct the operators $\mathbf{B}^{(\nu)}$ defined by their action on the basis vectors

$$\mathbf{B}^{(\nu)}|\ldots\psi_k^{(\nu)}\ldots\rangle = \sum_i B_{ik}|\ldots\psi_i^{(\nu)}\ldots\rangle.$$

In obvious notation we write this operator as

$$\mathbf{B}^{(\nu)} = \sum_{ij} B_{ij}|\psi_i\rangle\langle\psi_j|^{(\nu)} \tag{5.8}$$

which operates on the one-particle basis vector that occupies the ν-th position on a vector of the product basis $|\psi_a^{(1)}\psi_b^{(2)}\ldots\rangle$. The sum

$$\mathbf{B}_n = \sum_\nu \mathbf{B}_n^{(\nu)}$$

is already symmetric on $\mathbb{T}$, and its restriction to $\mathbb{F}_b$ is the operator we are looking for.

Because of (5.8) we need only look at the operators $|\psi_i\rangle\langle\psi_j|^{(\nu)}$ whose action is explicitly given by

$$|\psi_i\rangle\langle\psi_j|^{(\nu)}|\ldots\psi_k^{(\nu)}\ldots\rangle = \delta_{jk}|\ldots\psi_i^{(\nu)}\ldots\rangle.$$

Let us apply this operator to a basis vector in which $|\psi_j\rangle$ appears m times in positions $\nu_1, \nu_2, \ldots, \nu_m$ and then sum over "all particles", i.e., over all ν, we obtain for $j \neq i$

$$\sum_\nu |\psi_i\rangle\langle\psi_j|^{(\nu)}|\ldots\psi_j^{(\nu_1)}\ldots\psi_j^{(\nu_m)}\ldots\rangle$$
$$= |\ldots\psi_i^{(\nu_1)}\ldots\psi_j^{(\nu_m)}\ldots\rangle + \ldots + |\ldots\psi_j^{(\nu_1)}\ldots\psi_i^{(\nu_m)}\ldots\rangle. \tag{5.9}$$

There are m vectors on the right and in each of them $|\psi_j\rangle$ appears once less and $|\psi_i\rangle$ once more than in the vector on the left. For $j = i$ we obtain simply

$$\sum_\nu |\psi_i\rangle\langle\psi_i|^{(\nu)}|\ldots\psi_i^{(\nu_1)}\ldots\psi_i^{(\nu_m)}\ldots\rangle = m|\ldots\psi_i^{(\nu_1)}\ldots\psi_i^{(\nu_m)}\ldots\rangle. \tag{5.10}$$

We now symmetrize the above equations. Let $\mathbf{E}_S$ be the symmetrizing projection on $\mathbb{T}$. For $j \neq i$ we have by (3.11),

$$\mathbf{E}_S|\ldots\psi_j^{(\nu_1)}\ldots\psi_j^{(\nu_m)}\ldots\rangle = \frac{\sqrt{\ldots n_i!\ldots n_j!\ldots}}{\sqrt{n}}|\ldots n_i\ldots n_j\ldots\rangle,$$

$$n_j = m, \qquad n = \sum_j n_j.$$

Upon application of the $\mathbf{E}_S$ to each of the n_j vectors on the right of (5.9) we obtain

$$\frac{\sqrt{\ldots(n_i+1)!\ldots(n_j-1)!\ldots}}{\sqrt{n}}|\ldots n_i+1\ldots n_j-1\ldots\rangle$$

but by (5.3) and (5.4),

$$|\ldots n_i+1\ldots n_j-1\ldots\rangle = \frac{1}{\sqrt{n_j(n_i+1)}}b_i^\dagger b_j|\ldots n_i\ldots n_j\ldots\rangle$$

so that, after multiplying by the number n_j of vectors, the right member of (5.9) becomes

$$\frac{\sqrt{\ldots n_i!\ldots n_j!\ldots}}{\sqrt{n}}\, b_i^\dagger b_j|\ldots n_i\ldots n_j\ldots\rangle.$$

Thus, noting that the operator $\sum_\nu |\psi_i\rangle\langle\psi_j|^{(\nu)}$ is already symmetric and can therefore be regarded as an operator on $\mathbb{F}_b$, equation (5.9) yields

$$\sum_\nu |\psi_i\rangle\langle\psi_j|^{(\nu)} = b_i^\dagger b_j. \tag{5.11}$$

From (5.10) and (5.2) it can be seen without difficulty that this equation also applies in the case $i = j$. From (5.8) it follows immediately that the symmetric sum $\mathbf{B}$ on $\mathbb{F}_b$ of the single particle operators B on $\mathbb{H}$ is given by

$$\mathbf{B} = \sum_{ij} B_{ij} b_i^\dagger b_j. \tag{5.12}$$

An operator $\mathbf{C}$ on $\mathbb{F}_b$ which is the sum of two-particle observables is constructed next. We start with an observable for a pair of particles. We can think of it as a symmetric operator C on the tensor product of two single-particle spaces $\mathbb{H}\otimes\mathbb{H}$. We shall denote it by $C^{(\mu\nu)}$ when we need to label the position of the two copies of $\mathbb{H}$ in the tensor product $\mathbb{T}$, that is, $C^{(\mu\nu)}$ is an operator on $\mathbb{H}^{(\mu)}\otimes\mathbb{H}^{(\nu)}$.

In terms of the matrix elements $C_{ij,kl} = \langle\psi_i\psi_k|C|\psi_j\psi_l\rangle$, the operator $C^{(\mu\nu)}$ becomes

$$C^{(\mu\nu)} = \sum_{ijkl} C_{ij,kl}|\psi_i\rangle\langle\psi_j|^{(\mu)}\otimes|\psi_k\rangle\langle\psi_l|^{(\nu)}.$$

Our sum of two-particle operators is defined as the symmetric operator

$$\mathbf{C} = \sum_{\mu\neq\nu} C^{(\mu\nu)} \tag{5.13}$$

or, more precisely, its restriction to $\mathbb{F}_b$.

As before we only need to consider the operators

$$\sum_{\mu\neq\nu} |\psi_i\rangle\langle\psi_j|^{(\mu)}\otimes|\psi_k\rangle\langle\psi_l|^{(\nu)}$$
$$= \sum_{\mu\nu} |\psi_i\rangle\langle\psi_j|^{(\mu)}\otimes|\psi_k\rangle\langle\psi_l|^{(\nu)} - \delta_{jk}\sum_\mu |\psi_i\rangle\langle\psi_l|^{(\mu)}.$$

Reviewing the calculation leading to (5.11) one can see without difficulty that

$$\sum_{\mu\neq\nu} |\psi_i\rangle\langle\psi_j|^{(\mu)} \otimes |\psi_k\rangle\langle\psi_l|^{(\nu)} = b_i^\dagger b_j b_k^\dagger b_l - \delta_{jk} b_i^\dagger b_l. \tag{5.14}$$

Interchanging b_j and $b_k^\dagger$ in the right member with the help of the commutation relation (5.5), the preceding equation simplifies to

$$\sum_{\mu\neq\nu} |\psi_i\rangle\langle\psi_j|^{(\mu)} \otimes |\psi_k\rangle\langle\psi_l|^{(\nu)} = b_i^\dagger b_k^\dagger b_j b_l$$

and we get, therefore, the following expression for the operator (5.13)

$$\mathbf{C} = \sum_{ijkl} C_{ij,kl} b_i^\dagger b_k^\dagger b_j b_l. \tag{5.15}$$

Similar expressions are obtained for observables which are sums of operators defined on larger subsystems of particles.

The boson field. – Imagine that we consider an assembly of bosons consisting of particles with an internal degree of freedom that we think of as spin. A maximal observable on the single-particle space $\mathbb{H}$ is the composite observable $X = \{\mathbf{Q}, s_3\}$ formed by the position and a component of the spin. We denote a composite eigenvalue of this observable corresponding to the eigenvalue $\boldsymbol{\xi}$ of $\mathbf{Q}$ and μ of s_3 by $x = \{\boldsymbol{\xi}, \mu\}$ and the corresponding eigenket by $|x\rangle = |\boldsymbol{\xi}, \mu\rangle$. To simplify the writing further we introduce the following abbreviations

$$\int dx \cdot = \sum_\mu \int d^3\boldsymbol{\xi} \cdot, \qquad \psi(x) = \psi_\mu(\boldsymbol{\xi}) = \langle \boldsymbol{\xi}, \mu|\psi\rangle = \langle x|\psi\rangle$$

in terms of which, the spectral resolution of X on $\mathbb{H}$ becomes

$$X = \int dx\; x|x\rangle\langle x|.$$

Let us suppose that a single-particle observable B is given in the position-spin or X-representation and call $\hat{B}_x$ the corresponding operator in position-spin space,

$$\hat{B}_x\psi(x) = \langle x|B|\psi\rangle = (B\psi)(x)$$

(the subscript x is used to stress that $\hat{B}_x$ is an operator on a space of functions of x). Assume furthermore that the basis $\{|\psi_i\rangle\}$ in $\mathbb{H}$ that induces the creation and annihilation operators $b_i^\dagger, b_i$ is also given in the X-representation

$$\psi_i(x) = \langle x|\psi_i\rangle,$$

then, according to (5.12), the sum of the single-particle operators B is given by

$$\mathbf{B} = \sum_{ij} b_i^\dagger b_j \int dx\ \psi_i^*(x)\hat{B}_x\psi_j(x). \tag{5.16}$$

One may define on $\mathbb{F}_b$ the operators

$$\Psi^\dagger(x) = \sum_i b_i^\dagger \psi_i^*(x), \qquad \Psi(x) = \sum_i b_i\psi_i(x) \tag{5.17}$$

in terms of which the operator (5.16) becomes

$$\mathbf{B} = \int dx\ \Psi^\dagger(x)\hat{B}_x\Psi(x)$$

(note that $\hat{B}_x$ operates here on functions of x whose values are operators on $\mathbb{F}_b$).

An analogous formula follows from (5.15) for an operator $\mathbf{C}$ on $\mathbb{F}_b$ which is a sum of two-particle operators C on $\mathbb{H}\otimes\mathbb{H}$

$$\mathbf{C} = \iint dxdx'\ \Psi^\dagger(x)\Psi^\dagger(x')\hat{C}_{xx'}\Psi(x)\Psi(x'), \tag{5.18}$$

where

$$\hat{C}_{xx'}\psi(x,x') = \langle xx'|C|\psi\rangle = (C\psi)(x,x').$$

The operators $\Psi^\dagger(\mathrm{x})$ and $\Psi(\mathrm{x})$ are called boson creation and annihilation *field operators*. Rewriting (5.17) as

$$\Psi^\dagger(x) = \sum_i \langle\psi_i|x\rangle b_i^\dagger, \qquad \Psi(x) = \sum_i \langle x|\psi_i\rangle b_i$$

and comparing these expressions with (5.7) one can think of the field operators as the creation and annihilation operators induced by the one-particle position-spin "basis vectors" $|x\rangle = |\boldsymbol{\xi},\mu\rangle$.

From the definition (5.17) of the boson fields and the first commutation relation (5.5) one finds

$$[\Psi(x),\Psi^\dagger(x')] = \sum_i\sum_j [b_i, b_j^\dagger]\psi_i(x)\psi_j^*(x') = \sum_i \psi_i(x)\psi_i^*(x').$$

The last term is just $\delta(x-x') = \delta(\boldsymbol{\xi}-\boldsymbol{\xi}')\delta_{\mu\mu'}$, so that

$$[\Psi(x),\Psi^\dagger(x')] = \delta(x-x')$$

while the other commutation relations (5.5) yield

$$[\Psi(x),\Psi(x')] = [\Psi^\dagger(x),\Psi^\dagger(x')] = 0$$

These commutation relations characterize a boson field.

Although we started with the position-spin representation it is clear that everything that has been said applies to any other, e.g., the momentum-spin, representation.

6. An assembly of fermions

In the case of fermions the Fock space is

$$\mathbb{F}_f = \bigoplus_{n=0}^{\infty} \mathbb{A}_n$$

with basis vectors $|n_1 \ldots n_r \ldots\rangle$, $n_r = 0, 1$. The number operator $\mathbf{N}_r$, defined as in (4.3) by

$$\mathbf{N}_r |n_1 \ldots n_r \ldots\rangle = n_r |n_1 \ldots n_r \ldots\rangle, \tag{6.1}$$

has eigenvalues $n_r = 0, 1$. We attempt to write it again as a product

$$\mathbf{N}_r = a_r^\dagger a_r \tag{6.2}$$

of mutually adjoint creation and annihilation operators $a_r^\dagger$ and a_r. The subspace $\mathbb{M}_r \subset \mathbb{F}_f$ spanned by the basis vectors $|n_1 n_2 \ldots n_r \ldots\rangle$ with n_i, $i \neq r$, fixed is now two-dimensional and isomorphic with spin 1/2 space. The operators $a_r^\dagger$ and a_r correspond to the raising and lowering spin space operators s_+ and s_-. Therefore, if in $\mathbb{M}_r$ we represent the occupied ($n_r = 1$) and unoccupied ($n_r = 0$) states by $|1\rangle = \begin{pmatrix} 1 \\ 0 \end{pmatrix}$ and $|0\rangle = \begin{pmatrix} 0 \\ 1 \end{pmatrix}$, respectively, we have in terms of the Pauli matrices (II.3.5) $a^\dagger = \frac{1}{2}(\sigma_1 + i\sigma_2) = \begin{pmatrix} 0 & 1 \\ 0 & 0 \end{pmatrix}$, $a = \frac{1}{2}(\sigma_1 - i\sigma_2) = \begin{pmatrix} 0 & 0 \\ 1 & 0 \end{pmatrix}$, $a^\dagger a = \begin{pmatrix} 1 & 0 \\ 0 & 0 \end{pmatrix}$ (where the subscript has been dropped). From these expressions one finds that $a^\dagger|1\rangle = |0\rangle$, $a^\dagger|0\rangle = |1\rangle$, $a|1\rangle = |0\rangle$, $a|0\rangle = 0$ and $a^\dagger a|1\rangle = |1\rangle$, $aa^\dagger|0\rangle = 0$. Putting $n = 0, 1$, the last relations can be written $a^\dagger|n\rangle = (1-n)|1-n\rangle$, $a|n\rangle = n|1-n\rangle$, $a^\dagger a|n\rangle = n|n\rangle$, which leads us to define the fermion creation and annihilation operators on $\mathbb{F}_f$ by

$$a_r^\dagger |\ldots n_r \ldots\rangle = \gamma_r^*(1 - n_r)|\ldots n_r \ldots\rangle, \tag{6.3}$$

$$a_r |\ldots n_r \ldots\rangle = \gamma_r n_r |\ldots n_r \ldots\rangle, \tag{6.4}$$

where we have included a phase factor γ_r to be determined next in such a way that the relation (6.2) is satisfied. To this end, let us regard the vector $|\ldots n_r \ldots\rangle$ as a vector of the tensor product $\mathbb{T}_n$ of $n = \Sigma_r n_r$ single-particle spaces and write it as an $n \times n$ Slater determinant

$$|\ldots n_r \ldots\rangle = \frac{1}{\sqrt{n!}} \begin{vmatrix} |\psi_{i_1}\rangle^{(1)} & |\psi_{i_1}\rangle^{(2)} & \ldots & |\psi_{i_1}\rangle^{(n)} \\ |\psi_{i_2}\rangle^{(1)} & |\psi_{i_2}\rangle^{(2)} & \ldots & |\psi_{i_2}\rangle^{(n)} \\ \ldots & \ldots & \ldots & \ldots \\ |\psi_{i_n}\rangle^{(1)} & |\psi_{i_n}\rangle^{(2)} & \ldots & |\psi_{i_n}\rangle^{(n)} \end{vmatrix} \tag{6.5}$$

(remember that n_r equals one if r is one of the members of the set $\{i_j\}$ and is zero otherwise). It is assumed that a certain order of the single-particle basis vectors $|\psi_1\rangle, |\psi_2\rangle, \ldots$ has been chosen.

Consider the action of the operator $a_s^\dagger a_t$, $s \neq t$, on the above vector. If $s \in \{i_j\}$ or $t \notin \{i_j\}$ or both, then the result is the zero vector. In the remaining case the vector (6.5) is of the form $|\ldots 0_s \ldots 1_t \ldots\rangle$ and the action of $a_s^\dagger a_t$ on it results in the replacement of the row $|\psi_t\rangle^{(1)}, |\psi_t\rangle^{(2)}, \ldots$ by the row $|\psi_s\rangle^{(1)}, |\psi_s\rangle^{(2)}, \ldots$. If the number ν of indices between s and t in the set $\{i_j\}$ is odd (even), the parity of the permutation formed by the row indices in the final determinant is different from (the same as) the corresponding parity in the original determinant. Thus, $a_s^\dagger a_t|\ldots 0_s \ldots 1_t \ldots\rangle = (-1)^\nu |\ldots 1_s \ldots 0_t \ldots\rangle$. All cases can be included by writing

$$a_s^\dagger a_t|\ldots n_s \ldots n_t \ldots\rangle = (-1)^\nu (1-n_s) n_t |\ldots 1-n_s \ldots 1-n_t \ldots\rangle$$

with $n_r = 0, 1$. Comparing this result with (6.3) and (6.4) one finds that $\gamma_s^* \gamma_t = (-1)^\nu$. We can then take $\gamma_r = (-1)^{\nu_r}$ where ν_r is the number of n_i, $i < r$, different from zero or, equivalently,

$$\nu_r = \sum_{i=1}^{r-1} n_i.$$

We thus define the fermion creation and annihilation operators by

$$a_r^\dagger|\ldots n_r \ldots\rangle = (-1)^{\nu_r}(1-n_r)|\ldots 1-n_r \ldots\rangle,$$
$$a_r|\ldots n_r \ldots\rangle = (-1)^{\nu_r} n_r |\ldots n_r \ldots\rangle.$$

Defining the *anticommutator* of two operators A, B as

$$[A, B]_+ = AB + BA,$$

one can verify immediately the following anticommutation relations

$$[a_r, a_s^\dagger]_+ = \delta_{rs}, \qquad [a_r^\dagger, a_s^\dagger]_+ = [a_r, a_s]_+ = 0 \tag{6.6}$$

that characterize the fermion operators. Note that the transformation formulae analogous to (5.7) still hold, that is, if $a_r^\dagger, a_r$ are induced by the basis $\{|\psi_i\rangle\}$ in $\mathbb{H}$

and $\alpha_r^\dagger, \alpha_r$ by $\{|\phi_i\rangle\}$, then

$$\alpha_i^\dagger = \sum_j \langle\psi_j|\phi_i\rangle a_j^\dagger, \qquad \alpha_i = \sum_j \langle\phi_i|\psi_j\rangle a_j.$$

We turn now to the representation of observables on $\mathbb{F}_f$ in terms of $a_r^\dagger$ and a_r. Consider first an observable $\mathbf{A}$ which is the symmetrized sum of the single-particle observable A on $\mathbb{H}$ with matrix elements $A_{ij} = \langle\psi_i|A|\psi_j\rangle$. The argument leading to (5.12) can be repeated almost *verbatim* so that we have

$$\mathbf{A} = \sum_{ij} A_{ij} a_i^\dagger a_j. \tag{6.7}$$

In the case of the sum $\mathbf{D}$ of two-particle observables D on $\mathbb{H}\otimes\mathbb{H}$ with matrix elements $D_{ij,kl} = \langle\psi_i\psi_k|D|\psi_j\psi_l\rangle$, the argument leading to (5.15) can also be carried through. The difference in the commutation relations between the fermion and boson operators does not affect the result because in the present case the term with $j = k$ is not present in the sum analogous to (5.14). One thus gets

$$\mathbf{D} = \sum_{ijkl} D_{ij,kl} a_i^\dagger a_k^\dagger a_j a_l. \tag{6.8}$$

The fermion field. – As in the case of bosons, the fermion fields $\Phi^\dagger(x)$ and $\Phi(x)$ can also be introduced. Using the same notation as before they are defined by

$$\Phi^\dagger(x) = \sum_i a_i^\dagger \psi_i^*(x), \qquad \Phi(x) = \sum_i a_i \psi_i(x).$$

In terms of them the above operators $\mathbf{A}$ and $\mathbf{D}$ of equations (6.7) and (6.8) can be represented by formulae entirely analogous to (5.16) and (5.18).

The anticommutation relations obeyed by the Fermi field operators follow directly from (6.6),

$$[\Phi(x), \Phi^\dagger(x')]_+ = \delta(x - x'),$$
$$[\Phi(x), \Phi(x')]_+ = [\Phi^\dagger(x), \Phi^\dagger(x')]_+ = 0.$$

An important property, absent in the boson case, arises from the fact that the fermion anticommutation relations are symmetric in $a_r^\dagger$ and a_r [or $\Phi^\dagger(x)$ and $\Phi(x')$]. This symmetry makes it possible to regard the creation and annihilation operators respectively as annihilating and creating "holes".

In conclusion it should be pointed out that the discussion of Section V.7 shows that, were it not for the spin-statistics correlation, it would be possible to regard a system of bosons and a system of fermions of the same mass as a single system that obeys a superselection rule in the Hilbert space $\mathbb{T}$ defined by (4.2), confined to the coherent subspaces $\mathbb{F}_b$ and $\mathbb{F}_f$, respectively.

A FINITE-DIMENSIONAL VECTOR SPACES

A reminder on the nomenclature of mappings. – Throughout the mathematical appendices, and to a lesser extent in the main text of this book, we shall be referring to different kinds of mappings and it may be useful to remind the reader here of the usual nomenclature.

Given two sets $\mathcal{A}$ and $\mathcal{B}$, let $a \in \mathcal{A}$, $b \in \mathcal{B}$. A mapping T from $\mathcal{A}$ to $\mathcal{B} : a \mapsto b = Ta$, can be of the following types.

T is mapping of $\mathcal{A}$ *into* $\mathcal{B}$ if to each $a \in \mathcal{A}$ there corresponds a definite element $Ta \in \mathcal{B}$ (there may be elements of $\mathcal{B}$ that do not correspond to any element of $\mathcal{A}$ and different elements of $\mathcal{A}$ may correspond to the same element of $\mathcal{B}$). The *range* of the mapping is the subset $T\mathcal{A} \subset \mathcal{B}$ formed by those elements that correspond to some element of $\mathcal{A}$.

T is a mapping of $\mathcal{A}$ *onto* $\mathcal{B}$ if to each element of $\mathcal{A}$ there corresponds a definite element $Ta \in \mathcal{B}$ and to every element of $\mathcal{B}$ there corresponds at least one element of $\mathcal{A}$ (in this case $T\mathcal{A} = \mathcal{B}$).

A *one-to-one* mapping is one in which to distinct elements of $\mathcal{A}$ there correspond distinct elements of $\mathcal{B} : Ta \neq Tb$ if $a \neq b$.

It follows that if T is a one-to-one mapping from $\mathcal{A}$ *onto* $\mathcal{B}$, there exists an inverse one-to-one mapping T^{-1} from $\mathcal{B}$ onto $\mathcal{A}$. Such an inverse does not exist if T is a one-to-one mapping from $\mathcal{A}$ *into* $\mathcal{B}$ (an inverse mapping from $T\mathcal{B} \subset \mathcal{B}$ onto $\mathcal{A}$ does exist in this case).

1. Basic definitions

A vector space is defined with reference to a field. We consider the case where the field is the field $\mathbb{C}$ of complex numbers because this is the case of interest in quantum mechanics.

1.1 *Vector spaces.* – A **vector space** $\mathbb{H}$ is a set $\{f, g, h, \ldots\}$ whose elements are called vectors, such that

(i) An addition $f + g \in \mathbb{H}$ is defined which is commutative, $f + g = g + f$, and associative, $(f + g) + h = f + (g + h)$.

(ii) There is a zero element $0 : f + 0 = f$ (no distinction in notation is made between the zero vector and the complex number zero). For every vector f there is a vector $-f : f + (-f) = 0$. One writes $f - g = f + (-g)$.

(iii) Let $\alpha, \beta, \ldots$, be complex numbers. The product $\alpha f \in \mathbb{H}$ of a complex number times a vector is defined. It is distributive and associative: $(\alpha + \beta)f = \alpha f + \beta f$, $\alpha(\beta f) = (\alpha\beta)f$ – one writes simply $\alpha\beta f$ –. Moreover $1 \cdot f = f$, $0 \cdot f = 0$.

1.2 *Isomorphism.* – Two vector spaces $\mathbb{U}, \mathbb{V}$ (defined over the same field) are said to be **isomorphic** if there exists a one-to-one and onto correspondence $u \leftrightarrow v$ between every vector $u \in \mathbb{U}$ and every vector $v \in \mathbb{V}$ that preserves the linear operations, that is, $\alpha u_1 + \beta u_2 \leftrightarrow \alpha v_1 + \beta v_2$ whenever $u_1 \leftrightarrow v_1$ and $u_2 \leftrightarrow v_2$ for $u_1, u_2 \in \mathbb{U}$ and $v_1, v_2 \in \mathbb{V}$. Such a correspondence is called an **isomorphism**. If one is only interested in linear operations on vectors, two isomorphic spaces are indistinguishable from each other.

1.3 *Linear dependence.* – The vectors f_i, $i = 1, \ldots k$, are **linearly independent** if $\sum_i \alpha_i f_i = 0$ implies that every $\alpha_i = 0$. Otherwise they are **linearly dependent.** The largest number of linearly independent vectors is the **dimension** $\dim(\mathbb{H})$ of the space $\mathbb{H}$. If such a number does not exist the dimension is infinite.

In the remainder of this appendix we consider only finite-dimensional vector spaces.

1.4 *Subspaces.* – A subset of $\mathbb{H}$ containing all the linear combinations of any number of its vectors is a **subspace**. The smallest subspace that contains a set $\mathcal{S}$ of vectors is said to be **spanned** by $\mathcal{S}$. Note that the intersection $\mathbb{M} \cap \mathbb{N}$ – in the sense of set theory – of two subspaces $\mathbb{M}$ and $\mathbb{N}$ is a subspace but that in general their union $\mathbb{M} \cup \mathbb{N}$ is in not.

1.5 *Bases.* – If $\dim(\mathbb{H}) = m$, a set of m linearly independent vectors is called a **basis** on $\mathbb{H}$. The m vectors b_i, $i = 1, \ldots, m$, form a basis if and only if every vector f can be written uniquely as a linear combination $f = \sum_i \beta_i b_i$. Clearly a basis spans the whole of $\mathbb{H}$.

1.6 *Norm.* – The vector space $\mathbb{H}$ is **normed** if for every vector f there exist a real number $\|f\| \geq 0$, called the **norm** of f, such that $\|f\| = 0$ if and only if $f = 0$, $\|\alpha f\| = |\alpha|\|f\|$ and $\|f + g\| \leq \|f\| + \|g\|$.

1.7 *Dual space, scalar product.* – A **linear functional** on $\mathbb{H}$ is a complex valued function $a(f)$ such that $a(\alpha f + \beta g) = \alpha a(f) + \beta a(g)$. When the argument of the functional $a(f)$ is unimportant one simply omits it and denotes the functional by a. We define the sum $a + b$ of two linear functionals as $(a + b)(f) = a(f) + b(f)$, and the product αa of a number times a linear functional as $(\alpha a)(f) = \alpha^* a(f)$ (note the complex conjugation).[1] The linear functionals on $\mathbb{H}$ with the operations just defined form a vector space $\mathbb{H}^\dagger$ called the **dual space** of $\mathbb{H}$. Because both a and f are vectors (although in different spaces) it is often convenient to represent $a(f)$ by the more symmetric symbol $\langle a; f\rangle$ whose value is called the **scalar product** of a and f. The scalar product is clearly antilinear in a and linear in f.

1.8 *Inner product space.* – An inner product vector space is one in which an inner product is defined. An **inner product** $\langle f, g\rangle$ is a complex valued function of pairs of vectors with the properties $\langle f, g\rangle = \langle g, f\rangle^*$; $\langle f, f\rangle > 0$ for $f \neq 0$; $\langle f, g\rangle$ is linear in g, that is, $\langle f, g + h\rangle = \langle f, g\rangle + \langle f, h\rangle$, $\langle f, \alpha g\rangle = \alpha\langle f, g\rangle$. It follows that $\langle f, g\rangle$ is **antilinear** in f, that is, $\langle f + h, g\rangle = \langle f, g\rangle + \langle h, g\rangle$, $\langle \alpha f, g\rangle = \alpha^*\langle f, g\rangle$. In an inner product space we shall always work with the norm $\|f\| = \sqrt{\langle f, f\rangle}$, which obeys **Schwarz' inequality**, $|\langle f, g\rangle| \leq \|f\|\|g\|$. A **unitary** space is a finite-dimensional inner product vector space. Every finite-dimensional space can be made into a unitary space.

1.8.1. If $\mathbb{H}$ is a unitary space the inner product $\langle g, f\rangle$ for each fixed vector g is a linear functional a_g with value $a_g(f) = \langle g, f\rangle$. The properties of the inner product ensure that the addition and multiplication rules of 1.7 are obeyed if we set $a_g + a_h = a_{g+h}$ and $\alpha a_g = a_{\alpha g}$. We can then identify $\mathbb{H}^\dagger$ with $\mathbb{H}$ by identifying $a_g \in \mathbb{H}^\dagger$ with $g \in \mathbb{H}$. The terms scalar product and inner product can then be used synonymously. Except when the conceptual difference is important we shall use the first term, which, as will be seen, is more appropriate when using Dirac's notation

1.9 *Orthogonality.* – Two vectors f and g are **orthogonal** if $\langle f, g\rangle = 0$. A set of vectors $\{f_k\}$ is **orthonormal** if $\langle f_k, f_l\rangle = \delta_{kl}$ for all k, l. The expression of a vector f as a linear combination of the vectors b_i, $i = 1, \ldots, m$ of an orthonormal basis is particularly useful:

$$f = \sum_i \langle b_i, f\rangle b_i.$$

[1] In the mathematical literature the scalar product is usually defined as an antilinear functional. The convention used here has been chosen to make it consistent with the Dirac notation of bras and kets.

The coefficient $\langle b_i, f\rangle$ is the **component** of f along b_i. Given a basis $\{f_i\}$, $i = 1, ..., k$, on a subspace $\mathbb{M}$ one can construct an orthonormal basis $\{b_i\}$ on $\mathbb{M}$ by use of the **Gram-Schmidt** orthonormalization procedure: set $g_1 = f_1$, $b_1 = g_1/\|g_1\|$ and proceed with the recursion formula

$$g_i = f_i - \sum_{j=1}^{i-1} \langle b_j, f_i\rangle b_j, \qquad b_i = \frac{g_i}{\|g_i\|}, \qquad i = 2, ..., k.$$

1.9.1. Two subspaces $\mathbb{M}$ and $\mathbb{N}$ are orthogonal if every vector of one is orthogonal to every vector of the other. The subspace $\mathbb{M}^\perp$ consisting of all the vectors of $\mathbb{H}$ orthogonal to every vector of $\mathbb{M}$ is the **orthogonal complement** of $\mathbb{M}$.

1.10 *Direct sum.* – Given two unitary spaces $\mathbb{U}$ and $\mathbb{V}$, their **direct sum** $\mathbb{W} = \mathbb{U} \oplus \mathbb{V}$ is the unitary space consisting of the pairs $\{f, g\}$, written $f \oplus g$, $f \in \mathbb{U}$, $g \in \mathbb{V}$ with the operations defined by

$$\alpha_1 f_1 \oplus g_1 + \alpha_2 f_2 \oplus g_2 = (\alpha_1 f_1 + \alpha_2 f_2) \oplus (\alpha_1 g_1 + \alpha_2 g_2)$$

$$\langle f_1 \oplus g_1, f_2 \oplus g_2\rangle = \langle f_1, f_2\rangle + \langle g_1, g_2\rangle.$$

A definition that can obviously be extended to any finite number of spaces.

1.10.1. The space $\mathbb{H}$ of the previous paragraph can be regarded as the direct sum of $\mathbb{M}$ and its orthogonal complement $\mathbb{M}^\perp$. More generally, given a set of mutually orthogonal subspaces $\mathbb{M}_k$ whose vectors span the whole space $\mathbb{H}$ we can regard the latter as the direct sum of all the subspaces and write

$$\mathbb{H} = \bigoplus_k \mathbb{M}_k.$$

1.11 *Dirac's notation.* – In spite of the possibility of identifying $\mathbb{H}$ and $\mathbb{H}^\dagger$, we shall see that there are some formal advantages in not doing so. These advantages are exploited in **Dirac's notation.** in which the vectors of $\mathbb{H}$ are represented by **kets** $|f\rangle$ and the vectors of $\mathbb{H}^\dagger$ by **bras** $\langle g|$. A correspondence between bras and kets is established by writing $\langle f|$ for the vector of $\mathbb{H}^\dagger$ previously identified with $|f\rangle$. It follows that the bra corresponding to the ket $\alpha|f\rangle$ (also written $|f\rangle\alpha$) is $\alpha^*\langle f|$ (or $\langle f|\alpha^*$). The scalar product $\langle g, f\rangle$ is written $\langle g|f\rangle$.

2. Operators in unitary spaces

2.1 *Linear operators.* – An **operator** on a vector space $\mathbb{H}$ is a mapping $A : \mathbb{H} \to \mathbb{H}$; if g is the image of f under the mapping A we write $g = Af$. The **domain** of A is the set of vectors f for which Af is defined. Its **range** is the set of all vectors of the form Af. The **product** AB of two operators is defined by $(AB)f = A(Bf)$; we write it as ABf. The equality and algebraic operations among operators must be interpreted with due regard to their domain. Thus $A = B$ means that $Af = Bf$ for every vector f in the intersection of the domains of A and B.

2.1.1. The operator A is **linear** if, for all $f, g \in \mathbb{H}$, $A(f+g) = Af + Ag$, $A(\alpha f) = \alpha Af$. Linear operators form a linear space with operations $(A+B)f = Af + Bf$, $(\alpha A)f = \alpha Af$, $A = 0$ if and only if $Af = 0$ for all f. The **norm** $\|A\|$ of the linear operator A is defined as the least upper bound of $\|Af\|$ for all f such that $\|f\| \leq 1$. The **identity** operator I is defined by $If = f$.

2.1.2. We are concerned here with finite-dimensional vector spaces. In such spaces the domain and the range of a linear operator are subspaces. Moreover, by an extension of its definition, the domain of any given linear operator can always taken to be the whole of the space. Thus, in dealing with linear operators, domain considerations can be ignored.

From now on we shall be concerned mostly with linear operators. We shall drop the adjective "linear" so that unless otherwise explicitly specified the term "operator" is to be understood as "linear operator."

2.2 *Commutator.* – The **commutator** of A and B is $[A, B] = AB - BA$, it is linear and antisymmetric in its arguments

$$[\alpha A + \beta B, \gamma C] = \alpha\gamma[A, C] + \beta\gamma[B, C], \qquad [A, B] = -[B, A],$$

and satisfies the **Jacobi identity**:

$$\Big[[A, B], C\Big] + \Big[[B, C], A\Big] + \Big[[C, A], B\Big] = 0.$$

2.3 *Inverse.* – The operator A^{-1} is the **inverse** of A if $AA^{-1} = A^{-1}A = I$. The inverse of A exists if and only if the corresponding mapping is one-to-one. If A and B have inverses then so does AB and $(AB)^{-1} = B^{-1}A^{-1}$.

2.4 *Adjoint.* – Since $\langle f, Ag\rangle$ is linear in g and antilinear in f, one can define a linear operator $A^\dagger$ called the **adjoint** of A by $\langle f, A^\dagger g\rangle = \langle Af, g\rangle = \langle g, Af\rangle^*$, for all vectors f, g. The adjoint obeys the following properties: $\|A^\dagger\| = \|A\|$, $(A^\dagger)^\dagger = A$, $(A+B)^\dagger = A^\dagger + B^\dagger$, $(\alpha A)^\dagger = \alpha^* A^\dagger$, $(AB)^\dagger = B^\dagger A^\dagger$, $(A^{-1})^\dagger = (A^\dagger)^{-1}$.

2.5 *Hermitean operators.* – An operator H is **hermitean** (also called **self-adjoint**) if $H^\dagger = H$ or equivalently $\langle f, Hg\rangle = \langle Hf, g\rangle = \langle g, Hf\rangle^*$, for all vectors f, g. If

H, K are hermitean so are $H + K$, $i[H, K]$, αH for all real α, and $A^\dagger H A$ for any operator A; but HK is hermitean if and only if $[H, K] = 0$.

2.5.1. The hermitean operator H is **positive** if $\langle f, Hf \rangle \geq 0$ for all f; note that if the equality holds for all f then $H = 0$. If H is positive it obeys the inequality $|\langle f, Hg \rangle|^2 \geq \langle f, Hf \rangle \langle g, Hg \rangle$, called **Schwarz' inequality for positive operators**.

2.5.2. If $K^\dagger = -K$ then the operator K is called **antihermitean**. Clearly iK is hermitean. An arbitrary operator A can always be written as the sum $A = H_A + K_A$ of its hermitean part $H_A = \frac{1}{2}(A + A^\dagger)$ and its antihermitean part $K_A = \frac{1}{2}(A - A^\dagger)$.

2.6 *Unitary operators.* – A linear operator U is **unitary** if, for all f, $\|Uf\| = \|f\|$ or equivalently if U has an inverse $U^{-1} = U^\dagger$. Let $\{b_i\}$ be an orthonormal basis then $\{Ub_i\}$ is also an orthonormal basis. Conversely if $\{Ub_i\}$ is an orthonormal basis and U is linear, it is also unitary. Later on we shall meet operators that are not linear that also preserve orthonormal bases.

2.7 *Normal operators.* – Hermitean and unitary operators are particular types of normal operators. An operator is **normal** if it commutes with its adjoint. It follows that the hermitean and antihermitean parts of a normal operator commute.

2.8 *Projections.* – Let $\mathbb{M} \subset \mathbb{H}$ be a subspace and $\mathbb{M}^\perp$ its orthogonal complement. Every vector h can be written in an unique manner as $h = f + g$, with $f \in \mathbb{M}$ and $g \in \mathbb{M}^\perp$. f is called the **projection** of h on $\mathbb{M}$.[2] The linear operator E_M defined by $f = E_M h$ is called the projection operator on $\mathbb{M}$. Its domain is the whole space and its range the subspace $\mathbb{M}$. We say that E_M projects on $\mathbb{M}$. An operator which projects on some subspace is called a **projection operator**, or simply a projection when the context prevents confusion with a vector in its range. An operator E is a projection operator if and only if $E^2 = E$ and $E^\dagger = E$.

2.8.1. Let E_M an E_N be the projections on the subspaces $\mathbb{M}$ and $\mathbb{N}$ respectively. The product $E_M E_N$ is a projection if and only if both operators commute in which case $E_M E_N$ projects on the intersection of $\mathbb{M}$ and $\mathbb{N}$.[3] The sum $E_M + E_N$ is a projection if and only if $E_M E_N = 0$, which means that $\mathbb{M}$ and $\mathbb{N}$ are orthogonal. Then $E_M + E_N$ projects on $\mathbb{M} \oplus \mathbb{N}$. The difference $E_M - E_N$ is a projection operator if and only if $E_M E_N = E_N$ which is equivalent to $\mathbb{N} \subseteq \mathbb{M}$. In this case

[2] Properly speaking f is the orthogonal projection of h on $\mathbb{M}$. There are projections that are not orthogonal but they are not of interest to us.

[3] In general the projection on $\mathbb{M} \cap \mathbb{N}$ is given by $\lim_{n\to\infty}(E_M E_N)^n$ which becomes simply $E_M E_N$ when both projections commute.

$\|E_N f\| \leq \|E_M f\|$ for all $f \in \mathbb{H}$ and we write $E_N \leq E_M$.

2.8.2. The dimension of a projection operator E_M is the dimension of its range $\mathbb{M}$.

2.8.3. Any two vectors f, g determine an operator $f \times g^*$ defined by $(f \times g^*)h = \langle g, h\rangle f$. The notation includes the complex conjugation symbol as a reminder that the operator is a linear function of f and an antilinear function of g. We have $(f \times g^*)^\dagger = g \times f^*$. In particular $f \times f^*$ is hermitean and, if f is normalized, it is the one-dimensional projection whose range is f.

2.9 *Reducibility.* – A subspace $\mathbb{M}$ is **invariant** under an operator A if for all vectors $f \in \mathbb{M}$, we have $Af \in \mathbb{M}$. If, furthermore, $Ag \in \mathbb{M}^\perp$ for all $g \in \mathbb{M}^\perp$, that is, if both $\mathbb{M}$ and $\mathbb{M}^\perp$ are invariant under A, the subspace $\mathbb{M}$ is said to **reduce** the operator A. The statements "$\mathbb{M}$ reduces A" and "A commutes with E_M" are equivalent.

2.9.1. Let $\mathbb{M}$ be invariant under A. With f and g as above, we have $\langle A^\dagger g, f\rangle = \langle g, Af\rangle = 0$, therefore $A^\dagger g \in \mathbb{M}^\perp$ or, equivalently, $\mathbb{M}^\perp$ is invariant under $A^\dagger$.

2.9.2. From this result one sees immediately that if $\mathbb{M}$ is invariant under the hermitean operator H, then $\mathbb{M}$ reduces H. The same is true if U is unitary because then, from $\langle Uf, Ug\rangle = \langle f, g\rangle = 0$, we conclude that Ug, being orthogonal to Uf, must be in $\mathbb{M}^\perp$. A more general result holds: if $\mathbb{M}$ is invariant under a normal operator N, then $\mathbb{M}$ reduces N.

2.9.3. If no subspaces other than $\mathbb{H}$ itself and $\{0\}$ reduce every member of a set of operators then the set is called **irreducible**. It follows that a set of operators is irreducible if and only if the only operators that commute with every member of the set are multiples of the identity. If there is a subspace that reduces every operator of the set, then the set is said to be **reducible**.

2.10 *Matrix representation.* – Let $\{b_i\}$ be an orthonormal basis on an m-dimensional space $\mathbb{H}$. An arbitrary vector g can be written in terms of its components $\gamma_i = \langle b_i, g\rangle$, $i = 1, \ldots, m$,

$$g = \sum_i \gamma_i b_i.$$

Given an operator A, the vector $h = Ag$ has components

$$\eta_i = \langle b_i, Ag\rangle = \sum_j \gamma_j \langle b_i, Ab_j\rangle$$

One can identify the vectors g and h with the column matrices formed by their components,

$$g = \begin{pmatrix} \gamma_1 \\ \gamma_2 \\ \dots \\ \gamma_m \end{pmatrix}, \qquad h = \begin{pmatrix} \eta_1 \\ \eta_2 \\ \dots \\ \eta_m \end{pmatrix}.$$

Indeed the vector space of column matrices with the usual definitions of addition of matrices and multiplication by a number is isomorphic with $\mathbb{H}$. With these definitions the operator A can be identified with the square matrix whose elements are the m^2 numbers $A_{ij} = \langle b_i, Ab_j \rangle$,

$$A = \begin{pmatrix} A_{11} & A_{12} & \dots & A_{1m} \\ A_{21} & A_{22} & \dots & A_{2m} \\ \dots & \dots & \dots & \dots \\ A_{m1} & A_{m2} & \dots & A_{mm} \end{pmatrix}.$$

With these identifications the components η_i, can be rewritten

$$\eta_i = \sum_j A_{ij}\gamma_j.$$

An expression which is identical to the matrix equation

$$\begin{pmatrix} \eta_1 \\ \eta_2 \\ \dots \\ \eta_m \end{pmatrix} = \begin{pmatrix} A_{11} & A_{12} & \dots & A_{1m} \\ A_{21} & A_{22} & \dots & A_{2m} \\ \dots & \dots & \dots & \dots \\ A_{m1} & A_{m2} & \dots & A_{mm} \end{pmatrix} \begin{pmatrix} \gamma_1 \\ \gamma_2 \\ \dots \\ \gamma_m \end{pmatrix},$$

or more succintly $(\eta_i) = (A_{ij})(\gamma_j)$.

2.10.1. When the above identifications are made we speak of the **matrix representation of $\mathbb{H}$ with respect to the basis** $\{b_i\}$, or simply of the $\{b_i\}$-representation of $\mathbb{H}$.

2.10.2. If in a given representation $A = (A_{ij})$ and $A^\dagger = \left(A^\dagger_{ij}\right)$, then $A^\dagger_{ij} = A^*_{ji}$; thus, if H is hermitean, $H_{ij} = H^*_{ji}$. The representation of the identity operator with respect to any orthonormal basis is the identity matrix (δ_{ij}). The inverse A^{-1} exists if and only if the determinant $\det(A_{ij}) \neq 0$; then $A^{-1}_{ij} = \operatorname{cof}(A_{ji}) / \det(A_{ij})$. The matrix elements of a unitary operator U satisfy the relations $\sum_k U_{ik}U^*_{jk} = \delta_{ij}$.

2.10.3. If E_N is the projection on the n-dimensional subspace $\mathbb{N}$, one can select an orthonormal basis $\{b_i\}$ on $\mathbb{H}$, n of whose vectors $b_{i_1}, \dots, b_{i_n}$ form a basis on $\mathbb{N}$. In the corresponding representation E_N has the n diagonal matrix elements $b_{i_k i_{k'}}$, $k = 1, \dots, n$ equal to one and all the others equal to zero.

2.11 *Change of representation.* – Let $\{b_i\}$, $\{\bar{b}_i\}$ be two bases on $\mathbb{H}$, what is the relation between the representations of a vector g and an operator A with respect to the bases $\{b_i\}$ and $\{\bar{b}_i\}$?

2.11.1. Consider first the vector g and let $\gamma_i = \langle b_i, g\rangle$ and $\bar{\gamma}_i = \langle \bar{b}_i, g\rangle$ be its components along b_i and $\bar{b}_i$ respectively. Since $\bar{b}_i = \sum_j \langle b_j, \bar{b}_i\rangle b_j$, we have $\bar{\gamma}_i = \sum_j \langle b_j, \bar{b}_i\rangle^* \langle b_j, g\rangle$. Defining the matrix

$$S = (S_{ij}), \qquad S_{ij} = \left(\langle \bar{b}_i, b_j\rangle\right)$$

we can write $(\bar{\gamma}_i) = (S_{ij})(\gamma_j)$. A similar calculation for an operator A with matrix elements $A_{ij} = \langle b_i, Ab_j\rangle$ and $\bar{A}_{ij} = \langle \bar{b}_i, A\bar{b}_j\rangle$ yields $\left(\bar{A}_{ij}\right) = (S_{ik})(A_{kl})\left(S^*_{jl}\right)$.

2.11.2. The matrix S can be be verified to be unitary in the sense that $\sum_k S_{ik} S^*_{jk} = \delta_{ij}$. Instead of thinking of S as a matrix performing a change of bases we can think of it as a unitary operator that generates a **unitary transformation** of $\mathbb{H}$ onto itself given by the correspondence

$$\bar{f} = Sf, \qquad \bar{A} = SAS^{\dagger}.$$

For any vectors f, g and any operator A we then have

$$\langle \bar{f}, \bar{g}\rangle = \langle f, g\rangle, \qquad \langle \bar{f}, \bar{A}\bar{g}\rangle = \langle f, Ag\rangle.$$

2.12 *Trace.* – The **trace** $\mathrm{Tr}\,(A)$ of an operator A is defined by $\mathrm{Tr}\,(A) = \sum_i A_{ii}$. The definition is justified because the trace is independent of the representation:

$$\sum_i \bar{A}_{ii} = \sum_{ijk} S_{ij} A_{jk} S^*_i = \sum_{jk} A_{jk}\delta_{jk} = \sum_i A_{ii}.$$

It follows immediately from the definition that $\mathrm{Tr}\,(AB) = \mathrm{Tr}\,(BA)$, therefore, if the argument of the trace is a monomial product of operators, its factors can be permuted cyclically.

2.12.1. The trace of a projection operator equals its dimension. This can be seen by selecting a representation of the type introduced in 2.10.3.

2.13 *Direct sum.* – Let U, V be operators on two spaces $\mathbb{U}$ and $\mathbb{V}$, respectively. Their **direct sum** $C = U \oplus V$ is the operator on $\mathbb{U} \oplus \mathbb{V}$ defined by $C(f \oplus g) = (Uf) \oplus (Vg)$, $f \in \mathbb{U}$, $g \in \mathbb{V}$. . If a subspace $\mathbb{M}$ of $\mathbb{H}$ reduces an operator A then A can be regarded as a direct sum $A = A_M \oplus A_{M\perp}$, where A_M and $A_{M\perp}$ are the operators $E_M A = AE_M$ and $(I - E_M)A$ considered as operators on $\mathbb{M}$ and $\mathbb{M}^{\perp}$, respectively. In particular we can write $E_M = I_M \oplus 0_{M^\perp}$, where I_M is the identity on $\mathbb{M}$ and $0_{M^\perp}$ the zero operator on $\mathbb{M}^\perp$.

2.14 *Dirac's notation.* – In Dirac's notation the vector Af is written in the ket space $\mathbb{H}$ as $A|f\rangle$. Consider the dual space $\mathbb{H}^\dagger$, a vector $a \in \mathbb{H}^\dagger$, and an operator Q on $\mathbb{H}^\dagger$. As we know the vector a is denoted in Dirac's notation by the bra $\langle a|$. The vector $Qa \in \mathbb{H}^\dagger$ is denoted by $\langle a|Q$; the convention is that the operator operating on a vector is written to the left of kets but to the right of bras. The relation $\langle Af, h\rangle = \langle f, A^\dagger h\rangle$ shows that the bra that corresponds to the ket $A|f\rangle$ in the sense of 1.11 is $\langle f|A^\dagger$.

2.14.1. The scalar product $\langle g, Af\rangle$ is written in Dirac's notation as $\langle g|A|f\rangle$, while $\langle Af, g\rangle$ cannot be literally translated and must be first changed into one of the equivalent expressions $\langle g, Af\rangle^*$ or $\langle f, A^\dagger g\rangle$. Thus the defining relation in 2.4 for the adjoint of an operator becomes $\langle f|A^\dagger|g\rangle = \langle g|A|f\rangle^*$, while a hermitean operator H is characterized by $\langle f|H^\dagger|g\rangle = \langle g|H|f\rangle^*$.

2.14.2. In Dirac's notation the operator $f \times g^*$ of 2.8.3 is written as $|f\rangle\langle g|$ so that we have the formally pleasing relations $(|f\rangle\langle g|)\,|h\rangle = |f\rangle\langle g|h\rangle$, $\langle h|\,(|f\rangle\langle g|) = \langle h|f\rangle\langle g|$. If $|b\rangle$ is normalized, $|b\rangle\langle b|$ is the one-dimensional projection operator that projects on the subspace spanned by $|b\rangle$.

2.14.3. Let $\{|b_i\rangle\}$ be an orthonormal basis on $\mathbb{H}$. By computing its matrix elements one sees immediately that $\sum_i |b_i\rangle\langle b_i| = I$, the identity operator. Use of this expression of the identity simplifies many formal manipulations. For example, inserting it in each of the equations $|g\rangle = I|g\rangle$, $\langle g| = \langle g|I$ we obtain the expansions of $|g\rangle$ and $\langle g|$ in terms of the basis vectors

$$|g\rangle = \sum_i |b_i\rangle\langle b_i|g\rangle, \qquad \langle g| = \sum_i \langle g|b_i\rangle\langle b_i|.$$

A similar expansion for operators can be obtained by inserting the expression for the identity into the equation $A = IAI$,

$$A = \sum_i \sum_j |b_i\rangle\langle b_i|A|b_j\rangle\langle b_j|$$

or equivalently

$$A = \sum_i \sum_j |b_i\rangle A_{ij}\langle b_j|$$

with the conventional notation $|g\rangle\alpha\langle g| = \alpha|f\rangle\langle g|$.

2.14.4. In the $\{|b_i\rangle\}$-representation it is customary to represent the kets by column matrices. Thus, if $\gamma_i = \langle b_i|g\rangle$, one writes

$$|g\rangle = \begin{pmatrix} \gamma_1 \\ \gamma_2 \\ \cdots \\ \gamma_m \end{pmatrix}.$$

The corresponding bra is represented then by a row matrix with complex conjugate elements

$$\langle g| = (\,\gamma_1^* \quad \gamma_2^* \quad \cdots \quad \gamma_m^*\,).$$

The scalar product $\langle f|g\rangle$ is then given by the product of the matrices representing $\langle f|$ and $|g\rangle$.

2.14.5. Due to the convention on where to place an operator acting on a vector, the matrix representation of an operator is identical with the representation discussed in 2.10.

2.14.6. Let $\mathbb{N}$ be an n-dimensional subspace and $\{|a_1\rangle, \ldots, |a_n\rangle\}$ an orthonormal basis on it. A projection E_N on $\mathbb{N}$ can then be written as the sum of the one-dimensional projections along the basis vectors

$$E_N = \sum_{k=1}^{n} |a_i\rangle\langle a_i|;$$

a decomposition which is obviously not unique unless $n = 1$.

3. The spectral theorem

3.1 *The eigenvalue problem.* – If A is a linear operator and λ a complex number such that the equation $Af = \lambda f$, called the eigenvalue equation, has a solution $f \neq 0$, then λ is an **eigenvalue** and f an **eigenvector** of A. Since the eigenvalue equation is homogeneous any multiple αf of an eigenvector f is an eigenvector and we can take $\|f\| = 1$ without loss of generality.

3.1.1. In the m-dimensional space $\mathbb{H}$ choose a representation in which $A = (A_{ij})$, $g = (\gamma_i)$. Written in this representation, the eigenvalue equation $Ag = \lambda g$ becomes a system of m homogeneous linear equations in the γ_i. The eigenvalues λ are the roots of the algebraic equation of degree m

$$\det(A_{ij} - \lambda\delta_{ij}) = 0,$$

known as the secular equation, which expresses the condition for the homogeneous system to have non-trivial solutions. It follows from a well known result of algebra that, in an m-dimensional vector space, every operator has at least one and at most m distinct eigenvalues. If all the eigenvalues of an arbitrary operator A are simple then there are m linearly independent eigenvectors of A but there may be fewer if A has multiple eigenvalues.

3.1.2. Let λ be an eigenvalue of A with multiplicity s, from the linearity of the eigenvalue equation it follows that the eigenvectors of A belonging to λ constitute a subspace, called an **eigensubspace**, with dimension no greater than s. Its dimension is the **degeneracy**[4] of λ. One can easily see that the multiple of the identity αI has a single eigenvalue α of multiplicity $m = \dim(\mathbb{H})$ and that the corresponding eigensubspace is the whole of $\mathbb{H}$; thus, in this case the multiplicity and the degeneracy of α are equal. Also, by writing the secular equation for the operator $A^\dagger$ one sees immediately that λ^* is an eigenvalue of $A^\dagger$ with the same multiplicity as λ.

3.2 *Spectrum and resolvent operator.* – The set of eigenvalues of A is called the **spectrum** of A. The spectrum is simple if all the eigenvalues are. The operator $(A - \alpha I)^{-1}$ is called the **resolvent** of A. The discussion in (2.10) shows that the resolvent exists if and only if α does not satisfy the secular equation, that is, if and only if α is not in the spectrum of A.

3.3 *The eigenvalue problem for hermitean operators.* – Consider a *hermitean* operator H so that $H^\dagger = H$. Let λ, μ, be eigenvalues of H and f, g, corresponding eigenvectors. Taking the complex conjugate of the equation $\langle f, Hf\rangle = \lambda\langle f, f\rangle$ one finds $\langle Hf, f\rangle = \langle f, Hf\rangle = \lambda^*\langle f, f\rangle$ which shows that $\lambda = \lambda^*$: *the eigenvalues of a hermitean operator are real.*

3.3.1. Left-multiply the equation $Hf = \lambda f$ by g, right-multiply $Hg = \mu g$ by f and subtract; we obtain $(\lambda - \mu)\langle g, f\rangle = 0$. Therefore if $\lambda \neq \mu$ then $\langle g, f\rangle = 0$: *the eigenvectors of a hermitean operator belonging to different eigenvalues are orthogonal.* So are therefore the corresponding eigensubspaces and eigenprojections.

3.4 *The spectral theorem.* – Let H be a hermitean operator with r distinct eigenvalues λ_i, $i = 1, \ldots, r$. Call $\mathbb{M}_i$ and E_i the eigensubspace and eigenprojection belonging to the eigenvalue λ_i. Obviously, $\mathbb{M}_i$ is invariant under H; therefore, by 2.9.2 it reduces H or, equivalently, $E_iH = HE_i$. Evidently the subspace $\mathbb{M} = \bigoplus_i \mathbb{M}_i$ spanned by all the eigenvectors of H also reduces H and the corresponding pro-

[4] The degeneracy of an eigenvalue is also called its *geometric multiplicity* while its multiplicity in the sense used here is called its *algebraic multiplicity*.

jection $E = \sum_i E_i$ commutes with H. This result has the following important consequence: *the eigenvectors of a hermitean operator span the whole space*, that is, $E = I$ or $\mathbb{M} = \mathbb{H}$. To see this assume the statement not to be true. It follows that $H(I - E) \neq 0$ can be regarded as an operator on $\mathbb{M}^{\perp}$ that has at least one eigenvalue ν. There is therefore a vector $g \in \mathbb{M}^{\perp}$ which satisfies the eigenvalue equation $H(I - E)g = \nu g$. Since $Eg = 0$, we have $Hg = \nu g$ which means that $g \in \mathbb{M}$, in contradiction with the hypothesis.

3.4.1. Regarded as an operator on $\mathbb{M}_i$ the operator H multiplies every vector by the number λ_i; therefore it is equal to $\lambda_i I_i$ and the multiplicity of λ_i equals its degeneracy (cf. 3.1.2). One can then write H as the direct sum $\bigoplus_i \lambda_i I_i$ or equivalently as the sum $\sum_i \lambda_i E_i$ in terms of the eigenprojections E_i on $\mathbb{H}$. Collecting together the pertinent results we have the **spectral theorem** in a unitary space.

3.4.2. *To every hermitean operator H on an m-dimensional unitary space $\mathbb{H}$ there corresponds a unique family of non-zero projection operators, the eigenprojections of H, E_i, $i = 1, ..., r$, $r \leq m$ with the following properties:*

(i) The projections E_i are pairwise orthogonal,

$$E_i E_j = E_i \delta_{ij}.$$

(ii) The family of projections is complete,

$$\sum_{i=1}^{r} E_i = I.$$

(iii) There exists a unique set of r distinct real numbers λ_i, the eigenvalues of H, such that

$$H = \sum_{i=1}^{r} \lambda_i E_i.$$

This expression is called the **spectral resolution** of H.

3.4.3. The range of E_i is the eigensubspace $\mathbb{M}_i$ belonging to λ_i. Its dimension is the degeneracy or multiplicity s_i of λ_i. On it one can construct an orthonormal basis $\{b_i^r\}$, $r = 1, \ldots, s_i$. It follows from the completeness of the family E_i that the union of all those bases $\{b_i^r\}$, $r = 1, \ldots, s_i$, $i = 1, \ldots, m$, constitutes an orthonormal basis on $\mathbb{H}$. One often expresses this fact by saying that H possesses a **complete set** of eigenvectors.

3.4.4. It is an immediate consequence of (i) and (ii) that for every vector f we have $\langle f, E_i f \rangle \geq 0$ and $\sum_i \langle f, E_i f \rangle = 1$, that is, for each vector and each family of projections, the set of numbers $\mathcal{P}_i = \langle f, E_i f\rangle$ constitutes a discrete probability distribution.

3.4.5. Two hermitean operators $A = \sum_i \alpha_i E_i$ and $B = \sum_j \beta_j F_j$ commute if and only if every pair E_i, F_i, of their eigenprojections commute. The first implication follows trivially from the spectral resolutions. Assume next that $[A, B] = 0$, then $ABE_i = BAE_i = \alpha_i BE_i$. Thus, if f is any vector, not only $E_i f$ but also $BE_i f$ are eigenvectors of A belonging to the eigenvalue α_i, therefore $E_i BE_i f = BE_i f$, or $E_i BE_i = BE_i$. The adjoint of this equation reads $E_i BE_i = E_i B$; thus $[E_i, B] = 0$. Similarly we can prove that $[A, F_j] = 0$ so that in the first part of the proof F_j can replace B and we conclude, as stated, that $[E_i, F_j] = 0$.

3.4.6. From the orthogonality of the projections in the spectral resolution of H we have for every normalized vector f

$$\|Hf\|^2 = \langle Hf, Hf\rangle = \sum_{i=1}^{r} \lambda_i^2 \langle f, E_i f\rangle \leq \lambda_m$$

where λ_m is the largest eigenvalue and the equality holds for f in the range of E_m. It follows that the norm of H is $|\lambda_m|$.

3.5 *Functions of a hermitean operator.* – Using the spectral resolution of a hermitean operator $H = \sum_i \lambda_i E_i$ one can verify that for a non-negative integer n one has $H^n = \sum_i \lambda_i^n E_i$ (the formula is also valid for n a negative integer if all the λ_i are non-zero). This property suggests the following definition of function of a hermitean operator: *Let $F(x)$ be a complex valued function of the real variable x, the function $F(H)$ of the hermitean operator H is the operator*

$$F(H) = \sum_i F(\lambda_i) E_i.$$

It follows from this expression that $F(H)$ commutes with its adjoint $\sum_i F^*(\lambda_i) E_i$ and is, by definition, normal.

3.6 *Sets of commuting hermitean operators.* – The spectral theorem can be generalized to a set $H = \{H_r\}$, $r = 1, \ldots, n$ of hermitean operators that commute with each other. Let E_{ri}, λ_{ri}, $i = 1, \ldots, s_r$ be the eigenprojections and eigenvalues of H_r. Since all the E_{ri} commute with each other, for each set $\sigma = \{i_1, i_2, \ldots, i_n\}$,we can construct the projection

$$E_\sigma = E_{1i_1} E_{2i_2} \ldots E_{ni_n}$$

that projects on the intersection of eigensubspaces of the H_r. The E_σ are orthogonal and add to the identity. Setting

$$\lambda_\sigma = \{\lambda_{1i_1}, \lambda_{2i_2}, \ldots \lambda_{ni_n}\},$$

we can regard the set of operators H as a composite hermitean operator with composite eigenvalues λ_σ and with spectral resolution

$$H = \sum_\sigma \lambda_\sigma E_\sigma.$$

3.7 *Normal operators.* – Every normal operator can be written as the sum $N = A+iB$ of its commuting hermitean and antihermitean parts A and iB, respectively. Except for a trivial modification, it can be regarded, as the set of the two commuting hermitean operators A and B. From this property it is not difficult to see that the spectral theorem 3.4.2 applies almost *verbatim* to a normal operator N; the only change is that the eigenvalues of N are, in general complex. Moreover N and $N^\dagger$ have the same eigenprojections and their respective eigenvalues are complex conjugate of each other.

3.8 *Unitary operators.* – Let U be unitary and therefore normal and let its spectral resolution be $\sum_i \lambda_i E_i$. According to the previous paragraph, we have $U^\dagger = \sum_i \lambda_i^* E_i$; therefore, by the orthogonality of the E_i, the relation $UU^\dagger = I$ becomes $\sum_i |\lambda_i|^2 E_i = \sum_i E_i$ which implies that $|\lambda_i|^2 E_i = E_i$ from which it follows that $|\lambda_i|^2 = 1$. *The eigenvalues of a unitary operator have absolute value one.*

3.8.1. It follows that we can write $U = \sum_i e^{i\theta_i} E_i$, θ_i real, that is, $U = e^{iA}$ with $A = \sum_i \theta_i E_i$. *Every unitary operator is of the form e^{iA}, with A hermitean.*

3.9 *Operators with a simple spectrum.* – Consider together with a hermitean operator $H = \sum_i \lambda_i E_i$ any operator A that commutes with it and consequently with all the projections E_i. The operator A is reduced by all the eigensubspaces $\mathbb{M}_i$ of H and can therefore written as a direct sum $\bigoplus_i A_i$, where A_i is an operator on $\mathbb{M}_i$. If the spectrum of H is simple, the $\mathbb{M}_i$ are one-dimensional and, since in a one-dimensional space all linear operators are multiples of the identity, we have $A_i = \alpha_i I_i$ for some number α_i. Considered as an operator on $\mathbb{H}$, A has therefore the form $A = \sum_i \alpha_i E_i$ and is normal. We can define a function $F(\lambda)$ such that $F(\lambda_i) = \alpha_i$ and write $A = \sum_i F(\lambda_i) E_i$ or, equivalently, $A = F(H)$. If the spectrum of H is not simple then at least one of its eigensubspaces, say $\mathbb{M}_k$, has dimension greater than one. The corresponding operator A_k may be any operator on $\mathbb{M}_k$ and consequently cannot be determined by a function as described above. We have then shown that *if and only if the spectrum of a hermitean operator H is simple, is every linear operator that commutes with H a function of H.*

3.9.1. Suppose that $H = \sum_i \lambda_i E_i$ is a hermitean operator with a simple spectrum $\Sigma = \{\lambda_i\}$ and $\{b_i\}$ an orthonormal basis of eigenvectors of H. The matrix representation of $\mathbb{H}$ with respect to the basis $\{b_i\}$ is called the H-representation.[5]

3.10 *Function space representation.* – The following remarks add nothing new to what has been said in the previous paragraph but they offer another way of looking at the H-representation that will prove useful in the discussion of the general Hilbert spaces of Appendix B. Let $a = \sum_i \alpha_i b_i$ be a vector whose components α_i are all different from zero. For any function $F(H)$ of the operator H we have $F(H)a = \sum_i \alpha_i F(\lambda_i) b_i$. It is possible to write any vector $g = \sum_i \gamma_i b_i$ in the form $F(H)a$ by simply choosing the function $F(\lambda)$ to be given by $F(\lambda_i) = \gamma_i/\alpha_i$. This possibility hinges on the non-degeneracy of the eigenvalues λ_i. Vectors such as a are called **generating vectors**. Thus, every generating vector induces a one-to-one correspondence between the vectors of $\mathbb{H}$ and the functions $F(\lambda)$ of the discrete variable $\lambda \in \Sigma$. In particular to the basis vector b_i there corresponds the function $\delta_{\lambda\lambda_j}/\alpha(\lambda)$, where $\alpha(\lambda_j) = \alpha_j$. In general we denote by by $g^a = g(H)a$ the vector corresponding to the function $g(\lambda)$. The scalar product can be written

$$\langle h^a, g^a \rangle = \langle h(H)a, g(H)a \rangle = \langle a, h^\dagger(H) g(H) a \rangle.$$

Using the definition of a function of H, the last expression becomes

$$\sum_{\lambda \in \Sigma} h^*(\lambda) g(\lambda) \langle a, E(\lambda) a \rangle,$$

where $E(\lambda_i) = E_i$. Now, $\langle a, E(\lambda) a \rangle = \|E_i a\|^2 = |\alpha(\lambda_i)|^2$, which gives

$$\langle h^a, g^a \rangle = \sum_{\lambda \in \Sigma} |\alpha(\lambda)|^2 h^*(\lambda) g(\lambda).$$

The natural norm is therefore given by

$$\|g^a\|^2 = \sum_{\lambda \in \Sigma} |\alpha(\lambda)|^2 |g(\lambda)|^2.$$

3.10.1. The vector $Hg^a = Hg(H)a$ is evidently represented by $\lambda g(\lambda)$, that is, in the function space, the operator H becomes multiplication by λ. Similarly the

[5] This representation is unique except for arbitrary phase factors resulting from the fact that for any real φ_i, $\{e^{i\varphi_i} b_i\}$ is an orthonormal basis of eigenvectors of H.

representation of an operator $F(H)$ is the multiplication by the function $F(\lambda)$ and, in particular, the eigenprojection E_i is the multiplication by $\delta_{\lambda\lambda_i}$. To find the representation of an arbitrary operator A with matrix elements A_{ij} we observe that the component of g^a along b_i is $\langle b_i, g^a\rangle = \alpha(\lambda_i)g(\lambda_i)$. Therefore by the rule of matrix multiplication we have $(Ag^a)_i = \sum_j A_{ij}\alpha(\lambda_j)g(\lambda_j)$ so that, representing the function corresponding to the vector Ag^a by $(Ag^a)(\lambda)$, one has

$$(Ag^a)(\lambda) = \sum_{\lambda\in\Sigma} A^a(\lambda,\lambda')g(\lambda'), \qquad A^a(\lambda_i,\lambda_j) = \frac{\alpha(\lambda_j)}{\alpha(\lambda_i)}A_{ij}.$$

3.10.2. The usual H-representation is obtained by selecting as generating vector $u = \sum_i b_i$. In this case we have

$$\langle h^u, g^u\rangle = \sum_{\lambda\in\Sigma} h^*(\lambda)g(\lambda), \qquad A^u(\lambda_i,\lambda_j) = A_{ij}.$$

This function space is obviously identical with the vector space of column matrices with elements $g(\lambda_i) = \langle b_i, g^u\rangle$, that is, with the $\{b_i\}$-representation of $\mathbb{H}$. The vector represented by $f(\lambda)$ in this space is represented by $\alpha(\lambda)f(\lambda)$ in the function space of the previous two paragraphs. Thus, function spaces induced by different generating vectors differ by a simple rescaling.

3.10.3. Clearly two operators with a simple spectrum and the same family of eigenprojections determine function space representations that are essentially identical (they differ only in the labelling of the points of the spectrum).

3.11 *Analysis in vector spaces.* – In a vector space $\mathbb{H}$ one defines limits of sequences of vectors in terms of limits of sequences of numbers. There are two kinds of limits that are defined for a sequence $\{f_n\}$ of vectors:

(i) The *weak limit* of $\{f_n\}$ is a vector f such that for all vectors g one has $\lim_{n\to\infty}\langle g, f_n - f\rangle = 0$

(ii) The *strong limit* or *limit in the norm* of $\{f_n\}$ is a vector f such that $\lim_{n\to\infty}\|f_n - f\| = 0$.

In a finite-dimensional space both definitions coincide and moreover any Cauchy sequence of vectors has a limit. Neither of these assertions is true in an infinite-dimensional vector space.

3.11.1. Limits of sequences of operators are defined through the corresponding sequences of vectors. Thus we say that the sequence $\{A_n\}$ of operators tend to the limit A if $\lim_{n\to\infty}(A_n g) = Ag$ for all vectors g.

3.11.2. The derivative of an operator valued function $A(t)$ of a numerical variable t is defined in the usual way

$$\frac{dA}{dt} = A'(t) = \lim_{h\to 0} \frac{A(t+h) - A(t)}{h}.$$

With due regard for the lack of commutativity of operators, the usual rules for differentiation apply. Thus

$$\frac{dAB}{dt} = A'(t)B(t) + A(t)B'(t).$$

Note that, in general, $A'(t)$ does not commute with $A(t)$ so that

$$\frac{dA^2}{dt} = A'(t)A(t) + A(t)A'(t).$$

Also differentiating the expression $AA^{-1} = I$ one obtains $A'A^{-1} + A(A^{-1})' = 0$ which after multiplying by A^{-1} on the left gives

$$\frac{dA^{-1}}{dt} = -[A(t)]^{-1}A'(t)[A(t)]^{-1}.$$

If $A(t) = \sum_i \alpha_i(t)E_i$ *with* E_i *independent of* t, then $A(t)$ commutes with its derivatives and rules identical to the rules for numerical functions apply.

3.11.3. For all $f, g \in \mathbb{H}$, and $A = A(t)$, t real, we have

$$\left\langle f, \left(\frac{dA}{dt}\right)^{\dagger} g\right\rangle = \left\langle \left(\frac{dA}{dt}\right) f, g\right\rangle = \frac{d}{dt}\langle Af, g\rangle = \frac{d}{dt}\langle f, A^{\dagger}g\rangle = \left\langle f, \left(\frac{dA^{\dagger}}{dt}\right) g\right\rangle,$$

that is,

$$\left(\frac{dA^{\dagger}}{dt}\right) = \left(\frac{dA}{dt}\right)^{\dagger}.$$

3.12 *Dirac's notation.* – Let H have r distinct eigenvalues λ_i, $i = 1, \ldots, r$ and let s_i be the degeneracy of λ_i. In the eigensubspace $\mathbb{M}_i$ belonging to λ_i we select an arbitrary orthonormal basis with kets labelled in Dirac's notation by $|\lambda_i^k\rangle$, $k = 1, \ldots, s_i$. We then have for the eigenprojection belonging to λ_i

$$E_i = \sum_{k=1}^{s_i} |\lambda_i^k\rangle\langle\lambda_i^k|.$$

The spectral resolution of H can then be written

$$H = \sum_{i=1}^{r} \lambda_i \sum_{k=1}^{s_i} |\lambda_i^k\rangle\langle\lambda_i^k|.$$

We can, of course, label the vectors of the orthonormal basis on $\mathbb{H}$ formed by the union of the bases on all $\mathbb{M}_i$ by a single index $j = 1, \ldots, m$, and, in the understanding that some of the λ_j may be equal, write

$$H = \sum_{j=1}^{m} \lambda_j |\lambda_j\rangle\langle\lambda_j|,$$

a decomposition of H which is obviously not unique unless the spectrum of H is simple.

4. The tensor product.

4.1 *Definition.* – The tensor product of two vector spaces $\mathbb{U}, \mathbb{V}$ is the vector space $\mathbb{T} = \mathbb{U} \otimes \mathbb{V}$ formed by all linear combinations of ordered pairs $\{f, g\}$, written $f \otimes g$, $f \in \mathbb{U}$, $g \in \mathbb{V}$ such that

$$(\alpha' f' + \alpha'' f'') \otimes g = \alpha'(f' \otimes g) + \alpha''(f'' \otimes g),$$
$$f \otimes (\beta' g' + \beta'' g'') = \beta'(f \otimes g') + \beta''(f \otimes g'').$$

In particular we find that $f \otimes g = 0$ if an only if $f = 0$ or $g = 0$, or both. Note that the tensor product is not commutative.

4.1.1. The definition can obviously be extended to any finite number of spaces $\mathbb{H}_i$, $i = 1, \ldots, n$. We write

$$\mathbb{T} = \bigotimes_i \mathbb{H}_i.$$

4.2 *Product bases.* – Let $\{b_i\}$, $\{c_j\}$ be bases on $\mathbb{U}$ and $\mathbb{V}$, respectively, then every vector of $\mathbb{T}$ can be written as a linear combination of the vectors $b_i \otimes c_j \in \mathbb{T}$. Indeed a general vector $\boldsymbol{\Phi} \in \mathbb{T}$ is of the form

$$\boldsymbol{\Phi} = \sum_{r,s} \alpha^{rs} f^r \otimes g^s, \qquad f^r \in \mathbb{U}, g^s \in \mathbb{V}$$

so that, writing f^r and g^s as

$$f^r = \sum_i \beta_i^r b_i, \qquad g^s = \sum_j \gamma_j^s c_j,$$

we have

$$\mathbf{\Phi} = \sum_{rs} \alpha^{rs} \Big(\sum_i \beta_i^r b_i \Big) \otimes \Big(\sum_j \gamma_j^s c_j \Big) = \sum_{rs} \alpha^{rs} \sum_{ij} \beta_i^r \gamma_j^s b_i \otimes c_j$$

which is an expression of the desired form:

$$\mathbf{\Phi} = \sum_{ij} \varphi_{ij} b_i \otimes c_j, \qquad \varphi_{ij} = \sum_{rs} \alpha^{rs} \beta_i^r \gamma_j^s.$$

Since by 1.5 the components β_i^r, γ_j^s of f^r and g^s are unique so are the coefficients φ_{ij}. It follows that $\{b_i \otimes c_j\}$ is a basis on $\mathbb{T}$. By counting the number of basis vectors we find immediately that $\dim(\mathbb{T}) = \dim(\mathbb{U}) \dim(\mathbb{V})$. The basis $\{b_i \otimes c_j\}$ is called a **product basis** because its elements are **product vectors**, that is, vectors of the form $f \otimes g$.

4.3 *Scalar product.* – If scalar products $\langle f', f'' \rangle$, $\langle g', g'' \rangle$ are defined on $\mathbb{U}$ and $\mathbb{V}$, respectively, a scalar product is defined on the product vectors of $\mathbb{T}$ by

$$\langle f' \otimes g', f'' \otimes g'' \rangle = \langle f', f'' \rangle \langle g', g'' \rangle,$$

a definition which is extended by linearity to all vectors of $\mathbb{T}$.

4.4 *Linear operators.* – If B and C are linear operators on $\mathbb{U}$ and $\mathbb{V}$, respectively, a corresponding operator $\mathbf{A} = B \otimes C$ on $\mathbb{T}$ is defined by $\mathbf{A}(f \otimes g) = (Bf) \otimes (Cg)$. $\mathbf{A}$ is called the tensor product of B and C. It follows immediately that $B \otimes C$ is linear with respect to each factor

$$(\alpha' B' + \alpha'' B'') \otimes C = \alpha'(B' \otimes C) + \alpha''(B'' \otimes C),$$
$$B \otimes (\alpha' C' + \alpha'' C'') = \alpha'(B \otimes C') + \alpha''(B \otimes C''),$$

and that

$$(A' \otimes B')(A'' \otimes B'') = (A' A'') \otimes (B' B'').$$

Also, if B and C have inverses so does $B \otimes C$ and

$$(B \otimes C)^{-1} = B^{-1} \otimes C^{-1}.$$

Using the definition of the scalar product on $\mathbb{T}$ we find

$$(B \otimes C)^\dagger = B^\dagger \otimes C^\dagger.$$

Of course, an arbitrary linear operator on $\mathbb{T}$ is not in general a tensor product.

4.4.1. By applying $B \otimes C$ to the vectors of a product basis on $\mathbb{T}$ and using 4.1.1, one can easily see that $B \otimes C = 0$ if and only if $B = 0$ or $C = 0$, or both.

4.5 *Matrix representation.* – A representation of $\mathbb{T}$ with respect to a product basis is called a **product representation**. Suppose that the representations of $f \in \mathbb{U}$ and $g \in \mathbb{V}$ with respect to the bases $\{b_i\}$ and $\{c_j\}$, respectively, are

$$f = \begin{pmatrix} \beta_1 \\ \beta_2 \\ \dots \\ \beta_m \end{pmatrix}, \qquad g = \begin{pmatrix} \gamma_1 \\ \gamma_2 \\ \dots \\ \gamma_m \end{pmatrix}.$$

The corresponding product representation of $f \otimes g$, that is, its representation with respect to the basis $\{b_i \otimes c_j\}$ is the column matrix with $m \times n$ elements

$$f \otimes g = \begin{pmatrix} \beta_1\gamma_1 \\ \dots \\ \beta_1\gamma_n \\ \dots \\ \dots \\ \beta_m\gamma_1 \\ \dots \\ \beta_m\gamma_n \end{pmatrix}.$$

In an obvious and useful shorthand notation this matrix can be written

$$f \otimes g = \begin{pmatrix} \beta_1\,(\gamma_j) \\ \dots \\ \beta_m\,(\gamma_j) \end{pmatrix}.$$

4.5.1. Similarly if $B_{ij} = \langle b_i, Bb_j \rangle$, $C_{ij} = \langle c_i, Cc_j \rangle$ are the matrix elements of the operators B on $\mathbb{U}$ and C on $\mathbb{V}$, the matrix elements of the operator $\mathbf{A} = B \otimes C$ on $\mathbb{T}$ are $A_{ij,i'j'} = \langle b_i \otimes c_j, \mathbf{A} b_{i'} \otimes c_{j'} \rangle = B_{ii'} C_{jj'}$. The representation of $\mathbf{A}$ is thus the $(m \times n) \times (m \times n)$ matrix

$$B \otimes C = \begin{pmatrix} B_{11}C_{11} & \dots & B_{11}C_{1n} & \dots & \dots & B_{1m}C_{11} & \dots & B_{1m}C_{1n} \\ \dots & \dots & \dots & \dots & \dots & \dots & \dots & \dots \\ B_{11}C_{n1} & \dots & B_{11}C_{nn} & \dots & \dots & B_{1m}C_{n1} & \dots & B_{1m}C_{nn} \\ \dots & \dots & \dots & \dots & \dots & \dots & \dots & \dots \\ \dots & \dots & \dots & \dots & \dots & \dots & \dots & \dots \\ B_{m1}C_{11} & \dots & B_{m1}C_{1n} & \dots & \dots & B_{mm}C_{11} & \dots & B_{mm}C_{1n} \\ \dots & \dots & \dots & \dots & \dots & \dots & \dots & \dots \\ B_{m1}C_{n1} & \dots & B_{m1}C_{nn} & \dots & \dots & B_{mm}C_{n1} & \dots & B_{mm}C_{nn} \end{pmatrix},$$

or in shorthand notation

$$B \otimes C = \begin{pmatrix} B_{11}\,(C_{ij}) & \dots & B_{1m}\,(C_{ij}) \\ \dots & \dots & \dots \\ B_{m1}\,(C_{ij}) & \dots & B_{mm}\,(C_{ij}) \end{pmatrix}.$$

4.5.2. From the last matrix representation one finds

$$\mathrm{Tr}_T(B \otimes C) = \mathrm{Tr}_U(B)\mathrm{Tr}_V(C),$$

where, for clarity, a subscript has been added to indicate in which space the traces are performed.

4.6 *Conjugate tensor products.* – Along with the tensor product space $\mathbb{T} = \mathbb{U} \otimes \mathbb{V}$ one can define its **conjugate** $\mathbb{T}^{\ddagger} = \mathbb{V} \otimes \mathbb{U}$. The space $\mathbb{T}^{\ddagger}$ is spanned by vectors of the form $f \otimes g$. Clearly $(\mathbb{T}^{\ddagger})^{\ddagger} = \mathbb{T}$. We say that the vectors $f \otimes g \in \mathbb{T}$ and $g \otimes f \in \mathbb{T}^{\ddagger}$ are conjugate. The correspondence between conjugate vectors establishes an isomorphism between the two spaces. A general vector of $\mathbb{T}^{\ddagger}$ is written $\boldsymbol{\Psi}^{\ddagger} = \sum_{ij} \psi^{\ddagger}_{ji} c_j \otimes b_i$. Under the above isomorphism $\boldsymbol{\Psi}^{\ddagger}$ is conjugate to the vector $\boldsymbol{\Phi} = \sum_{ij} \varphi_{ij} b_i \otimes c_j$ of $\mathbb{T}$ if $\psi^{\ddagger}_{ji} = \varphi_{ij}$, we denote it then by $\boldsymbol{\Phi}^{\ddagger}$. Thus, $\boldsymbol{\Phi}^{\ddagger} = \sum_{ij} \varphi_{ij} c_j \otimes b_i$. The matrices that represent conjugate vectors and linear operators in both spaces differ only in the order of their elements.

4.7 *The tensor product as a set of transformations.* – With each product vector $f \otimes g \in \mathbb{T}$, let us associate a transformation $\mathbf{T}[f \otimes g]$ from $\mathbb{U}$ into $\mathbb{V}$ such that $\mathbf{T} = \mathbf{T}[f \otimes g]f' = \langle f', f\rangle g$. We then have

$$\mathbf{T}(f' + f'') = \langle f' + f'', f\rangle g = \langle f', f\rangle g + \langle f'', f\rangle g = \mathbf{T}f' + \mathbf{T}f'',$$
$$\mathbf{T}(\alpha f') = \langle \alpha f', f\rangle g = \alpha^* \langle f', f\rangle g = \alpha^* \mathbf{T}f'.$$

From the definition we also find immediately,

$$\mathbf{T}[(\alpha' f' + \alpha'' f'') \otimes g] = \alpha' \mathbf{T}[f' \otimes g] + \alpha'' \mathbf{T}[f'' \otimes g],$$
$$\mathbf{T}[f \otimes (\beta' g' + \beta'' g'')] = \beta' \mathbf{T}[f \otimes g'] + \beta'' \mathbf{T}[f \otimes g''],$$

that is, $\mathbf{T}[f \otimes g]$ is bilinear in f and g. This property permits us to extend the definition of $\mathbf{T}$ to all vectors of $\mathbb{T}$ and to verify that, if $\boldsymbol{\Phi}, \boldsymbol{\Phi}' \in \mathbb{T}$, then $\mathbf{T}[\alpha\boldsymbol{\Phi} + \alpha'\boldsymbol{\Phi}'] = \alpha\mathbf{T}[\boldsymbol{\Phi}] + \alpha'\mathbf{T}[\boldsymbol{\Phi}']$. It follows from all these properties that the set of transformations $\mathbf{T}[\boldsymbol{\Phi}]$, constitutes a vector space which, with the scalar product definition $\langle \mathbf{T}[\boldsymbol{\Phi}], \mathbf{T}[\boldsymbol{\Phi}']\rangle = \langle \boldsymbol{\Phi}, \boldsymbol{\Phi}'\rangle$, is isomorphic to $\mathbb{T}$. Consequently we can identify $\boldsymbol{\Phi}$ with $\mathbf{T}[\boldsymbol{\Phi}]$, and shall do so whenever is convenient.

4.7.1. With the identification just mentioned we can write the definition of the transformation $f \otimes g$ as follows,

$$(f \otimes g)f' = \langle f', f \rangle g.$$

Let $\{b_i\}$, $\{c_j\}$ be bases on $\mathbb{U}$ and $\mathbb{V}$, respectively, and $\boldsymbol{\Phi} = \sum_{ij} \varphi_{ij} b_i \otimes c_j$ an arbitrary vector in $\mathbb{T}$. We have

$$\boldsymbol{\Phi} f = \sum_{ij} \varphi_{ij}(b_i \otimes c_j) f = \sum_{ij} \varphi_{ij} \langle f, b_i \rangle c_j.$$

If $f = \sum_i \alpha_i b_i$, then

$$\boldsymbol{\Phi} f = \sum_{ij} \varphi_{ij} \alpha_i^* c_j.$$

From this expression it is easy to verify that

$$\boldsymbol{\Phi}(f' + f'') = \boldsymbol{\Phi} f' + \boldsymbol{\Phi} f'', \qquad \boldsymbol{\Phi}(\alpha f) = \alpha^* \boldsymbol{\Phi} f.$$

4.7.2. Let $\boldsymbol{\Phi}' = \sum_{ij} \varphi'_{ij} b_i \otimes c_j$. If in the above paragraph we replace f by the basis vector b_i, one has

$$\boldsymbol{\Phi} b_i = \sum_j \varphi_{ij} c_j, \qquad \boldsymbol{\Phi}' b_i = \sum_j \varphi'_{ij} c_j.$$

The scalar product of these two vectors of $\mathbb{V}$ is

$$\langle \boldsymbol{\Phi} b_i, \boldsymbol{\Phi}' b_i \rangle = \sum_j \varphi_{ij}^* \varphi'_{ij}.$$

Summing over i the right member yields the scalar product $\langle \boldsymbol{\Phi}, \boldsymbol{\Phi}' \rangle$ expressed in terms of the components along the basis $\{b_i \otimes c_j\}$. Thus, one gets the useful relation

$$\langle \boldsymbol{\Phi}, \boldsymbol{\Phi}' \rangle = \sum_i \langle \boldsymbol{\Phi} b_i, \boldsymbol{\Phi}' b_i \rangle.$$

If $\boldsymbol{\Phi}' = f \otimes g$, this relation yields

$$\langle \boldsymbol{\Phi}, f \otimes g \rangle = \langle \boldsymbol{\Phi} f, g \rangle.$$

4.7.3. Let B and C be operators on $\mathbb{U}$ and $\mathbb{V}$, respectively. Then $B \otimes C \boldsymbol{\Phi}$ is a vector of $\mathbb{T}$. In the particular case that $\boldsymbol{\Phi}$ is a product $f \otimes g$ we have

$$(B \otimes C)(f \otimes g)f' = (Bf \otimes Cg)f' = \langle f', Bf \rangle Cg = C\langle B^\dagger f', f \rangle g = C(f \otimes g)B^\dagger f'.$$

Expanding an arbitrary $\mathbf{\Phi} \in \mathbb{T}$ with respect to a product basis and using linearity, one can generalize the above relation to read $B \otimes C\mathbf{\Phi} f' = C\mathbf{\Phi} B^{\dagger} f'$. Since this relation holds for arbitrary f', it can be written

$$B \otimes C\mathbf{\Phi} = C\mathbf{\Phi} B^{\dagger}.$$

Note that $C\mathbf{\Phi}$ and $\mathbf{\Phi} B$ are vectors in $\mathbb{T}$. In fact $C\mathbf{\Phi} = I_U \otimes C\mathbf{\Phi}$ and $\mathbf{\Phi} B = B^{\dagger} \otimes I_V \mathbf{\Phi}$, where I_U and I_V are the identities in $\mathbb{U}$ and $\mathbb{V}$, respectively.

4.7.4. The above considerations can be repeated for the conjugate space $\mathbb{T}^{\ddagger} = \mathbb{V} \otimes \mathbb{U}$. A vector $\mathbf{\Psi}^{\ddagger} \in \mathbb{T}^{\ddagger}$ can be regarded as a transformation from $\mathbb{V}$ into $\mathbb{U}$ such that

$$\mathbf{\Psi}^{\ddagger}(g' + g'') = \mathbf{\Psi}^{\ddagger} g' + \mathbf{\Psi}^{\ddagger} g'', \qquad \mathbf{\Psi}^{\ddagger}(\alpha g) = \alpha^* \mathbf{\Psi}^{\ddagger} g.$$

In particular if $\mathbf{\Psi}^{\ddagger} = g \otimes f$, we have

$$(g \otimes f) g' = \langle g', g \rangle f.$$

By expansion of the vectors in terms of conjugate product bases one sees easily that $\mathbf{\Phi} \in \mathbb{T}$ and $\mathbf{\Phi}^{\ddagger} \in \mathbb{T}^{\ddagger}$ are conjugate in the sense of 4.6 if and only if one of the two equivalent relations

$$\langle g, \mathbf{\Phi} f \rangle = \langle f, \mathbf{\Phi}^{\ddagger} g \rangle, \qquad \langle \mathbf{\Phi} f, g \rangle = \langle \mathbf{\Phi}^{\ddagger} g, f \rangle$$

is satisfied.

4.7.5. It follows from the previous paragraph that, if $\mathbf{\Phi} \in \mathbb{T}$ and $\mathbf{\Psi}^{\ddagger} \in \mathbb{T}^{\ddagger}$, then $\mathbf{\Phi}\mathbf{\Psi}^{\ddagger}$ is a linear operator on $\mathbb{V}$ and $\mathbf{\Psi}^{\ddagger}\mathbf{\Phi}$ is one on $\mathbb{U}$. In particular if $\mathbf{\Phi}$ and $\mathbf{\Psi}^{\ddagger}$ are product vectors we have

$$(f' \otimes g')(g'' \otimes f'') g = \langle g'', g \rangle \langle f'', f' \rangle g'.$$

Using the last relations of 4.7.4 we obtain

$$\langle f, \mathbf{\Psi}^{\ddagger}\mathbf{\Phi} f' \rangle = \langle \mathbf{\Phi} f', \mathbf{\Psi} f \rangle = \langle \mathbf{\Phi}^{\ddagger}\mathbf{\Psi} f, f' \rangle,$$

that is,

$$(\mathbf{\Psi}^{\ddagger}\mathbf{\Phi})^{\dagger} = \mathbf{\Phi}^{\ddagger}\mathbf{\Psi}.$$

In particular, the operator $\mathbf{\Phi}^{\ddagger}\mathbf{\Phi}$ on $\mathbb{U}$ is hermitean. Obviously the same is true of $\mathbf{\Phi}\mathbf{\Phi}^{\ddagger}$ on $\mathbb{V}$.

4.7.6. The last equality in 4.7.2 allows us to write another useful expression for the scalar product on $\mathbb{T}$,

$$\langle \mathbf{\Phi}, \mathbf{\Psi} \rangle = \sum_i \langle \mathbf{\Phi} b_i, \mathbf{\Psi} b_i \rangle = \sum_i \langle b_i, \mathbf{\Psi}^{\ddagger}\mathbf{\Phi} b_i \rangle = \mathrm{Tr}_U(\mathbf{\Psi}^{\ddagger}\mathbf{\Phi}).$$

If $\Psi = B \otimes C\Phi = C\Phi B^\dagger$, it is easily seen that $\Psi^\ddagger = C \otimes B\Psi^\ddagger = B\Psi^\ddagger C^\dagger$, and the above equation becomes

$$\langle \Phi, B \otimes C\Phi \rangle = \mathrm{Tr}_U \left(B(\Psi^\ddagger C^\dagger \Phi) \right) = \mathrm{Tr}_U \left((\Phi^\ddagger C \Psi) B^\dagger \right).$$

4.7.7. In the general case if $\Phi = \sum_{ij} \varphi_{ij} b_i \otimes c_j$ and $\Psi^\ddagger = \sum_{ij} \psi^\ddagger_{ji} c_j \otimes b_i$, then the matrix elements of $\Phi\Psi^\ddagger$ and $\Psi^\ddagger\Phi$ with respect to the bases $\{b_i\}$ and $\{c_j\}$ are respectively

$$\langle c_{j'}, \Phi\Psi^\ddagger c_j \rangle = \sum_k \varphi_{kj'} \psi^{\ddagger *}_{jk}, \qquad \langle b_{i'}, \Psi^\ddagger \Phi b_i \rangle = \sum_k \psi^\ddagger_{ki'} \varphi^*_{ik}.$$

In particular, for conjugate vectors

$$\langle c_{j'}, \Phi\Phi^\ddagger c_j \rangle = \sum_k \varphi_{kj'} \varphi^*_{kj}, \qquad \langle b_{i'}, \Phi^\ddagger \Phi b_i \rangle = \sum_k \varphi_{i'k} \varphi^*_{ik}.$$

4.8 *Dirac's notation.* – In Dirac's notation, the product ket and bra corresponding to $g \otimes f \in \mathbb{U} \otimes \mathbb{V}$ are written as $|f\rangle|g\rangle$ and $\langle g|\langle f|$, respectively (for a formal reason that will become apparent in the next paragraph, the kets are in the same order as the spaces while the bras are in reverse order). However, when it is not necessary to stress that a vector of $\mathbb{T}$ is a product, it is customary to write $|f, g\rangle = |f\rangle|g\rangle$ and $\langle f, g| = \langle g|\langle f|$. We shall keep the notation $B \otimes C$ for product operators, so that the ket and bra corresponding to $(B \otimes C)(f \otimes g)$ are, respectively,

$$B \otimes C|f, g\rangle = B \otimes C|f\rangle|g\rangle = B|f\rangle C|g\rangle,$$
$$\langle f, g|(B \otimes C)^\dagger = \langle g|\langle f|B^\dagger \otimes C^\dagger = \langle g|C^\dagger \langle f|B^\dagger.$$

4.8.1. When the vectors of $\mathbb{T}$ are regarded as transformations from $\mathbb{V}$ into $\mathbb{U}$ the equation $(f \otimes g)f' = \langle f', f\rangle g$ can be written in the two formally pleasing equivalent forms

$$\langle f'| \left(|f\rangle|g\rangle \right) = \langle f'|f\rangle|g\rangle, \qquad \left(\langle g|\langle f| \right) |f'\rangle = \langle g|\langle f|f'\rangle.$$

These formulae suffice to do all the calculations when the vectors of $\mathbb{T}$ are written as sums of product vectors. For example, it is easy to verify that the operator on $\mathbb{U}$ given by $T = (f' \otimes g')(g'' \otimes f'')$ becomes, in agreement with 4.7.5,

$$T = |g'\rangle\langle f''|f'\rangle\langle g''|.$$

4.8.2. When dealing with general vectors of $\mathbb{T}$, it is often more convenient not to work with the expansions. We shall write, somewhat inconsistently,[6] $|\mathbf{\Phi} f\rangle$ and $\langle f\mathbf{\Phi}|$ for the ket and bra corresponding to $\mathbf{\Phi} f$, and keep the notation $\mathbf{\Psi}^{\ddagger}\mathbf{\Phi}$ for the operator on $\mathbb{U}$. Thus, the ket and bra corresponding to $\mathbf{\Psi}^{\ddagger}\mathbf{\Phi} f \in \mathbb{U}$ are, respectively, $\mathbf{\Psi}^{\ddagger}\mathbf{\Phi}|f\rangle$ and $\langle f|\mathbf{\Phi}^{\ddagger}\mathbf{\Psi}$.

4.8.3. In the product space we write $|\mathbf{\Phi}\rangle$ and $\langle\mathbf{\Phi}|$ for the ket and bra corresponding to $\mathbf{\Phi}$. Thus, the last equation in 4.7.6 reads

$$\langle\mathbf{\Phi}|B\otimes C|\mathbf{\Phi}\rangle = \mathrm{Tr}_U\left(B(\mathbf{\Psi}^{\ddagger}C^{\dagger}\mathbf{\Phi})\right) = \mathrm{Tr}_U\left((\mathbf{\Phi}^{\ddagger}C\mathbf{\Psi})B^{\dagger}\right).$$

5. Antilinear operators

The only type of non-linear operators that are important in quantum mechanics are the antilinear operators. We give here the basic definitions and properties pertaining to them.

5.1 *Definition of antilinear operators.* – An operator X on the vector space $\mathbb{H}$ is antilinear if for all $f, g \in \mathbb{H}$, $X(f+g) = Xf + Xg$, $X(\alpha f) = \alpha^* Xf$.

5.1.2. If X and Y are antilinear we have $XY(f+g) = X(Yf+Yg) = XYf + XYg$, while $XY(\alpha f) = X(\alpha^* Yf) = \alpha(XY)f$. Thus, the product of two antilinear operators is linear. A similar calculation shows that if A is linear then both the products AX and XA are antilinear. The sum $X+Y$ of two antilinear operators is defined by $(X+Y)f = Xf + Yf$. One can verify immediately that $X+Y$ is antilinear.

5.2 *Inverse.* – If X defines a one to one mapping of $\mathbb{H}$ onto itself then an inverse X^{-1} such that $X^{-1}X = XX^{-1} = I$ exists and is antilinear.

5.3 *Adjoint.* – Since the scalar product $\langle f, Xg\rangle$ is antilinear both in f and in g, so is $\langle Xg, f\rangle^*$. One defines then the adjoint $X^{\dagger}$ of X as the antilinear operator such that $\langle X^{\dagger}f, g\rangle^* = \langle f, Xg\rangle$, that is, $\langle g, X^{\dagger}f\rangle = \langle f, Xg\rangle$, for all vectors f, g. It follows that $(V^{\dagger})^{\dagger} = V$.

[6] In the general case it is possible, if one insists, to use a consistent bra and ket notation. Denote the bra and ket corresponding to Φ by $\langle\!\langle\Phi\|$ and $\|\Phi\rangle\!\rangle$. The equation $g = \Phi f$ can then be written in the two alternative forms $\langle\!\langle\Phi\|f\rangle = \langle g|$, $\langle f\|\Phi\rangle\!\rangle = |g\rangle$ (the unpaired "$\langle$" or "$\rangle$" indicates whether the expression is a bra or a ket in $\mathbb{U}$). However, calculations with this notation become somewhat cumbersome. For example, the expression for $\langle f, \Psi^{\ddagger}\Phi f'\rangle$ becomes $\langle f|\langle\!\langle\Phi\|f'\rangle\|\Psi^{\ddagger}\rangle\!\rangle$, and the ket and bra corresponding to $C\Phi B^{\dagger}f$ are written $C\langle f|B\|\Phi\rangle\!\rangle$ and $\langle\!\langle\Phi\|B^{\dagger}|f\rangle C^{\dagger}$, respectively.

5.4 *Antiunitary operators.* – An antilinear operator V is **antiunitary** if, for all vectors f, $\|Vf\| = \|f\|$ or equivalently if V has an inverse $V^{-1} = V^\dagger$. It follows that $\langle Vf, Vg\rangle = \langle f, g\rangle^*$. Therefore if $\{b_i\}$ is an orthonormal basis so is $\{Vb_i\}$.

5.4.1. An antiunitary operator V generates a **antiunitary transformation** of $\mathbb{H}$ onto itself with the correspondence

$$\bar{f} = Vf, \qquad \bar{A} = VAV^\dagger$$

between vectors and between operators. We then have for any pair of vectors f, g, and any operator A

$$\langle \bar{f}, \bar{g}\rangle = \langle f, g\rangle^*, \qquad \langle \bar{f}, \bar{A}\bar{g}\rangle = \langle f, Ag\rangle^*.$$

It is clear from the first of these equations that, along with unitary transformations, antiunitary transformations preserve the absolute value of the scalar product, $|\langle \bar{f}, \bar{g}\rangle| = |\langle f, g\rangle|$, and therefore change orthonormal bases into orthonormal bases. The converse of this statement is true in a certain sense given by a theorem due to Wigner that we study in the next section.

6. Wigner's theorem

As discussed in the main body of this book, the states of systems can be represented in many ways which are equivalent in the sense that they yield the same experimental predictions. Wigner's theorem establishes how the different equivalent representations are related to each other. The theorem plays such an important role in quantum mechanics that it is desirable to study it in some detail.

6.1 *Notation.* – A one-dimensional subspace of a Hilbert space $\mathbb{H}$ is called a **ray**; it consists of all the multiples of a given vector. Here we are concerned with a one-to-one mapping of the set of rays of $\mathbb{H}$ onto itself. Since to each ray there corresponds a one-dimensional projection operator and vice versa, the mapping can also be regarded as a one to one mapping of the set $\mathcal{R}$ of one-dimensional projections onto itself. We shall use the following notation: R_μ are *one*-dimensional projections distinguished by the index, the corresponding rays are their ranges $\mathbb{R}_\mu$. Given $\mathbb{R}_\mu$ (or R_μ) we denote by u_μ, u'_μ *unit* vectors and by r_μ, r'_μ vectors of arbitrary norm in $\mathbb{R}_\mu$. Thus, in our notation $R_\mu = u_\mu \times u^*_\mu = u'_\mu \times u'^*_\mu$ and

$$\mathrm{Tr}\,(R_\mu R_\nu) = |\langle u_\mu, u_\nu\rangle|^2 = |\langle u'_\mu, u'_\nu\rangle|^2.$$

If $f \in \mathbb{H}$ is given we shall denote the ray it determines by $\mathbb{R}_f$ and the corresponding projection by R_f. Therefore,

$$\mathrm{Tr}\,(R_f R_g) = (\|f\|\|g\|)^{-2}\,|\langle f, g\rangle|^2.$$

Wigner's theorem can now be stated as follows.

6.2 *Wigner's theorem. – Let* $\mathbf{V}$ *be a one-to-one mapping of* $\mathcal{R}$ *onto itself:* $\bar{R} = \mathbf{V}R$, *such that for every pair of one-dimensional projections* $R_\mu, R_\nu \in \mathcal{R}$ *we have*

$$\mathrm{Tr}\,(\bar{R}_\mu \bar{R}_\nu) = \mathrm{Tr}\,(R_\mu R_\nu).$$

The mapping $\mathbf{V}$ *induces a one-to-one transformation* $f \mapsto \bar{f}$ *of the Hilbert space onto itself, generated by an operator* V, *determined up to a phase factor, which is either unitary or antiunitary. That is,*

$$\begin{aligned} \bar{f} &= Vf, \\ \bar{f} + \bar{f}' &= V(f + f'), \\ \langle \bar{f}, \bar{f}'\rangle &= \eta\left(\langle f, f'\rangle\right), \end{aligned}$$

where either $\eta(\alpha) = \alpha$ *or* $\eta(\alpha) = \alpha^*$.

6.3 *Outline of the proof.* – The theorem is proven by constructing an operator V that satisfies the conditions of the theorem. It follows from 6.1 that the correspondence $u \mapsto \bar{u} = Vu$ between unit vectors must be such that

$$|\langle \bar{u}_1, \bar{u}_2\rangle| = |\langle u_1, u_2\rangle|.$$

Since V is either unitary or antiunitary, a similar expression holds for for any pair of vectors f, g, of arbitrary norm

$$|\langle \bar{f}, \bar{g}\rangle| = |\langle f, g\rangle|.$$

with $\bar{f} = Vf \in \mathbb{R}_f$ and $\bar{g} = Vg \in \mathbb{R}_g$. It is therefore permissible to include this condition among those that the operator V must satisfy. It implies that if u_i, $i = 1, \ldots, k$, are the vectors of an orthonormal set, then the set of vectors $\bar{u}_i$ is also orthonormal and, moreover, that if $f = \sum_i \alpha_i u_i$ then the corresponding vector is of the form $\bar{f} = \sum_i \bar{\alpha}_i \bar{u}_i$, with $|\alpha_i| = |\bar{\alpha}_i|$.

6.3.1. Let $\mathbb{R}_0$ be a given ray and $\mathbb{R}_1$ an arbitrary ray orthogonal to $\mathbb{R}_0$. To construct V we start by making correspond to a given $u_0 \in \mathbb{R}_0$ an arbitrary unit vector $\bar{u}_0 = Vu_0$ in $\bar{\mathbb{R}}_0 = \mathbf{V}\mathbb{R}_0$. Next we construct V for the vectors of the form

$$f = u_0 + r_1.$$

If $r_1 = \|r_1\|u_1$, then any $\bar{f}' \in \bar{\mathbb{R}}_f = \mathbf{V}\mathbb{R}_f$ that satisfies the conditions of 6.3 must be of the form $\bar{\alpha}\bar{u}_0 + \bar{\beta}\bar{u}_1$ with $|\bar{\alpha}| = 1, |\bar{\beta}| = \|r_1\|$. We define $\bar{f} = Vf$ as the unique $\bar{f}'$ for which $\bar{\alpha} = 1$,

$$\bar{f} = Vf = \bar{u}_0 + \bar{\gamma}_1\bar{u}_1, \qquad \bar{\gamma}_1 = \bar{\alpha}^{-1}\bar{\beta}, \qquad |\bar{\gamma}_1| = \|r_1\|$$

and, correspondingly, we define an operator W, whose domain is the subspace $\mathbb{H}^\perp \subset \mathbb{H}$ orthogonal to $\mathbb{R}_0$ and whose range the subspace $\bar{\mathbb{H}}^\perp \subset \mathbb{H}$ orthogonal to $\bar{\mathbb{R}}_0$, by

$$Wr_1 = \bar{r}_1 = Vf - Vu_0 = \bar{\gamma}_1\bar{u}_1.$$

6.3.2. Next we study the operator W and prove that it satisfies the theorem. To this end consider another ray $\mathbb{R}_2 \subset \mathbb{H}^\perp$. Then if $g = u_0 + r_2$, we have $\bar{g} = Vg \in \bar{\mathbb{R}}_g$ given by

$$\bar{g} = Vg = \bar{u}_0 + \bar{\gamma}_2\bar{u}_2, \qquad |\bar{\gamma}_2| = \|r_2\|,$$

where, as before, $r_2 = \|r_2\|u_2$. Therefore,

$$|\langle\bar{r}_2, \bar{r}_1\rangle| = |\bar{\gamma}_2||\bar{\gamma}_1||\langle\bar{u}_2, \bar{u}_1\rangle| = \|r_2\|\|r_1\||\langle u_2, u_1\rangle| = |\langle r_2, r_1\rangle|.$$

Moreover according to 6.3 we must also have $|\langle\bar{f}, \bar{g}\rangle| = |\langle f, g\rangle|$, that is,

$$|\langle\bar{u}_0 + \bar{r}_1, \bar{u}_0 + \bar{r}_2\rangle| = |\langle u_0 + r_1, u_0 + r_2\rangle|.$$

Expanding the products in this expression and taking the previous one into account we find

$$\mathrm{Re}\langle\bar{r}_2, \bar{r}_1\rangle = \mathrm{Re}\langle r_2, r_1\rangle, \qquad \mathrm{Im}\langle\bar{r}_2, \bar{r}_1\rangle = \pm\mathrm{Im}\langle r_2, r_1\rangle.$$

6.3.3. Evidently this expression also holds if the vector r_2 is replaced by a vector $r_1' = \gamma_1' u_1$ in the ray $\mathbb{R}_1$. In this case the preceding equation yields

$$\mathrm{Re}(\bar{\gamma}_1'^*\bar{\gamma}_1) = \mathrm{Re}(\gamma_1'^*\gamma_1), \qquad \mathrm{Im}(\bar{\gamma}_1'^*\bar{\gamma}_1) = \pm\mathrm{Im}(\gamma_1'^*\gamma_1).$$

If we write $r_1' = \alpha r_1$, $\alpha = \gamma_1'/\gamma_1$, $\bar{r}_1' = \bar{\alpha}\bar{r}_1$, $\bar{\alpha} = \bar{\gamma}_1'/\bar{\gamma}_1$, and note that $\bar{\gamma}_1^*\bar{\gamma}_1 = \gamma_1^*\gamma_1$ is real, the equations read

$$\mathrm{Re}(\bar{\alpha}) = \mathrm{Re}(\alpha), \qquad \mathrm{Im}(\bar{\alpha}) = \pm\mathrm{Im}(\alpha).$$

Since $\mathbb{R}_1$ is any ray orthogonal to $\mathbb{R}_0$, we conclude that for any vector $r \in \mathbb{H}^\perp$, $W(\alpha r) = \eta(\bar{\alpha})Wr$, where either $\eta(\alpha) = \alpha$ or $\eta(\alpha) = \alpha^*$. It follows immediately that

$$\eta(\alpha + \beta) = \eta(\alpha) + \eta(\beta),$$

$$\eta(\alpha\beta) = \eta(\alpha)\eta(\beta),$$

$$[\eta(\alpha)]^* = \eta(\alpha^*).$$

That $\eta(\alpha)$ is independent of the vector r will be shown in the next paragraph.

6.3.4. Consider now the vector $r = \sum_i \alpha_i u_i$, $i = 1, 2$. It follows from 6.3 that, if $\bar{r} = Wr$ and $\bar{u}_i = Wu_i$, then $\bar{r} = \sum_i \bar{\alpha}_i \bar{u}_i$ with $\langle \bar{u}_i, \bar{u}_j \rangle = \delta_{ij}$ and $|\bar{\alpha}_i| = |\alpha_i|$. From 6.3.2 we obtain

$$\mathrm{Re}(\bar{\alpha}_i) = \mathrm{Re}(\langle \bar{u}_i, \bar{r} \rangle) = \mathrm{Re}(\langle u_i, r \rangle) = \mathrm{Re}(\alpha_i),$$

$$\mathrm{Im}(\bar{\alpha}_i) = \mathrm{Im}(\langle \bar{u}_i, \bar{r} \rangle) = \pm\mathrm{Im}(\langle u_i, r \rangle) = \pm\mathrm{Im}(\alpha_i).$$

That is, $\bar{\alpha}_i = \eta_i(\alpha_i)$, where as above either $\eta_i(\alpha) = \alpha$ or $\eta_i(\alpha) = \alpha^*$. In fact, $\eta_i(\alpha)$ is independent of i: consider the case in which $r = u_1 + u_2$. Since both coefficients are real, $\bar{r} = \bar{u}_1 + \bar{u}_2$; then, from $\alpha r = \alpha u_1 + \alpha u_2$, and 6.3.3 we have

$$W(\alpha r) = \eta_1(\alpha)\bar{u}_1 + \eta_2(\alpha)\bar{u}_2 = \eta(\alpha)\bar{r}.$$

Whence, $\eta(\alpha)(\bar{u}_1 + \bar{u}_2) = \eta_1(\alpha)\bar{u}_1 + \eta_2(\alpha)\bar{u}_2$, or $\eta_1(\alpha) = \eta_2(\alpha) = \eta(\alpha)$.

6.3.5. Let r and r' be orthogonal to $\mathbb{R}_0$ and let $\{u_1, u_2\}$ be an orthonormal basis in the subspace spanned by r and r' so that

$$r = \beta_1 u_1 + \beta_2 u_2, \qquad r' = \gamma_1 u_1 + \gamma_2 u_2.$$

From the properties of $\eta(\beta)$ at the end of 6.3.3 it follows

$$W(r + r') = \eta(\beta_1 + \gamma_1)\bar{u}_1 + \eta(\beta_2 + \gamma_2)\bar{u}_2 = Wr + Wr',$$

$$W(\alpha r) = \eta(\alpha\beta_1)\bar{u}_1 + \eta(\alpha\beta_2)\bar{u}_2 = \eta(\alpha)Wr,$$

$$\langle Wr, Wr' \rangle = \eta(\beta_1^*\gamma_1) + \eta(\beta_2^*\gamma_2) = \eta\left(\langle r, r' \rangle\right).$$

Thus, the operator W constructed in 6.3.1 and defined for the vectors $r \in \mathbb{H}^\perp$ satisfies the theorem and, as can be verified immediately, so does the operator V defined there on vectors of the form $u_0 + r$.

6.3.6. It remains to construct the operator for a general vector, which can always be written in the form $f = \alpha u_0 + r$. It suffices to write $f' = \alpha^{-1} f = u_0 + \alpha^{-1} r$ and define

$$Vf = \eta(\alpha)Vf' = \eta(\alpha)\bar{u}_0 + Wr$$

which can be easily verified to satisfy the theorem.

6.3.7. The operator V is completely determined once the unit vector $\bar{u}_0 \in \bar{\mathbb{R}}_0$ of 6.3.1 is chosen. One can show that two operators V and V' resulting from different choices of unit vectors differ only by a phase factor: $V' = e^{i\theta}V$, θ real.

B THE HILBERT SPACE

A Hilbert space is a generalization of a unitary space to infinite dimensions. Most of the results of Appendix A that make no reference to the dimension can be transferred to the infinite-dimensional Hilbert space. In this appendix we give those results that need modification, they arise mostly as a result of the existence of unbounded operators.

1. Basic definitions, unbounded operators

1.1 *Definition of Hilbert Space.* – A Hilbert space is a vector space $\mathbb{H}$ such that

(i) An inner product $\langle f, g\rangle, f, g \in \mathbb{H}$, and the natural norm $\|f\| = \langle f, f\rangle^{1/2}$ are defined on $\mathbb{H}$.

(ii) $\mathbb{H}$ is **complete**: every sequence $\{f_n\}$ of vectors with the Cauchy property, $\lim_{m,n\to\infty} \|f_n - f_m\| = 0$, converges in the norm (cf. A.3.11) to a unique limit $f \in \mathbb{H}$.

(iii) $\mathbb{H}$ is **separable**: there exists a sequence $\{g_n\}$ of vectors which is dense on $\mathbb{H}$, that is, given any $g \in \mathbb{H}$ and any positive number ϵ one can find a g_n such that $\|g_n - g\| < \epsilon$.[1]

1.1.1. The definition of a Hilbert space is tailored to maintain the main properties of the unitary spaces. That (i) must be required is obvious. In unitary spaces (ii) and (iii) follow from (i) but here they must be introduced separately. Axiom (ii) ensures the validity of the notion of continuity; a similar axiom is used, for example,

[1] Properly speaking a Hilbert space is defined by (i) and (ii) while a space that in addition obeys (iii) is called a *separable* Hilbert space. Since in quantum mechanics only the latter one is of interest we shall omit the qualifier.

to extend the rational into the real numbers. The importance of (iii) is explained below.

1.1.2. The dual space $\mathbb{H}^\dagger$ of the Hilbert space $\mathbb{H}$ is the vector space formed by all the *continuous*, linear functionals $a(f)$ on $\mathbb{H}$. Continuity means that if $\|f_n - f\| \to 0$ then $a(f_n) \to a(f)$. As in the finite-dimensional case $\mathbb{H}^\dagger = \mathbb{H}$. This means that the continuous linear functionals $a(f)$ are the inner products $\langle a, f\rangle$. Following the convention adopted in A.1.8.1 we shall use for $\langle a, f\rangle$ the synonymous term scalar product.

1.2 *Bases.* – A **basis** on $\mathbb{H}$ is a set $\{a_i\}$ of vectors such that every $f \in \mathbb{H}$ can be written uniquely in the form $f = \sum_i \eta_i a_i$. The necessary and sufficient condition for a countable orthonormal basis to exist on a complete, inner product space is that it be separable. That is why (iii) was included in the definition. If $\{b_i\}$ is an orthonormal basis, then $f = \sum_i \alpha_i b_i$, $\alpha_i = \langle b_i, f\rangle$, and $\|f\| = \sum_i |\alpha_i|^2 < \infty$.

1.3 *Examples of Hilbert spaces.* – Obviously unitary spaces are Hilbert spaces. The following concrete examples of infinite-dimensional Hilbert spaces are extensively used in quantum mechanics.

1.3.1. The set of all countable sequences $\{\alpha_k\}$ of complex numbers such that $\sum_k |\alpha_k|^2 < \infty$ is a Hilbert space with respect to the following operations, if $f = \{\alpha_k\}$, $g = \{\beta_k\}$, then $\mu f + \nu g = \{\mu\alpha_k + \nu\beta_k\}$, $\langle f, g\rangle = \sum_k \alpha_k^* \beta_k$. This space is called the *l^2-space.* An orthonormal basis on it is formed by the sequences $b_n = \{\delta_{nk}\}$, $n = 1, 2, \ldots$.

1.3.2. The set of all complex valued functions $f(x)$ of a real variable x, defined on a (finite or infinite) interval (a, b) and such that the Lebesgue integral[2]

$$\int_a^b dx\ |f(x)|^2 < \infty,$$

constitutes a Hilbert space if the operations are defined as follows: if $f = f(x)$, $g = g(x)$, then $\mu f + \nu g = \mu f(x) + \nu g(x)$,

$$\langle f, g\rangle = \int_a^b dx\ f^*(x) g(x).$$

[2] We need not define here the Lebesgue integral. Suffice it to say that there are many more functions integrable in the Lebesgue sense that in the ordinary Riemann sense and that if a function is Riemann integrable it is also Lebesgue integrable and both integrals are equal.

This space is called the $L^2(a,b)$-*space.* An example of an orthogonal basis on this space, for (a,b), finite is the set of trigonometric functions

$$b_n(x) = \frac{1}{\sqrt{b-a}} \exp\left(\frac{2\pi i}{b-a} nx\right), \qquad n = 0, \pm 1, \pm 2, \ldots,$$

The expression $f = \sum_n \langle b_n, f\rangle b_n$ yields in this case the Fourier series expansion of the function $f(x)$.

1.4 *Subspaces.* – The definition of subspace given in the finite-dimensional case must be slightly refined to include not only all vectors f that are linear combinations of a (finite or infinite) subset of $\mathbb{H}$ but also the limits of any Cauchy sequence of the vectors f (otherwise we would have subspaces that are not themselves Hilbert spaces).

1.5 *Linear operators.* – The definitions and properties of linear operators, or operators for short, given for unitary spaces are valid in Hilbert spaces with some modifications which we give below.

1.6 *Boundedness and continuity.* – An operator A is **bounded** if there exists a positive number α such that, for every $f \in \mathbb{H}$, $\|Af\| \leq \alpha\|f\|$. It is **continuous** if it is defined for the limit f of any convergent sequence $\{f_n\}$ in its domain and $\|A(f_n - f)\| \to 0$. An operator is continuous if and only if it is bounded. In a unitary space every operator is bounded but this is not so in an infinite-dimensional space.

1.6.1. As an example of an unbounded operator consider the operator Q defined on $L^2(-\infty, \infty)$ by $(Qf)(x) = xf(x)$, for $f(x)$ such that $\int dx\ |xf(x)|^2 < \infty$. Consider the unit vectors $h_m(x) = \sqrt{m - \frac{1}{2}}(1 + |x|)^{-m}$, $m > 1/2$. For $m = (3/2) + \epsilon$, $\epsilon > 0$ we have $\|Qf\|^2 = \int dx\ |xh_m(x)|^2 = [\epsilon(1 + 2\epsilon)]^{-1}$ so that the vectors h_m, $m > 3/2$ are in the domain of Q. Since, for sufficiently small ϵ, $\|Qf\|$ exceeds any given number, Q is unbounded. One can verify directly that Q is not continuous, for Q is defined for all vectors $f_n = h_{3/2+1/n}$, $n = 1, 2, \ldots$, but not for their limit as $n \to \infty$.

1.7 *Closure.* – Although the operator Q is not continuous it however has the property that if $\{g_n\}$ is a convergent sequence in its domain with limit g *and* $\{Qg_n\}$ converges then $\|Q(g_n - g)\| \to 0$. Operators with such a property are said to be **closed**. A closed operator defined on the whole of $\mathbb{H}$ is continuous and therefore bounded. By appropriately enlarging their domain many operators which are not closed can be extended in a unique manner to a closed operator.

1.7.1. Here we are only concerned with closed operators that furthermore have dense domains in the sense of 1.1, (iii).

1.8 *Algebra with operators.* – Contrary to what is the case in unitary spaces one has to be careful with the domains of operators that enter in an algebraic expression. For example, the operator $A + B$ has a domain which is the intersection of the domains of A and B; if that intersection is the zero vector, $A + B$ has a trivial meaning.

1.9 *Adjoint.* – If A has a dense domain, its **adjoint** $A^\dagger$ is an operator whose domain is the set of all vectors g such that $\langle A^\dagger g, f\rangle = \langle g, Af\rangle$ for every f in the domain of A. Note that because of the different domains it is not in general the case that A is the adjoint of $A^\dagger$.

1.10 *Hermitean operators.* – An operator S is **symmetric** if its domain is a subset of the domain of $S^\dagger$ and $S^\dagger f = Sf$ for every f in the domain of S. An operator H is **hermitean** if it is symmetric and its domain coincides with the domain of $H^\dagger$.

1.10.1. As an example of a symmetric operator which is not hermitean consider the operator P defined on $L^2(a, b)$ by $(Pf)(x) = -if'(x)$ and whose domain is the set of functions that have a square integrable first derivative such that $f(a) = f(b) = 0$. The operator P is symmetric. Indeed, *regardless* of the value of $g(x)$ at the end points of the interval we have

$$\langle g, Pf\rangle = -i\int_a^b dx\ g^*(x)f'(x) = +i\int_a^b dx\ g'^*(x)f(x)$$

because in integrating by parts the boundary term $g^*(x)f(x)\ |_a^b$ vanishes on account of the condition on $f(x)$. If we put $(P^\dagger g)(x) = -ig'(x)$ the integral can be written $\langle P^\dagger g, f\rangle$ and the equality $\langle P^\dagger g, f\rangle = \langle g, Pf\rangle$ holds for the set of all functions with square integrable derivative. This set is therefore the domain of the adjoint $P^\dagger$ of P. Clearly $P^\dagger$ has a larger domain than P and coincides with P in the domain of the latter. Therefore P is symmetric but not hermitean.

1.10.2. The result of the previous paragraph holds also if one of the end points of the interval (a, b) is infinite. However the operator P defined as above on $L^2(-\infty, \infty)$ is hermitean because since all the functions of such a space vanish at infinity, the boundary conditions that restrict the domain of P are superfluous. The operator Q of 1.6.1. is also hermitean.

2. The spectral theorem

Rather than attempting to sketch the mathematical reasoning that generalizes to the infinite dimensional Hilbert space the spectral theorem given in Appendix A, it is more useful to give an intuitive discussion based on an analogy, which can in fact be made precise, between the family of projections belonging to a hermitean operator and a probability distribution.

2.1 *A reminder about probability.* – As we saw in A.3.4.4, for every vector f the family of projections E_i belonging to a normal operator determines a probability distribution on a discrete finite sample space with the probabilities given by $\langle f, E_i f\rangle$. In probability theory the passage to a discrete but infinite sample space offers no difficulty: the sums that yield averages are simply replaced by convergent series. The difficulties appear in the case where the sample space is continuous. One cannot then construct the probabilities for every possible set of outcomes from the probability of the points of the space (it is in general zero). Instead one must consider the probabilities of appropriate elementary sets which, in the one-dimensional sample space in which we are interested here, can be taken to be intervals of the real line. To discuss such a case it is convenient to introduce the probability distribution function $\mathcal{D}(\alpha)$ defined as the probability that the outcome of a trial be less than α.

2.1.1. In the discrete case, let $\mathcal{P}_i$ be the probability that the outcome of a trial yields the value ξ_i and let us order the indices in such a way that $\xi_i < \xi_j$ for $i < j$. Written in terms of the $\mathcal{P}_i$ the distribution function is

$$\mathcal{D}(\alpha) = \sum_{i=-\infty}^{i_\alpha} \mathcal{P}_i,$$

where i_α is the largest index such that $\xi_i \leq \alpha$ (we have taken the sample space to be infinite but the finite case is included if one puts the appropriate $\mathcal{P}_i = 0$). The function $\mathcal{D}(\alpha)$ is clearly a non-decreasing ladder function with the properties $\mathcal{D}(-\infty) = 0$, $\mathcal{D}(\infty) = 1$. We have chosen the upper limit in the summation so as to satisfy the convention that $\mathcal{D}(\alpha)$ is continuous on the right.

2.1.2. One can imagine a pure continuous sample space as the limit of a discrete one when the differences $\xi_{i+1} - \xi_i$ tend to zero. The distribution function becomes then the limit of a ladder function with the properties just mentioned. It is, therefore, a continuous non decreasing function $\mathcal{D}(\alpha)$ such that $\mathcal{D}(-\infty) = 0$, $\mathcal{D}(\infty) = 1$. In case $\mathcal{D}(\alpha)$ is everywhere differentiable, as is the case in most physical problems, the probability density $\pi(\alpha)$ is defined by $\pi(\alpha) = \mathcal{D}'(\alpha)$ and the average of the random variable $f(\alpha)$ is given by $\langle f(\alpha)\rangle = \int d\alpha\, f(\alpha)\pi(\alpha)$.

2.1.3. In the general case the sample space is continuous but with a non-zero probability $\mathcal{P}_i$ concentrated at a countable number of points α_i. In this case $\mathcal{D}(\alpha)$

will have a countable number of discontinuities of size $\mathcal{P}_i$. If $\mathcal{D}(\alpha)$ is differentiable everywhere else and $\pi(\alpha)$ is the corresponding probability density, we have

$$\langle f(\alpha)\rangle = \int_{-\infty}^{\infty} d\alpha\ f(\alpha)\pi(\alpha) + \sum_i f(\alpha_i)\mathcal{P}_i.$$

Alternatively one may use Dirac's δ-function and write

$$\langle f(\alpha)\rangle = \int_{-\infty}^{\infty} d\alpha\ f(\alpha)\pi_d(\alpha),$$

where $\pi_d(\alpha)$ is the derivative of $\mathcal{D}(\alpha)$ in the sense of a distribution (see Appendix C) that is,

$$\pi_d(\alpha) = \pi(\alpha) + \sum_i \mathcal{P}_i\delta(\alpha - \alpha_i).$$

2.1.4. The average can also be conveniently written by using the *Stieltjes integral* defined for a function $g(\alpha)$ with a countable number of discontinuities by

$$\int_a^b f(\alpha)\ dg(\alpha) = \lim \sum_{i=1}^{n} f(\alpha_i)[g(\alpha_i) - g(\alpha_{i-1})],$$

where the a_i, $a = a_0 \le a_1 \le \ldots \le a_n = b$, form a subdivision of the interval (a, b) and the limit is taken over a sequence of ever finer subdivisions. In terms of the Stieltjes integral we have then

$$\langle f(\alpha)\rangle = \int_{-\infty}^{\infty} f(\alpha)\ d\mathcal{D}(\alpha).$$

2.1.5. In a unitary space the family of projections E_i belonging to a hermitean operator $A = \sum_i \xi_i E_i$ can be thought of as constituting a probability distribution on a finite sample space of operators. We shall take the eigenvalues to be ordered in ascending order. For every vector f the "operator valued probability" E_i generates the ordinary probability distribution $\mathcal{P}_i(f) = \langle f, E_i f\rangle$, which in quantum mechanics, as discussed in Chapter II, gives the probability of the outcome ξ_i upon measurement of A on a system in state f. In analogy with ordinary probability one can construct a corresponding "operator valued probability distribution function"

$$E(\alpha) = \sum_{i=-\infty}^{i_\alpha} E_i.$$

A formula that we would expect to be also applicable in the case of the infinite-dimensional Hilbert space if the operator A corresponds to an observable which yields upon measurement a countable set of possible values. From the properties of the family E_i it follows that $E(\alpha)$ is a projection operator, that $E(\alpha)$ and $E(\alpha')$ commute, that in the sense of the inequality relation defined in A.2.8.1 we have $E(\alpha') \geq E(\alpha)$ for $\alpha' \geq \alpha$ and that $E(-\infty) = 0,\ E(\infty) = I$. $E(\alpha)$ is an operator valued "ladder" function whose jumps at the discontinuities are given by $E(\xi_i) - E(\xi_i-) = E_i$.

2.1.6. The practice of quantum mechanics also compels us to consider operators on the Hilbert space that correspond to observables that yield upon measurement a continuous range of possible values. As one would expect such operators are associated with "operator valued probability distribution functions" analogous to the continuous distribution functions of ordinary probability, that is, with a family of projections $E(\alpha)$ depending continuously on the parameter α such that $E(\alpha') \geq E(\alpha)$ for $\alpha' \geq \alpha$ and that $E(-\infty) = 0,\ E(\infty) = I$. In the most general case corresponding to the continuous sample space with points ξ_i of concentrated probability we would expect the family $E(\alpha)$ to have discontinuities at ξ_i of the form $E(\xi_i) - E(\xi_i-) = E_i$ where E_i is the "operator valued probability" concentrated at the point ξ_i.

The tentative conclusions of this intuitive discussion can, in fact, proven to be correct. We formulate them precisely in the following paragraphs.

2.2 *Spectral family.* – A **spectral family** is a one parameter family of commuting projection operators $E(\alpha)$ depending on the real parameter α that satisfies the following properties

(i) $E(\alpha)$ is increasing: $E(\alpha) \leq E(\alpha'),\ \alpha \leq \alpha'$.

(ii) $E(\alpha)$ is continuous on the right: $E(\alpha') \to E(\alpha),\ \alpha' \to \alpha,\ \alpha' > \alpha$.

(iii) $E(\alpha)$ is complete: $E(-\infty) = 0,\ E(\infty) = I$.

It follows that $E(\alpha)E(\alpha') = E(\alpha')E(\alpha) = E(\alpha)$ for $\alpha \leq \alpha'$.

2.3 *The spectral theorem.* – As in Appendix A we give first the spectral theorem for hermitean operators and extended it later to normal operators.

2.3.1. *To every hermitean operator H there corresponds a unique spectral family $E_H(\alpha)$ that commutes with H such that for every f and every g in the domain of H*

$$(f, Hg) = \int_{-\infty}^{\infty} \alpha \, d(f, E_H(\alpha)g),$$

where the integral is a Stieltjes integral.[3] An expression which is written in short as

$$H = \int_{-\infty}^{\infty} \alpha \, dE_H(\alpha).$$

and is called the **spectral resolution** of H. We have

$$E_H(\alpha) = \int_{-\infty}^{\alpha} dE_H(\alpha')$$

and therefore

$$\int_{-\infty}^{\infty} dE_H(\alpha) = I.$$

2.3.2. To every interval $\Delta = (\alpha_1, \alpha_2]$ of the real line there corresponds a projection operator

$$E_H[\Delta] = \int_{\alpha_1}^{\alpha_2} dE_H(\alpha) = E_H(\alpha_2) - E_H(\alpha_1),$$

a definition that can be extended to a set of intervals. It follows from the properties of the spectral family $E_H(\alpha)$ that $E_H[\Delta]E_H[\Delta'] = E_H[\Delta \cap \Delta']$ and $E_H[\Delta] + E_H[\Delta'] = E_H[\Delta \cup \Delta']$. The quantity $dE_H(\alpha)$ can be thought of as the projection $E_H[d\alpha]$ corresponding to an infinitesimal interval of length $d\alpha$ centred at α.

2.3.3. By the definition of the Stieltjes integral one has

$$H = \lim_{n \to \infty} \sum_{\Delta_j \in S_n} \alpha_j E_H[\Delta_j],$$

where $\{S_n\}$ is a sequence of subdivisions of the real line, such that S_n becomes infinitely fine as $n \to \infty$. The sum runs over all the intervals $\Delta_j^{(n)} = \alpha_j^{(n)} - \alpha_{j-1}^{(n)}$ of the subdivision S_n. If the spectral family $E_K(\alpha)$ belonging to a hermitean operator K is constant except for a countable number of isolated discontinuities at the points α_i, of size $E_K(\alpha_i) - E_K(\alpha_i -) = E_i$, then K has a spectral resolution entirely similar to the one in a unitary space

$$K = \sum_i \alpha_i E_i,$$

[3] The Stieltjes integral defined in 2.1.4 is the Riemann-Stieltjes integral which is sufficient in this case. To discuss the most general functions of an operator, defined below, it is necessary to use the more general Lebesgue-Stieltjes integral.

although in this case the number of terms in the sum may be infinite. A general operator H can be regarded as the limit of a sequence

$$H_n = \sum_{\Delta_j \in S_n} \alpha_j E_H[\Delta_j]$$

of operators of the type K.

2.3.4. Let A and B be two operators with spectral families $E_A(\alpha), E_B(\beta)$. Suppose that A and B commute in the sense that $[A,B]g = 0$ for every g in the domain of $[A,B]$. Then one can show that $[E_A(\alpha), E_B(\beta)]g = 0$. However, unless the domain of $[A,B]$ coincides with the whole space, this relation does not imply that the spectral families commute. For this reason when we deal with two operators A and B that possess a spectral family we shall use a stronger definition of commutativity: *A and B commute if their spectral families commute.*

2.4 *The eigenvalue problem and the spectrum.* – The eigenvalue problem is defined in the same way as in the finite-dimensional case. For the operator K of the previous paragraph let f_i be any vector in the range of E_i then, since the E_i are orthogonal, we have $Kf_i = \alpha_i f_i$ so that f_i is an eigenvector of K belonging to the eigenvalue α_i and E_i is the corresponding eigenprojection.

2.4.1. For a general hermitean operator we have $H = \lim_{n\to\infty} H_n$ as discussed in 2.3.3. Applying H_n to a vector $f_j^{(n)}$ in the range of $E_H[\Delta_j^{(n)}]$ we have $H_n f_j^{(n)} = E_H[\Delta_j^{(n)}]f_j^{(n)}$, where we have denoted by the superscript n the intervals belonging to the subdivision S_n. The completeness of the space ensures that the vectors of this sequence tend to a limit. Let us choose the subdivisions S_n in such a way that the point $\alpha_j^{(n)}$ remains fixed at λ. If λ is a point of discontinuity of $E_H(\alpha)$ the limit will be a vector f_λ in the range of $E_\lambda = \lim_{n\to\infty}[E_H(\lambda) - E_H(\lambda-)]$ and we have $Hf_\lambda = \lambda f_\lambda$. Thus f_λ is an eigenvector of H belonging to the eigenvalue λ and E_λ is the corresponding eigenprojection. If on the other hand λ is a point where $E_H(\alpha)$ is continuous then the limit would be the zero vector and does not yield a solution of the eigenvalue problem. These considerations lead to the following result.

2.4.2. *The eigenvalue problem $Hf = \lambda f$ has solutions if and only if the spectral family $E_H(\alpha)$ is discontinuous. The eigenvalues λ_i are the points of discontinuity of $E_H(\alpha)$. The eigenprojection E_i belonging to λ_i is the discontinuity of $E(\alpha)$ at $\lambda_i : E_i = E_H(\lambda_i) - E_H(\lambda_i-)$.* It follows that the eigenprojections are orthogonal and that (for hermitean H) the eigenvalues are real.

2.4.3. The **spectrum** of H consists of the set Σ of points of the real axis where the spectral family $E_H(\alpha)$ is increasing. If the spectral family is constant except for a (necessarily countable) number of discontinuities H is said to have a **pure discrete** spectrum. If the spectral family is everywhere continuous H is said to have a **pure continuous** spectrum. Otherwise the spectrum is said to be **mixed**. If and only if H is bounded (positive) is its spectrum bounded (positive). Since the points of constancy of $E_H(\alpha)$ do not contribute to the integral in 2.3.1, the region of integration can be restricted to the spectrum and we can write

$$H = \int_{\Sigma} \alpha \, dE_H(\alpha).$$

2.4.4. Suppose that an operator H has a mixed spectrum Σ. The subspace $\mathbb{M} \subset \mathbb{H}$ spanned by its eigenvectors reduces H so that we can write $H = H_d \oplus H_c$, where H_d and H_c are operators on $\mathbb{M}$ and $\mathbb{M}^{\perp}$, respectively. The set of eigenvectors of H coincides with the pure discrete spectrum Σ_d, of H_d, while the spectrum Σ_c of H_c is pure continuous. We call Σ_d the **discrete spectrum** and Σ_c the **continuous spectrum** of H.

2.4.5. Note that $\Sigma = \Sigma_d \cup \Sigma_c$ but that Σ_d and Σ_c may have points in common, that is, there may be eigenvalues embedded in the continuous spectrum.[4] Separating out the discrete spectrum, the spectral resolution of H can be written

$$H = \sum_i \alpha_i E_i + \int_{\Sigma_c} \alpha \, dE_c(\alpha), \qquad \alpha_i \in \Sigma_d.$$

Clearly the terms on the right are, respectively, H_d and H_c when regarded as operators on $\mathbb{H}$.

2.5 *The spectral family of Q.* – Consider the multiplicative operator Q defined on $L^2(-\infty, \infty)$ by $(Qf)(x) = xf(x)$ introduced in 1.6.1. Let $E_Q(x)$ be its spectral family and let us put $(E_Q(\alpha)f)(x) = g(\alpha, x)$. The spectral resolution of Q requires that

$$xf(x) = \int_{-\infty}^{\infty} \alpha \, dg(\alpha, x).$$

A solution of this equation is obtained if we set $dg(\alpha, x) = \delta(\alpha - x)f(x)d\alpha$. We get then

$$(E_Q(\alpha)f)(x) = g(\alpha, x) = \int_{-\infty}^{\alpha} dg(\alpha', x) = \chi(\alpha - x)f(x).$$

[4] The definitions of discrete and continuous spectrum used here are not universal. They are, however, the ones most commonly used in physics.

where $\chi(x)$ is the Heaviside function $\chi(x) = 1$ for $0 \leq x$ and $\chi(x) = 0$ otherwise. According to this calculation $E_Q(\alpha)$ is the multiplicative operator $\chi(\alpha - x)$. We still have to verify that the solution that we have constructed gives the unique spectral family belonging to Q.

2.5.1. First we show that our solution is a spectral family, that is, that it has the properties required by the definition in 2.2. Since the $E_Q(\alpha)$ are multiplicative operators it is clear that they commute. According to A.2.8.1, property (i) is equivalent to $E(\alpha)E(\alpha') = E(\alpha)$, $\alpha \leq \alpha'$, a property that is obviously satisfied by the multiplicative operator $\chi(\alpha - x)$. Property (ii) holds because $\chi(x)$ was defined to be continuous on the right, while properties (iii) are immediate.

2.5.2. Next we have to verify that the spectral family found does indeed belong to Q. According to the spectral theorem we must have

$$\int_{-\infty}^{\infty} dx\ f^*(x)xg(x) = \int_{-\infty}^{\infty} dx \int_{-\infty}^{\infty} d\alpha\ \alpha f^*(x)\delta(a - x)g(x).$$

An equality which is obviously satisfied.

2.5.3. The spectral family $E_Q(\alpha)$ is continuous everywhere and increasing in $(-\infty, \infty)$. The spectrum of Q consists therefore of the whole of the real axis and is pure continuous. The projection $E_Q[\Delta]$ is the characteristic function $\chi(\Delta)$ of the interval Δ, that is, a function of x equal to one if $x \in \Delta$ and zero otherwise.

2.6 *The spectral family of P.* – In 1.10.1 we noted that the operator P defined on $L^2(-\infty, \infty)$ by $(Pf)(x) = -if'(x)$ is hermitean. To find its spectral family we put $(E_p(\alpha)f)(x) = h(\alpha, x)$ and solve the equation

$$-if'(x) = \int_{-\infty}^{\infty} \alpha\ dh(\alpha, x).$$

The solution is most easily found by Fourier transformation. Defining the Fourier transform $\tilde{f}(k)$ of $f(x)$ by

$$\tilde{f}(k) = \frac{1}{\sqrt{2\pi}} \int_{-\infty}^{\infty} dx\ e^{-ikx} f(x),$$

one finds the equation

$$k\tilde{f}(k) = \int_{-\infty}^{\infty} \alpha\ d\tilde{h}(\alpha, k)$$

whose solution we already know to be

$$\tilde{h}(\alpha, k) = \chi(\alpha - k)\tilde{f}(k)$$

so that

$$h(\alpha, x) = \frac{1}{\sqrt{2\pi}} \int_{-\infty}^{\infty} dk\, e^{ikx} \chi(\alpha - k)\tilde{f}(k)$$

and the spectral family is given by

$$(E_P(\alpha)f)(x) = \frac{1}{2\pi} \int_{-\infty}^{\alpha} dk \int_{-\infty}^{\infty} dx'\, e^{ik(x-x')} f(x').$$

One can verify that $E_P(\alpha)$ is continuous and increasing from $-\infty$ to ∞ and that therefore P has a pure continuous spectrum that extends to the whole real axis.

2.7 *Normal operators.* – If N is normal, its hermitean part A and antihermitean part iB commute. Let $E_A(\alpha)$ and $E_B(\beta)$ be the commuting spectral families of the hermitean operators A and B, respectively. By substituting into $N = A + iB$ the spectral resolutions of A and B we have

$$N = \int_{-\infty}^{\infty} \int_{-\infty}^{\infty} (\alpha + i\beta)\, dE_A(\alpha) dE_B(\beta).$$

For any region Δ of the complex plane $\mathbb{C}$ one can define a projection operator

$$E_N[\Delta] = \iint_{\Delta} dE_A(\alpha)\, dE_B(\beta)$$

which, in terms of the complex variable $\zeta = \alpha + i\beta$ we write simply as

$$E_N[\Delta] = \int_{\Delta} E_N[d\zeta].$$

These operators generalize to the complex plane the projections introduced in 2.3.2. They constitute a family of commuting projections depending on the region Δ with the following properties

(i) $E_N[\Delta]E_N[\Delta'] = E_N[\Delta \cap \Delta']$.

(ii) $E_N[\Delta] + E_N[\Delta'] = E_N[\Delta \cup \Delta']$.

(iii) $E_N[\mathbb{C}] = I$.

The quantity $E_N[d\zeta]$ can be thought of as the projection corresponding to an infinitesimal rectangle of area $dxdy$ centred at $\zeta = x + iy$. In the probability analogy of 2.1, the projection $E_N[\Delta]$ is the analogue of the probability of a region Δ of a two-dimensional sample space.

2.7.1. The spectral theorem for normal operators states that *to every normal operator N there corresponds a unique family $E(\Delta)$ of projections with the above properties such that*

$$N = \int_{\mathbb{C}} \zeta \, E_N[d\zeta].$$

If N is bounded and such that for any unit vector f, $\mu \leq \|Nf\| \leq \nu$, then the integration can be restricted to the annular region of the complex plane $\mu \leq |\zeta| \leq \nu$.

2.8 *Unitary operators.* – Since a unitary operator U is normal, the preceding discussion applies and since $\|U\| = 1$ the integration is limited to the circle $|\zeta| = 1$. Putting $\zeta = re^{i\phi}$ we can then write the spectral resolution of U as

$$U = \int_0^{2\pi} e^{i\phi} \, dE_U(\phi),$$

where the spectral family $E(\phi)$ is defined on $(0, 2\pi)$ and is such that

$$\int_0^{2\pi} dE_U(\phi) = I.$$

2.9 *Functions of a hermitean operator.* – Generalizing the definition in A.3.5, if $F(x)$ is a complex valued function of the real variable x the function $F(H)$ of the hermitean operator $H = \int \alpha \, dE_H(\alpha)$ is defined as the normal operator

$$F(N) = \int F(\alpha) \, dE_H(\alpha).$$

A definition that can obviously be extended to functions of normal operators.

2.10 *Sets of commuting hermitean operators.* – Suppose that $\{H_i\}$, $i = 1, \ldots, n$ is a set of n hermitean operators that commute with each other and such that H_i has spectral family $E_i(\alpha)$. Generalizing the discussion in 2.7, for every region Δ of an n-dimensional space one can construct a projection operator $E[\Delta]$ by the formula

$$E[\Delta] = \int \ldots \int_{\Delta} dE_1(\alpha_1) \ldots dE_n(\alpha_n)$$

which we write simply as

$$E[\Delta] = \int_{\Delta} E[d^n\alpha].$$

As in A.3.6, the set of operators can be considered as a single composite operator $A = \{A_i\}$ whose spectrum is a region Σ of n-dimensional space. Putting $\alpha = \{\alpha_1, \ldots, \alpha_n\}$ we write for the spectral resolution of A

$$A = \int_{\Sigma} \alpha E[d^n\alpha].$$

As in the finite-dimensional case it follows that a normal operator can be regarded as a set of two hermitean operators. From now on we restrict the discussion to hermitean operators. With obvious modifications the results can be applied to any normal operator, in particular to unitary ones.

2.10.1. As an example consider the space $\mathbf{L}^2$ of scalar functions $f(\mathbf{x})$, where $\mathbf{x} = \{x_1, x_2, x_3\}$ is a vector in the three-dimensional euclidean space $\mathbb{E}_3$, such that $\int d^3x \, |f(\mathbf{x})|^2 < \infty$. One can show that $\mathbf{L}^2$ is a Hilbert space. On this space consider the commuting operators Q_i, $i = 1, 2, 3$, defined by $(Q_i f)(\mathbf{x}) = x_i f(\mathbf{x})$. One can regard the three operators Q_i as a single "vector operator" $\mathbf{Q} = \{Q_1, Q_2, Q_3\}$ and write

$$\mathbf{Q} = \int \alpha E_Q[d^3\alpha].$$

It follows from 2.5 that the projections

$$E_Q[\Delta] = \int_{\Delta} E_Q[d^3\alpha]$$

are multiplicative operators, functions of the vector $\mathbf{x}$, equal to one in the region Δ of $\mathbb{E}_3$ and to zero outside it, and that the spectrum of $\mathbf{Q}$ is pure continuous and consists of the whole of $\mathbb{E}_3$.

2.11 *Degeneracy of the spectrum.* – As discussed in A.3.2, in a unitary space the spectrum of a hermitean operator is said to be simple – or non-degenerate – if all its eigenvalues are simple. Such a definition is not appropriate in the infinite-dimensional case because in general the spectrum is larger than the set of eigenvalues of the operator. Instead, guided by the result of A.3.9, the following definition that applies to both the infinite- and finite-dimensional cases is given: *The spectrum of a hermitean operator H is simple if every operator that commutes with H is a function of H.*

3. Function space representation

We saw in A.3.10 that every hermitean operator on a unitary space determines a representation of the space as a space of functions defined over its spectrum. In this section we discuss the generalization of the results of A.3.10 to the infinite-dimensional case.

3.1 *Discrete spectrum.* – If K is a hermitean operator on $\mathbb{H}$ with a pure discrete simple spectrum Λ, eigenvalues λ_i, and eigenprojections E_{Ki},

$$K = \sum_i \lambda_i E_{Ki},$$

the generalization of the results of A.3.10 is straightforward. However, because Λ is a countable infinite set instead of a finite one, one must take care to restrict the functions so as to ensure the existence of a norm. We can thus state that a generating vector $a = \sum_i \alpha_i b_i$, $\sum_i |\alpha_i|^2 < \infty$, where $\{b_i\}$ is a basis of eigenvectors of K, determines a function space representation of $\mathbb{H}$ through the correspondence $f^a = f(K)a \leftrightarrow f(\lambda)$, $\lambda \in \Lambda$, such that

$$\sum_{\lambda\in\Lambda} |\alpha(\lambda)|^2 |f(\lambda)|^2 < \infty,$$

where we have put $\alpha(\lambda_i) = \alpha_i$, so that $|\alpha(\lambda_i)|^2 = \langle a, E_{Ki} a\rangle$. The scalar product is given by

$$\langle f^a, g^a\rangle = \sum_{\lambda\in\Lambda} |\alpha(\lambda)|^2 f^*(\lambda) g(\lambda).$$

As in A.3.10.1 the operator $F(K)$ is represented by multiplication by $F(\lambda)$ and in particular the eigenprojection E_{Ki} by multiplication by $\delta_{\lambda\lambda_i}$. An operator A defined on the whole of $\mathbb{H}$ is given by

$$(Af^a)(\lambda) = \sum_{\lambda'\in\Lambda} A^a(\lambda,\lambda') f(\lambda'), \qquad A^a(\lambda_i,\lambda_j) = \frac{\alpha(\lambda_j)}{\alpha(\lambda_i)} \langle b_i, A b_j\rangle.$$

3.1.1. Although in this case a generating vector analogous to u in A.3.10.2 does not exist we can, as in that paragraph, perform a rescaling and agree to represent the vector f^a by the function $\alpha(\lambda) f(\lambda)$ obtaining a new function space, called the K-representation of $\mathbb{H}$, in which the scalar product is given by

$$\langle f, g\rangle = \sum_{\lambda\in\Lambda} f^*(\lambda) g(\lambda).$$

Here we have denoted by f the vector which, in this new function space, is represented by $f(\lambda)$ which must be such that

$$\sum_{\lambda \in \Lambda} |f(\lambda)|^2 < \infty.$$

The function of two variables corresponding to the operator A of the last paragraph is simply given by $A(\lambda_i, \lambda_j) = \langle b_i, Ab_j \rangle$. In this case we can write a decomposition of A similar to the one in A.2.14.3

$$A = \sum_i \sum_j A_{ij} b_i \times b_j^*, \qquad A_{ij} = \langle b_i, Ab_j \rangle.$$

3.1.2 The K-representation of $\mathbb{H}$ is identical with the l^2-space in 1.3.1 of sequences $f = \{f(\lambda_i)\}$, which can also be thought of as the space of infinite column matrices whose elements are the components $f(\lambda_i) = \langle b_i, f \rangle$ of f.

3.2 *Continuous spectrum.* – For a hermitean operator H with a pure continuous simple spectrum Σ and spectral family $E_H(\alpha)$,

$$H = \int_\Sigma \alpha \, dE_H(\alpha),$$

the generalization is not so straightforward because a basis of eigenvectors of H does not exist; it can be accomplished by a limiting process similar to the one used in 2.3 and 2.4. Thus, using the notation of 2.3.3, we regard H as the limit of a sequence of operators

$$H_n = \sum_{\Delta_j \in S_n} \alpha_j E_H[\Delta_j],$$

which leads us to define a generating vector a by the condition $\langle a, E_H[\Delta]a \rangle \neq 0$. Each generating vector a determines then a representation of $\mathbb{H}$ as a space of functions of the continuous variable $\alpha \in \Sigma$ through the correspondence $f^a = f(H)a \leftrightarrow f(\alpha)$. To find the expression of the scalar product we note that $\langle f^a, g^a \rangle = \langle f(H)a, g(H)a \rangle = \langle a, f^\dagger(H)g(H)a \rangle$ so that, writing the last expression in terms of the spectral resolution of H, we have

$$\langle f^a, g^a \rangle = \int_\Sigma f^*(\alpha) g(\alpha) \, d\langle a, E_H(\alpha)a \rangle.$$

In this function space representation an operator $F(H)$ becomes multiplication by $F(\alpha)$ and in particular the projection $E_H[\Delta]$ becomes multiplication by the characteristic function $\chi_\Delta(\alpha)$ of Δ.

3.2.1. The operators H of interest to us, to which we shall henceforth restrict the discussion, will be assumed to have an **absolutely continuous spectrum**, which means that the derivative $w(\alpha) = d\langle a, E_H(\alpha)a\rangle/d\alpha$ exists everywhere. For such operators the preceding relation becomes

$$\langle f^a, g^a\rangle = \int_\Sigma d\alpha\ w(\alpha)f^*(\alpha)g(\alpha).$$

Note that because $\langle a, E_H(\alpha)a\rangle$ is increasing everywhere in Σ, the "weight function" $w(\alpha)$ is positive. Note also that the existence of the norm implies the restriction

$$\int_\Sigma d\alpha\ w(\alpha)|f(\alpha)|^2 < \infty$$

on the functions of the space.

3.2.2. As in 3.1.1 we can perform a rescaling consisting in representing the vector f^a by the function $\sqrt{w(\alpha)}f(\alpha)$ obtaining a new equivalent function space in which the vectors f and g represented by the functions $f(\alpha)$ and $g(\alpha)$ have scalar product

$$\langle f, g\rangle = \int_\Sigma d\alpha\ f^*(\alpha)g(\alpha)$$

and the functions of the space are restricted by

$$\int_\Sigma d\alpha\ |f(\alpha)|^2 < \infty.$$

Obviously this function space, that we shall call the H**-space representation** (or simply the H-representation) of $\mathbb{H}$, is precisely the $L^2(\Sigma)$-space of square integrable functions on Σ of the example 1.3.2. The function space in which the operators $x\cdot$ and $-id\cdot/dx$ of 2.5 and 2.6 were studied is nothing other than the Q-representation of an abstract Hilbert space on which an operator Q with a simple spectrum consisting of the whole real axis is defined.

C DIRAC'S FORMULATION IN THE HILBERT SPACE

When dealing with unitary spaces, the Dirac formalism is just a notation that has some mnemonic advantages for the manipulation of vectors and operators. Such is not the case of its extension to infinite-dimensional spaces where it constitutes a simple and powerful method applicable to most, if not all, problems in quantum mechanics. It is most remarkable that it took several decades to uncover the mathematical basis of so successful a formulation. In this appendix we present Dirac's formulation as well as a brief sketch of its connexion with the theory of Hilbert spaces.

1. Tempered distributions

To understand how Dirac's formulation fits into the general theory of the Hilbert space and to give rules for the determination of the spectral resolution of most operators of interest in quantum mechanics it is necessary to be acquainted with the theory of distributions. We give in this section a sketch of the theory of a particular type of distributions called tempered distributions, which are the ones most commonly found when solving problems in quantum mechanics. Other types of distributions will be mentioned in Section 2.

1.1 *The space of test functions.* – Consider the subset of the the Hilbert space $L^2(-\infty, \infty)$ formed by the infinitely differentiable functions $\psi(x)$ of the real variable $x \in (-\infty, \infty)$ that decrease at infinity faster than any negative power of x. These functions, which will be called test functions, constitute a vector space $S(-\infty, \infty)$ which cannot be made into a Hilbert space because there are in S sequences $\{\psi_n\}$, such that $\|\psi_m - \psi_n\| \to 0$, that do not have a limit in S (cf. B.1.1.(ii)). Therefore, although $S \subset L^2$, S is *not* a subspace of L^2.

1.1.1. It is possible to introduce the notion of convergence in the space S in such a way as to make it complete. Basically, without going into technical details, convergence in S of the sequence $\{\psi_n\}$ to the limit ψ means that, at each x, $\psi_n(x) \to \psi(x)$ and all the derivatives $\psi_n^{(m)}(x) \to \psi^{(m)}(x)$. It is important to distinguish between the convergence just defined of $\{\psi_n\}$ regarded as a sequence of elements of S and the convergence in the norm of the same sequence when its members are regarded as elements of L^2. One can see that convergence in S implies convergence in L^2 but that the converse is not true.

1.1.2. Note for future reference that there is an infinity of test functions that vanish everywhere outside a given open interval.

1.2 *Definition of tempered distribution.* – A tempered **distribution** is a complex valued linear continuous functional $\Phi(\psi)$ on $S(-\infty, \infty)$. That is, $\Phi(\alpha_1\psi_1+\alpha_2\psi_2) = \alpha_1\Phi(\psi_1) + \alpha_2\Phi(\psi_2)$, $\psi_1, \psi_2 \in S$, and if $\{\psi_n\}$ is a sequence of test functions that converges *in* S to a limit ψ, then, $\Phi(\psi_n) \to \Phi(\psi)$. The tempered distributions form a vector space $S^\dagger$, the dual of S.

1.2.1. In the interest of brevity we shall refer in this section to tempered distributions simply as distributions. We shall do the same throughout the book whenever there is no danger of confusion

1.3 *Examples of distributions.* – We refer to a function which is integrable over every finite interval as a locally integrable function. Every locally integrable function $\gamma(x)$ that increases at infinity no faster than a power of x defines a distribution $\Gamma(\psi)$ through the formula

$$\Gamma(\psi) = \int_{-\infty}^{\infty} dx\ \gamma^*(x)\psi(x).$$

One can identify the function $\gamma(x)$ with the functional $\Gamma(\psi)$. We shall do so and write $\Gamma(\psi)$ as the scalar product $\langle \gamma; \psi \rangle$, $\gamma \in S^\dagger$, $\psi \in S$, where we have used the semicolon to distinguish it from the scalar product $\langle f, g \rangle$, $f, g \in \mathbb{H}$ defined on $\mathbb{H}$.

1.3.1 For every $f \in L^2$ and every $\psi \in S$ the integral

$$\langle f; \psi \rangle = \int_{-\infty}^{\infty} dx\ f^*(x)\psi(x)$$

exists and defines a continuous linear functional on S. This means that the function $f(x)$ can be regarded as an element of $S^\dagger$. Thus the space L^2 is a subset of $S^\dagger$. There are, however, elements in $S^\dagger$ that are not in L^2, at least those functions

$\gamma(x)$ of the previous paragraph that increase at infinity. Thus we have the relation, $S \subset L^2 \subset S^\dagger$.

1.3.2. The distributions of the type presented in the last two paragraphs do not exhaust the elements of $S^\dagger$. As is well known the Dirac "δ-function" is not a function. In the present context, it is, rather, the continuous linear functional which, using the above notation, is defined by

$$\langle \delta_a; \psi \rangle = \psi(a).$$

In accordance with a universal convention one writes *symbolically*

$$\langle \delta_a; \psi \rangle = \int_{-\infty}^{\infty} dx\ \delta(x-a)\psi(x).$$

1.4. *Notation.* – In general we shall refer to the linear functional denoted by $\langle u; \cdot \rangle$ as the distribution u and write

$$\langle u; \psi \rangle = \int_{-\infty}^{\infty} dx\ u^*(x)\psi(x)$$

even if, as in the case of the δ-function, the integral can only be regarded as a symbolic notation (minor variations are universally used as when one writes $\delta(x-a)$ instead of δ_a to exhibit the behaviour of the functional under a translation of the coordinate x). Of course, if the functional is generated by a function, so that the integral has a meaning in its own right, we shall feel free to refer to the functional $\langle u; \psi \rangle$ as the function $u(x)$. Sometimes, when the general usage sanctions it, we shall use the word function even if the functional is not generated by a function. We did that in the last paragraph.

1.4.1. Although, as is clear from the definition, it is meaningless to speak about the value of a distribution at a point, it is possible to define the equality of two distributions on an open interval. Two distributions u and v are said to be **equal** in an open interval Δ if $\langle u; \psi \rangle = \langle v; \psi \rangle$ for every test function $\psi(x)$ that vanishes outside Δ (cf. 1.1.2). Thus, $\delta(x)$ equals zero on any open interval that does not include the origin.

1.5. *Change of variables.* – Consider a function $y = h(x)$ that has a derivative different from zero everywhere. It defines a change of variable of a function $f(x)$ through the correspondence $f(x) \mapsto \bar{f}(y) = f\left(h^{-1}(y)\right)$. Since the value of a distribution at a point is a meaningless concept one cannot define a change of variables for distributions in the same manner as for functions. The definition is possible but it is somewhat more complicated. To guide ourselves let us start with an ordinary function $\gamma(x)$ that defines a distribution. The usual formula for the change of integration variable tells us that

$$\int_{-\infty}^{\infty} dy\, \gamma^* \left(h^{-1}(y)\right) \psi(y) = \int_{-\infty}^{\infty} dx\, |h'(x)| \gamma^*(x) \psi\left(h(x)\right).$$

If furthermore $h(x)$ is such that, along with $\psi(x)$, the function $\bar{\psi}(x) = |h'(x)|\psi\left(h(x)\right)$ is a test function, the second integral defines a distribution which we denote by $\gamma(h^{-1})$. The definition can obviously be generalized to an arbitrary distribution u through the formula

$$\langle u(h^{-1}); \psi \rangle = \langle u; \bar{\psi} \rangle, \qquad \bar{\psi}(x) = |h'(x)|\psi\left(h(x)\right).$$

For example, if u is the δ-function we have

$$\langle \delta_0(h^{-1}); \psi \rangle = \langle \delta_0; \bar{\psi} \rangle = |h'(0)|\psi\left(h(0)\right).$$

Putting $h^{-1}(x) = k(x)$ this formula becomes

$$\int_{-\infty}^{\infty} dx\, \delta\left(k(x)\right) \psi(x) = \frac{1}{|k'(x_0)|}\psi(x_0),$$

where x_0 is the (necessarily simple) zero of $k(x)$. More succintly one writes

$$\delta\left(k(x)\right) = \frac{1}{|k'(x_0)|}\delta(x - x_0).$$

1.5.1. Because the value of the functional $\langle \delta_0(k); \psi \rangle$ depends only on the value of the function at the point x_0, the preceding equation can be generalized as follows. Let $g(x)$ be a function that possesses a countable number of isolated *simple* zeros x_m and has at every zero derivatives of all orders, then

$$\delta\left(g(x)\right) = \sum_m \frac{1}{|g'(x_m)|}\delta(x - x_m).$$

1.6 *Derivatives of distributions.* – The **derivative** $u^{(n)} = d^{(n)}u/dx^{(n)}$ of order n of a distribution u is the distribution

$$\langle u^{(n)}; \psi \rangle = (-1)^n \langle u; \psi^{(n)} \rangle.$$

A definition which is consistent with the formal integration by parts

$$\int_{-\infty}^{\infty} dx\ u'^*(x)\psi(x) = -\int_{-\infty}^{\infty} dx\ u^*(x)\psi'(x).$$

Two important properties follow from the definition:

(i) Every distribution is infinitely differentiable.

(ii) Every distribution is the derivative of some order of a locally integrable continuous function.

As an example of (ii), $\delta(x)$ is the second derivative of the continuous function $F(x) = 0,\ x < 0$ and $F(x) = x,\ x \geq 0$, whose first derivative is the Heaviside function $\chi(x)$.

1.7 *Some useful distributions.* – According to the above definition the n-th derivative $\delta_a^{(n)}$ of the δ-function δ_a is given by $\langle\delta_a^{(n)};\psi\rangle = (-1)^n\langle\delta_a;\psi^{(n)}\rangle = (-1)^n\psi^{(n)}(a)$. One writes

$$\langle\delta_a^{(n)};\psi\rangle = \int_{-\infty}^{\infty} dx\ \delta^{(n)}(x-a)\psi(x) = (-1)^n\psi^{(n)}(a).$$

1.7.1. Another set of useful distributions is formed by the derivatives of the continuous function $x\ln|x| - x$. The first derivative is the discontinuous function $\ln|x|$. According to the definition in 1.6, we have

$$\left\langle\frac{d\ln|x|}{dx};\psi\right\rangle = -\left\langle\ln|x|;\frac{d\psi}{dx}\right\rangle.$$

The last expression can be written

$$-\int_{-\infty}^{\infty} dx\ \ln|x|\psi'(x) = -\lim_{\epsilon\to 0}\int_{\epsilon}^{\infty} dx\ \ln x[\psi'(x)+\psi'(-x)]$$

or integrating by parts and remembering the definition of the Cauchy principal value of an integral,

$$\left\langle\frac{d\ln|x|}{dx};\psi\right\rangle = \lim_{\epsilon\to 0}\left\{\int_{-\infty}^{-\epsilon} + \int_{\epsilon}^{\infty}\right\} dx\ \frac{\psi(x)}{x} = \mathcal{P}\int_{-\infty}^{\infty} dx\ \frac{1}{x}\psi(x).$$

Although, as is common practice, we shall denote this distribution by $1/x$, the reader should be warned that it is different from the ordinary function $1/x$ although both are equal in any open interval that does not contain the origin.

1.7.2. Generalizing this notation we define the distribution $1/x^n$ by the formula

$$\frac{1}{x^n} = \frac{(-1)^{(n-1)}}{(n-1)!}\frac{d^n \ln|x|}{dx^n},$$

Which gives in particular

$$\langle 1/x^2; \psi\rangle = \int_0^\infty dx\, \frac{1}{x^2}[\psi(x) + \psi(-x) - 2\psi(0)].$$

1.8 *Products of distributions.* – It is not possible to define in general the product of two distributions. Suppose, however, that $\gamma(x)$ is an infinitely differentiable function that grows at infinity no faster than a power of x. Then its product $\gamma(x)\psi(x)$ by a test function is also a test function. We define then the product $u\gamma = \gamma u$ of an arbitrary distribution u times the distribution γ by

$$\langle u\gamma; \psi\rangle = \langle u; \gamma^*\psi\rangle.$$

For example, the distribution $\gamma(x)\delta(x-a)$ is defined, and $\langle \gamma\delta_a; \psi\rangle = \gamma^*(a)\psi(a)$.

1.8.1. In particular we have $x\delta(x) = 0$. In fact, the general solution of the equation $xu = 0$, where u is an unknown distribution, is $u = c\delta(x)$ with c an arbitrary constant. This can be seen by rewriting the equation as $\langle u; x\psi\rangle = 0$ and noting that $\phi(x) = x\psi(x)$ is a test function such that $\phi(0) = 0$ but is otherwise arbitrary.

1.8.2. From this result we can find the general solution of the corresponding inhomogeneous equation $xu = \gamma$, where $\gamma(x)$ is as above, to be the family of distributions

$$u = \frac{\gamma(x)}{x} + c\delta(x).$$

1.8.3. It is immediately verified that $u = \delta'(x)$ is a particular solution of the equation $xu = \delta(x)$. It follows from 1.8.1 that the general solution of this equation is

$$u = \delta'(x) + c\delta(x).$$

1.9 *Limits and integrals of distributions.* – A sequence of distributions $\{u_n\}$ is said to tend to a limit if there is a distribution u such that $\lim\langle u_n; \psi\rangle = \langle u; \psi\rangle$ for all ψ. One writes simply $u = \lim u_n$.

1.9.1. The following useful expressions give the δ-function as a limit of sequences of simple functions

$$\delta(x) = \lim_{n\to\infty} \sqrt{\frac{n}{\pi}}\, e^{-nx^2} = \lim_{n\to\infty} \frac{\sin nx}{\pi x} = \lim_{n\to\infty} \frac{1}{\pi} \frac{n}{1+n^2x^2}.$$

As an example we prove the first one. We have

$$|\langle u_n; \psi\rangle - \langle \delta_0; \psi\rangle| = \sqrt{\frac{n}{\pi}} \left| \int_{-\infty}^{\infty} dx\, e^{-nx^2} [\psi(x) - \psi(0)] \right|,$$

now, the right member is bounded from above by

$$2\sqrt{\frac{n}{\pi}}\, \max|\psi'(x)| \int_0^{\infty} dx\, x e^{-nx^2} = \frac{\max|\psi'(x)|}{\sqrt{\pi n}}$$

which tends to zero as $n \to \infty$.

1.9.2. Another useful distribution denoted by $\delta_+(x)$ is defined by the limit of the sequence $u_n = -(i/2\pi)d\ln(x - i/n)/dx$ as $n \to \infty$. We have

$$2\pi i \langle u_n; \psi\rangle = \int_{-\infty}^{\infty} dx\, \frac{d\ln|x - i/n|}{dx} \psi(x) - i \int_{-\infty}^{\infty} dx\, \frac{d\tan^{-1}(1/nx)}{dx} \psi(x).$$

By 1.7.2 the first integral tends to the limit $\langle 1/x; \psi\rangle$, while performing the differentiation the second integral becomes

$$-\int_{-\infty}^{\infty} dx\, \frac{n}{1+n^2x^2} \psi(x)$$

which, as we have just seen, tends to $-\pi\langle \delta_0; x\rangle$. We, thus, have

$$\delta_+(x) = \frac{1}{2\pi i x} + \tfrac{1}{2}\,\delta(x).$$

The complex conjugate of δ_+ is denoted by δ_-:

$$\delta_-(x) = \frac{i}{2\pi} \lim_{n\to\infty} \frac{d\ln(x + i/n)}{dx} = -\frac{1}{2\pi i x} + \tfrac{1}{2}\,\delta(x).$$

1.9.3. If u_y is a distribution depending on the parameter y and the integral $\int dy\ \langle u_y; \psi\rangle$ exists for all values of the functional, then this integral is a distribution w which is defined as the integral of u_y. One writes $w = \int dy\ u_y$. If $u_y(x)$

is an ordinary function the relation $\int dy\ \langle u_y;\psi\rangle = \langle \int dy\ u_y;\psi\rangle$ amounts to a statement of the possibility of interchanging the integrals over y and x. Some simple examples of integrals that can be easily verified are

$$\int_{-\infty}^{\infty} d\lambda\ \lambda^n e^{i\lambda x} = 2\pi(-i)^n \delta^{(n)}(x), \qquad \int_a^b d\lambda\ \delta(x-\lambda) = \chi(b-x) - \chi(a-x).$$

where we have used our notation $\chi(x)$ for the Heaviside function.

1.10 *Fourier transforms.* – One can easily see that the Fourier transform $\tilde{\psi}(k)$ of a test function $\psi(x)$

$$\tilde{\psi}(k) = \frac{1}{\sqrt{2\pi}} \int_{-\infty}^{\infty} dx\ e^{-ikx}\psi(x)$$

exists and is a test function of the variable k. The inverse transformation

$$\psi(x) = \frac{1}{\sqrt{2\pi}} \int_{-\infty}^{\infty} dk\ e^{ikx}\tilde{\psi}(k)$$

also exists and transforms test functions into test functions. The Fourier transformation is therefore an isomorphism of the space S onto itself.

1.10.1. It follows that, given a distribution u on the space of the functions $\psi(x)$, we can define its Fourier transform as the distribution $\tilde{u} = \mathcal{F}[u]$ on the space of the test functions $\tilde{\psi}(k)$ such that

$$\langle \tilde{u};\tilde{\psi}\rangle = \langle u;\psi\rangle.$$

A formula that, if $\tilde{u}$ is given also defines its inverse Fourier transform u. Therefore the Fourier transform is also an isomorphism of the space $S^\dagger$ of distributions onto itself.[1]

1.10.2. If u is an ordinary function that has a Fourier transform, it satisfies Parseval's relation

$$\int_{-\infty}^{\infty} dk\ \tilde{u}^*(k)\tilde{\psi}(k) = \int_{-\infty}^{\infty} dx\ u^*(x)\psi(x)$$

which, if u is interpreted as a distribution, is just the definition 1.10.1. Thus, the generalized definition of Fourier transform agrees in this case with the ordinary one.

[1] This property is not true of distributions that are not tempered and is one of the reasons for the usefulness of tempered distributions in many branches of physics.

1.10.3. The following important Fourier transforms can be readily verified

$$\mathcal{F}[\delta(x)] = \frac{1}{\sqrt{2\pi}},$$

$$\mathcal{F}[1/x] = -i\sqrt{\frac{\pi}{2}}\,\operatorname{sgn} k,$$

$$\mathcal{F}[\chi(x)] = \sqrt{2\pi}\,\delta_+(k).$$

1.10.4. It is not difficult to see that the usual formulae regarding the Fourier transforms of the derivatives, or the effect on the transforms of a linear change of the independent variable apply to distributions. If the product of two distributions is defined the convolution theorem applies and the corresponding convolution integral of their transforms is defined.

1.11 *Functions of several variables.* – The foregoing discussion can be generalized to a Hilbert space of functions $f(x_1, \ldots, x_n)$ of n real variables. In this case the test functions are required to decrease at infinity faster than any negative power of $(x_1^2 + \ldots + x_n^2)^{1/2}$. As before one writes for a distribution

$$\langle u; \psi\rangle = \int_{-\infty}^{\infty} \ldots \int_{-\infty}^{\infty} dx_1 \ldots dx_n\; u^*(x_1, \ldots, x_n)\psi(x_1, \ldots, x_n).$$

The definition of derivative is extended in the obvious manner

$$\Big\langle \frac{\partial^m u}{\partial x_1^{m_1} \ldots \partial x_n^{m_n}}; \psi\Big\rangle = (-1)^m \Big\langle u; \frac{\partial^m \psi}{\partial x_1^{m_1} \ldots \partial x_n^{m_n}}\Big\rangle.$$

It follows from the definition that the order of differentiation can always be interchanged.

1.12 *Linear operators on the space of test functions.* – We shall see now that there are classes of operators on the Hilbert space L^2 which are distributions on the space of functions of two variables. Suppose that A is an operator on L^2 such that when applied to a test function it yields another test function. We can think of it as an operator on S and, when we do so, we shall denote it by A^o and call it the restriction of A to S. Suppose furthermore that A^o is continuous (by this we mean, naturally, continuous in the sense of 1.2). The scalar product $\langle\phi; A^o\psi\rangle$ (which of course is identical to $\langle\phi, A\psi\rangle$ when A is regarded as an operator on L^2) is then a continuous functional bilinear in $\psi(x)$ and $\phi^*(x)$ and can, in fact, be regarded as a distribution A^* defined on the space $S^* \otimes S$ of test functions of two variables by the formula

$$\langle \phi; A^o \psi \rangle = \langle A^*; \phi^* \otimes \psi \rangle,$$

where we have represented by $\phi^* \otimes \psi$ the function $\phi^*(x)\psi(y) \in S^* \otimes S$. Use of a notation for $\langle A^o; \phi^* \otimes \psi \rangle$ similar to the one in 1.4 yields

$$\langle A^*; \phi^* \otimes \psi \rangle = \langle \phi; A^o \psi \rangle = \langle \phi, A\psi \rangle = \int_{-\infty}^{\infty} \int_{-\infty}^{\infty} dxdy \; A(x,y)\phi^*(x)\psi(y).$$

We call $A(x,y)$ the distribution (of two variables) that represents the operator A. Correspondingly one writes

$$(A\psi)(x) = \int_{-\infty}^{\infty} dy \; A(x,y)\psi(y).$$

One can see, in particular, that multiplication by the function $F(x)$ is represented by $F(x)\delta(x - x')$, the identity by $\delta(x - x')$ and the differentiation d^n/dx^n by $(-1)^n \delta^{(n)}(x - x')$.

1.12.1. The considerations of the preceding paragraph can be extended to a linear operator Γ (not defined, in general, on L^2) that maps continuously test functions into distributions, for, if $w = \Gamma^* \phi$, the functional $\langle w; \psi \rangle$ is again bilinear in ϕ^* and ψ so that Γ^* can also be regarded as a distribution on $S^* \otimes S$. As an important example let u and v be two distributions on S. We can then define an operator from S to $S^\dagger$, which we denote by $u \times v^*$, by the formula

$$\langle (u^* \times v)\phi; \psi \rangle = \langle u^* \otimes v; \phi^* \otimes \psi \rangle = \langle u; \phi \rangle^* \langle v; \psi \rangle$$

which can also be written

$$(u \times v^*)\psi = \langle v; \psi \rangle u.$$

In our usual symbolic notation we have

$$\langle (u^* \times v)\phi; \psi \rangle = \int_{-\infty}^{\infty} \int_{-\infty}^{\infty} dxdy \; u(x)v^*(y)\phi^*(x)\psi(y)$$

so that the representation of $v \times u^*$ is $v(x)u^*(y)$. In terms of it we can write

$$((u \times v^*)\psi)(x) = \int_{-\infty}^{\infty} dy \; u(x)v^*(y)\psi(y).$$

For example the representation of the operator $\delta_a \times \delta_b$ is $\delta(x - a)\delta(y - b)$ and its action on the test function ψ(x) is the distribution $\delta(x - a)\psi(b)$.

1.12.2. If $f, g \in L^2$, then the operator $f \times g^*$ can be regarded as an operator on L^2 since $(f \times g^*)\psi = \langle g; \psi\rangle f = (g, \psi) f \in L^2$. This operator is entirely analogous to the the operator introduced in A.2.8.3 which was written $|f\rangle\langle g|$ in Dirac's notation. In particular if f is normalized then $f \times f^*$ is just the projection on the one-dimensional subspace spanned by f. Its representation is the function of two variables $f(x)f^*(y)$.

1.12.3. Suppose now that the operator Γ_λ depending on the parameter λ is of the type that we have been discussing. Since it can be regarded as a distribution, its integral $\int d\lambda\, \Gamma_\lambda$, can be defined in the same way as for a distribution. For example, if Γ_λ has representation $\lambda^n e^{i\lambda(x-y)}$, we have

$$\int_{-\infty}^{\infty} d\lambda\, \lambda^n e^{i\lambda(x-y)} = 2\pi(-i)^n \delta^{(n)}(x-y) = 2\pi i^n \delta^{(n)}(y-x).$$

1.12.4. We return now to discuss operators with a continuous restriction to S of the type introduced in 1.12. For the sake of brevity we shall henceforth refer to them as "good" operators. Let A be a good operator. Given a distribution u one can define a corresponding distribution $A^\ddagger u$ by $\langle A^\ddagger u; \psi\rangle = \langle u; A^o \psi\rangle$. This formula defines a linear operator $A^\ddagger$ on the space $S^\dagger$ which coincides with the adjoint $A^\dagger$ of A on L^2 when u is a square integrable function regarded as a distribution, in the same way that, as we already saw, A coincides with A^o when the test function ψ is regarded as an element of L^2.

1.12.5. Suppose that the good operator H on L^2 is hermitean; then $H^\ddagger$ coincides with H when u is a square integrable function and with H^o when u is a test function. Thus $H^\ddagger$ can be thought of the extension of H (or of H^o) to the space $S^\dagger$ of distributions. In fact this manner of thinking lies behind the definition of derivative of a distribution in 1.6. Indeed, if one considers the hermitean operator $-id/dx$ on L^2 and uses the same notation for its restriction to S and its extension to $S^\dagger$, the equation $\langle H^\ddagger u; \psi\rangle = \langle u; H^o \psi\rangle$ becomes $i\langle du/dx; \psi\rangle = -i\langle u; d\psi/dx\rangle$ in agreement with the definition of derivative.

1.13 *The generalized eigenvalue problem.* – The generalized eigenvalue problem for a good operator A on L^2 is said to have a solution if there exists a complex number μ such that the equation

$$\langle u_\mu; A^o \psi\rangle = \mu^* \langle u_\mu; \psi\rangle$$

or, equivalently,

$$\langle A^\ddagger u_\mu; \psi\rangle = \mu \langle u_\mu; \psi\rangle$$

is satisfied by a distribution u_μ different from zero for all $\psi \in S$. If such is the case u_μ is the **generalized eigenvector** of A belonging to **the generalized eigenvalue** μ.

1.13.1. In a finite-dimensional space the problem of finding the spectral family of a hermitean or unitary operator is equivalent to the solution of the eigenvalue problem. Such is also the case in the L^2-space when we deal with operators that have a pure discrete spectrum. The two problems are equivalent because for each eigenvalue λ_i of degeneracy s_i there exist s_i orthonormal eigenvectors f_i^r, $r = 1, \ldots, s_i$ from which the corresponding eigenprojection $E_i = \sum_r f_i^r \times f_i^{r*}$ can be constructed and, moreover, the eigenprojections thus constructed are complete in the sense that $\sum_i E_i = I$. An analogous situation exists for all "good" hermitean or unitary operators on L^2. We shall restrict the following discussion to hermitean operators; the results can be applied to unitary operators with minor modifications that will be mentioned later. Also, for the sake of clarity, it is convenient to state the results for an operator with a pure simple spectrum before discussing the general case of an operator with a mixed degenerate spectrum.

1.14 *Operators with pure continuous spectra.* – Let H be a good hermitean operator on L^2 with a pure continuous spectrum Σ. As pointed out in B.3.2.1 we shall assume henceforth that our operators are not only "good" but also have an absolutely continuous spectrum. As is customary, we denote the extension $H^\ddagger$ of the hermitean operator H on L^2 to the space of distributions by the same symbol H and write the eigenvalue equation as $\langle Hu_\alpha; \psi\rangle = \alpha\langle u_\alpha; \psi\rangle$ or, in abbreviated notation

$$Hu_\alpha = \alpha u_\alpha.$$

In this case, for each value of α in the spectrum Σ there exists a unique set of linearly dependent solutions of the eigenvalue problem of the form cu_α which, as we know from B.2.4.2, cannot be elements of L^2. We can choose the u_α to form a family of distributions which is **complete** in the sense that for any pair of test functions $\phi(x)$, $\psi(x)$ we have

$$\langle \phi, \psi\rangle = \int_\Sigma d\alpha \, \langle u_\alpha; \phi\rangle^* \langle u_\alpha; \psi\rangle.$$

1.14.1. Using the definition of 1.12.1 this completeness relation can be rewritten as the resolution of the identity

$$I = \int_\Sigma d\alpha \, u_\alpha \times u_\alpha^*$$

which, with the replacement of the summation by an integration, is formally similar to the completeness relation for an operator with a pure discrete spectrum.

1.14.2. When this relation is applied to the vector $\psi \in S$ it yields

$$\psi = \int_{\Sigma} d\alpha \, u_\alpha \langle u_\alpha ; \psi \rangle .$$

which is again similar to the expansion of a vector as a linear combination of a basis of eigenvectors of an operator with a pure simple spectrum.

1.14.3. Forming the scalar product of the last relation with the generalized eigenvector u_α we obtain

$$\langle u_\alpha ; \psi \rangle = \left\langle u_\alpha ; \left(\int_{\Sigma} d\alpha' \, u_{\alpha'} \langle u_{\alpha'} ; \psi \rangle \right) \right\rangle$$

which, succumbing to the obvious temptation, we would like to write as

$$\langle u_\alpha ; \psi \rangle = \int_{\Sigma} d\alpha' \, \langle u_{\alpha'} ; \psi \rangle \langle u_\alpha ; u_{\alpha'} \rangle .$$

Unfortunately our temptation, as is too often the case, leads us to transgress the law, for as we have just seen the u_α are distributions which are *not* elements of L^2 and can therefore form scalar products only with test functions. Nevertheless, it is customary to keep the illegal symbol and satisfy the last equation by putting

$$\langle u_\alpha ; u_{\alpha'} \rangle = \delta(\alpha - \alpha')$$

and refer to this expression as the "orthonormality relation" for the generalized eigenvectors. The usefulness of this practice stems from the fact that very often the generalized eigenvectors can be regarded as distributions w_x on a space of test functions of the independent variable α that depend on the parameter x in which case the illegal symbol can be regarded as an operator of the form $w_x \times w_x^*$ on that space. The reader is however warned that in the general case the "orthonormality relation" means no more and no less than the expansion of ψ as an an integral over the eigen-distributions or any of the other two equivalent statements of the completeness relation given in the previous paragraph.

1.14.4. In terms of the representation of the operators $u_\alpha \times u_\alpha^*$ the resolution of the identity of 1.14 becomes

$$\delta(x - x') = \int_{\Sigma} d\alpha \, u_\alpha(x) u_\alpha^*(x')$$

and the expansion of ψ is written

$$\psi(x) = \int_{\Sigma} d\alpha\, u_{\alpha}(x) \int_{-\infty}^{\infty} dx'\, u_{\alpha}^{*}(x')\psi(x).$$

which is equivalent to the "orthonormality relation"

$$\int_{-\infty}^{\infty} dx\, u_{\alpha}^{*}(x) u_{\alpha'}(x) = \delta(\alpha - \alpha')$$

if the latter is meaningful in its own right.

1.14.5. If the expression for the identity in 1.14.1 is substituted into the equality $H = HI$ we obtain

$$H = \int_{\Sigma} d\alpha\, H u_{\alpha} \times u_{\alpha}^{*}$$

which, taking into account the eigenvalue equation in 1.14, gives the following form for the spectral resolution of H

$$H = \int_{\Sigma} d\alpha\, \alpha\, u_{\alpha} \times u_{\alpha}^{*}.$$

1.14.6. When this expression is compared with the corresponding expression in B.2.4.3, one finds the following relation between the projections $E_H(\alpha)$ and the operators $u_{\alpha} \times u_{\alpha}^{*}$

$$u_{\alpha} \times u_{\alpha}^{*} = \frac{dE_H(\alpha)}{d\alpha}$$

which means that for any pair of test functions ϕ, ψ, we have

$$\langle u_{\alpha}; \phi \rangle \langle u_{\alpha}^{*}; \psi \rangle^{*} d\alpha = d\langle \phi, E_H(\alpha)\psi \rangle.$$

If we set $u_{\alpha} = 0$ outside the spectrum, we can write

$$E_H(\alpha) = \int_{-\infty}^{\alpha} d\alpha'\, u_{\alpha'} \times u_{\alpha'}^{*}.$$

1.15 *Operators with a general spectrum.* – In the general case the operator H has a mixed spectrum with a discrete part consisting of real eigenvalues α_i and a real continuous spectrum that can be decomposed into a non-degenerate part Σ_1, a doubly degenerate part Σ_2, etc. As we know, for each eigenvalue α_{ki} of degeneracy k

there exists an orthonormal set of eigenvectors $f^r_{ki} \in L^2$, $r = 1, \ldots, k$. Similarly in each region Σ_k of the continuous spectrum there exists a set of generalized eigenvectors, $u^r_{k\alpha}$, $r = 1, \ldots, k$ such that the set formed by all these eigenvectors together with the f^r_{ki} is **complete**. What completeness means is that, for any pair of test functions ϕ, ψ, we have

$$\langle \phi, \psi \rangle = \sum_k \sum_{r=1}^{k} \Big(\sum_i \langle f^r_{ki}; \phi \rangle^* \langle f^r_{ki}; \psi \rangle + \int_{\Sigma_k} d\alpha \, \langle u^r_{k\alpha}; \phi \rangle^* \langle u^r_{k\alpha}; \psi \rangle \Big).$$

1.15.1. In operator notation this expression yields the following resolution of the identity,

$$I = \sum_k \sum_{r=1}^{k} \Big(\sum_i f^r_{ki} \times f^{r*}_{ki} + \int_{\Sigma_k} d\alpha \, u^r_{k\alpha} \times u^{r*}_{k\alpha} \Big).$$

1.15.2. An equivalent relation is given by the expansion

$$\psi = \sum_k \sum_{r=1}^{k} \Big(\sum_i f^r_{ki} \langle f^r_{ki}; \psi \rangle + \int_{\Sigma_k} d\alpha \, u^r_{k\alpha} \langle u^r_{k\alpha}; \psi \rangle \Big)$$

of an arbitrary test function.

1.15.3. As in 1.14.3, the completeness relation is often expressed in terms of the "orthonormality relations"

$$\langle f^r_{ki}; f^{r'}_{k'i'} \rangle = \delta_{kk'} \delta_{ii'} \delta_{rr'}, \qquad \langle u^r_{k\alpha}; u^{r'}_{k\alpha'} \rangle = \delta_{rr'} \delta(\alpha - \alpha'),$$

$$\langle u^r_{k\alpha}; f^r_{ki} \rangle = 0, \qquad \langle u^r_{k\alpha}; u^{r'}_{l\alpha'} \rangle = 0, \ k \neq l.$$

The first of these relations is defined if we interpret the symbol $\langle f^r_{ki}; f^{r'}_{ki'} \rangle$ as $\langle f^r_{ki}, f^{r'}_{ki'} \rangle$ and is just the usual orthonormality relation between eigenvectors. The other three relations are not defined, in general (obviously $\langle u^r_{k\alpha}; f^r_{ki} \rangle$ is defined if the eigenvector f^r_{ki} happens to be a test function), but they are useful for the reason given in 1.14.3. Needless to say that, unless the spectrum is simple, there exist infinitely many complete sets of generalized eigenvectors.

1.15.4. The reader should find no difficulty in rewriting the previous relations in the notation of 1.14.4. For example, the resolution of the identity becomes

$$\delta(x - x') = \sum_k \sum_{r=1}^{k} \Big(\sum_i f^r_{ki}(x) f^{r*}_{ki}(x') + \int_{\Sigma_k} d\alpha \, u^r_{k\alpha}(x) u^{r*}_{k\alpha}(x') \Big).$$

1.16 *The spectral resolution.* – On inserting the resolution of the identity in 1.15.1 into the equality $H = HI$ and using the eigenvalue equation for H we obtain after a minor rearrangement

$$H = \sum_k \Big(\sum_i \alpha \sum_{r=1}^{k} f_{ki}^r \times f_{ki}^{r*} + \int_{\Sigma_k} d\alpha\, \alpha \sum_{r=1}^{k} u_{k\alpha}^r \times u_{k\alpha}^{r*} \Big).$$

An expression which is obviously not unique if the spectrum is degenerate. Let us define the following operators from S to $S^\dagger$ (cf. 1.12.1),

$$E_{Hki} = \sum_{r=1}^{k} f_{ki}^r \times f_{ki}^{r*}; \qquad \Gamma_H(\alpha) = \sum_{r=1}^{k} u_{k\alpha}^r \times u_{k\alpha}^{r*}, \qquad \alpha \in \Sigma_k.$$

The operator E_{Hki} can also be regarded as a projection on L^2 but $\Gamma_H(\alpha)$ cannot be so regarded. With these definitions, a unique expression for the spectral resolution of H is obtained.

$$H = \sum_k \sum_i \alpha_{ki} E_{Hki} + \int_{\Sigma_c} d\alpha\, \alpha \Gamma_H(\alpha),$$

where $\Sigma_c = \bigcup_k \Sigma_k$ is the continuous spectrum of H. Comparing this expression with the corresponding one in B.2.4.5, we see that it gives the decomposition of H into the sum $H_d + H_c$ of two operators with a pure discrete and a pure continuous spectrum, respectively. Furthermore, if we denote by $E_c(\alpha)$ the spectral family of H_c, we have

$$\Gamma_H(\alpha) d\alpha = dE_c(\alpha).$$

The meaning of this formula is that for any pair ϕ, ψ, of test functions the equality

$$\langle \Gamma_H^*(\alpha)\phi; \psi \rangle d\alpha = d\langle \phi, E_c(\alpha)\psi \rangle$$

holds. If we replace $\Gamma_H(\alpha)$ by its definition in terms of the generalized eigenvectors, the equality becomes

$$\sum_{r=1}^{k} u_{k\alpha}^r \times u_{k\alpha}^{r*} d\alpha = dE_c(\alpha), \qquad \alpha \in \Sigma_k$$

or more explicitly

$$\sum_{r=1}^{k} \langle u_{k\alpha}^r; \phi \rangle \langle u_{k\alpha}^{r*}; \psi \rangle^* d\alpha = d\langle \phi, E_c(\alpha)\psi \rangle, \qquad \alpha \in \Sigma_k.$$

If we set $\Gamma_H(\alpha) = 0$ for $\alpha \notin \Sigma_c$, the spectral family of H_c is obviously given by

$$E_c(\alpha) = \int_{-\infty}^{\alpha} d\alpha' \, \Gamma_H(\alpha')$$

and the spectral family of H is obtaining by adding to $E_c(\alpha)$ the projection valued ladder function with discontinuities E_{Hki} at the eigenvalues α_{ki}.

1.17 *Unitary operators.* – We consider now a good unitary operator U on L^2. In this case $U^\ddagger$ coincides with U^{-1} for a distribution in L^2 and can be thought of as its extension to $S^\dagger$. Similarly $(U^{-1})^\ddagger$ can be thought as the extension of U to S. As above we use the same symbol for $U^\ddagger, U$ and U^o, and write the eigenvalue equation simply as $Uv_\beta = \beta v_\beta$. All the preceding results hold except that the generalized eigenvalues are complex numbers of modulus one. The spectral resolution of U can, then, be written

$$H = \sum_j e^{i\theta_j} E_{Uj} + \int d\theta \, e^{i\theta} \Gamma_U(\theta),$$

where the eigenvalues $\beta_j = e^{i\theta_j}$, $0 \leq \theta_j < 2\pi$, form the discrete spectrum and the integration is performed over those range of values of θ in $(0, 2\pi)$ such that $\beta = e^{i\theta}$ lies in the continuous spectrum. The operator $\Gamma_U(\theta)$ is related to the generalized eigenvectors u_β in the manner described above for a hermitean operator.

1.18 *Examples.* – As a first example consider the operator Q of B.2.5, namely the operator of multiplication by x on L^2. Although this example may seem rather trivial, it is useful to examine it in the context of the general theory. It illustrates some of the concepts we have been discussed and exhibits, in particular, several different meanings of the δ-function. The generalized eigenvalue equation $xu_\xi = \xi u_\xi$ becomes in this case

$$(x - \xi)u_\xi = 0$$

whose general solution by 1.8.1 is the distribution $c\delta(x - \xi)$ which exists for all real ξ. Thus the spectrum is pure continuous and consists of the whole of the real axis. The "orthonormality" relation of 1.15.3 requires that

$$|c| \int_{\infty}^{\infty} dx \; \delta(x - \xi)\delta(x - \xi') = \delta(\xi - \xi').$$

As mentioned in 1.14.3 the integrand can be regarded as the representation of the operator $\delta_x \times \delta_x^*$ on the space of test functions of the variable ξ, so that the integral has a meaning in its own right and equals $\delta(\xi - \xi')$. We can then satisfy the equation by choosing $c = 1$ and taking $u_\xi = \delta(x-\xi)$. That the completeness relation is satisfied can be immediately verified:

$$\int_{-\infty}^{\infty} d\xi \; u_\xi \times u_\xi^* = \int_{-\infty}^{\infty} d\xi \; \delta(x-\xi)\delta(x-\xi') = \delta(\xi-\xi').$$

The operator $\Gamma_Q(\xi) = u_\xi \times u_\xi^*$ has then a representation

$$\Gamma_{Q\xi}(x,x') = \delta(x-\xi)\delta(x'-\xi),$$

which gives the spectral resolution of Q:

$$Q(x,x') = \int_{-\infty}^{\infty} d\xi \; \xi\delta(x-\xi)\delta(x'-\xi) = x\delta(x-x'),$$

as befits the multiplicative operator x. The representation of the spectral family $E_Q(\xi)$ is then

$$E_{Q\xi}(x,x') = \int_{-\infty}^{\xi} d\xi' \; \delta(x-\xi')\delta(x'-\xi');$$

by use of the integral of the δ-function in 1.9.3 we obtain

$$\left(E_Q(\xi)\psi\right)(x) = \int_{-\infty}^{\xi} d\xi' \; \delta(x-\xi') \int_{-\infty}^{\infty} dx' \; \delta(x'-\xi')\psi(x') = \chi(\xi - x)\psi(x)$$

in agreement with our previous calculation in B.2.5.1.

1.18.1. We consider now the differential operator $P = -id\cdot/dx$ of B.2.6. Although this example can be reduced to the previous one by Fourier transformation, it is instructive to proceed directly. The eigenvalue equation

$$-iv'_\eta(x) = \mu v_\eta(x)$$

has, for each $\eta \in (-\infty, \infty)$, as general solution the distribution $v_\eta(x) = ce^{i\eta x}$ which is not in L^2. Thus, the pure continuous spectrum is the real axis. Completeness is satisfied if the constant c is chosen in such a way that

$$\int_{-\infty}^{\infty} dx \; v_\eta^*(x)v_{\eta'}(x) = \delta(\eta-\eta').$$

Since, as we saw in 1.9.3,

$$\int_{-\infty}^{\infty} dx\ e^{-i(\eta-\eta')x} = 2\pi\delta(\eta-\eta'),$$

we can choose $c = 1/\sqrt{2\pi}$ which yields

$$v_\eta = \frac{1}{\sqrt{2\pi}}\, e^{i\eta x}.$$

The operator $\Gamma_P(\eta)$ has representation

$$\Gamma_{P\eta}(x, x') = \frac{1}{2\pi}\, e^{i\eta(x-x')}$$

Therefore, the spectral resolution is given by

$$P(x, x') = \frac{1}{2\pi}\int_{-\infty}^{\infty} d\eta\ \eta e^{i\eta(x-x')} = i\delta'(x'-x),$$

where we have used the expression for the integral in 1.12.3. Consequently,

$$(P\psi)(x) = i\int_{-\infty}^{\infty} d\eta\ \delta'(x'-x)\psi(x') = -i\psi'(x).$$

as it should. The representation of the spectral family $E_P(\eta)$ is

$$E_{P\eta}(x, x') = \frac{1}{2\pi}\int_{-\infty}^{\eta} d\eta'\ e^{i\eta'(x-x')};$$

whence, its action on a test function

$$\left(E_P(\eta)\psi\right)(x) = \frac{1}{2\pi}\int_{-\infty}^{\eta} d\eta' \int_{-\infty}^{\infty} dx'\ e^{i\eta'(x-x')}\psi(x')$$

is in agreement with the result of B.2.6.

1.18.2 As a third example we study the differential operator $T = P^2 = -d^2\cdot/dx^2$. The generalized eigenvalue equation is

$$-w_\lambda''(x) = \lambda w_\lambda(x)$$

which for each real λ has a general solution that can be written

$$w_\lambda(x) = ae^{i\sqrt{\lambda}x} + be^{-i\sqrt{\lambda}x}.$$

By inspection we can deduce directly from this solution that the spectrum of T is pure continuous, for there is no λ for which a square integrable solution different from zero exists. Moreover, for negative λ the solution is not a (tempered) distribution; therefore the spectrum consists of the non-negative real axis. Since for each λ there are two linearly independent solutions, the spectrum is doubly degenerate. To satisfy the completeness relation we seek two families of solutions w_λ^1, w_λ^2 that satisfy the "orthonormality" conditions of 1.15.3, that is,

$$\int_{-\infty}^{\infty} dx\ w_\lambda^{1*}(x) w_{\lambda'}^2(x) = 0,$$

$$\int_{-\infty}^{\infty} dx\ w_\lambda^{1*}(x) w_{\lambda'}^2(x) = \int_{-\infty}^{\infty} dx\ w_\lambda^{2*}(x) w_{\lambda'}^2(x) = \delta(\lambda - \lambda').$$

This can be achieved by appropriately selecting the constants a and b of the general solution. For positive λ we calculate the following integrals with the help of 1.5.1

$$\int_{-\infty}^{\infty} dx\ e^{i(\sqrt{\lambda}+\sqrt{\lambda'})x} = 0,$$

$$\int_{-\infty}^{\infty} dx\ e^{i(\sqrt{\lambda}-\sqrt{\lambda'})x} = 2\pi\delta(\sqrt{\lambda} - \sqrt{\lambda'}) = 4\pi\sqrt{\lambda}\delta(\lambda - \lambda').$$

We can see, then, that an appropriate pair of linearly independent solutions is

$$w_\lambda^1(x) = (4\pi\sqrt{\lambda})^{-1/2} e^{i\sqrt{\lambda}x}, \qquad w_\lambda^2(x) = (4\pi\sqrt{\lambda})^{-1/2} e^{-i\sqrt{\lambda}x}.$$

As $\lambda \to 0$ they both approach a constant distribution. According to 1.16, the operator $\Gamma_T(\lambda)$ is $w_\lambda^1 \times w_\lambda^{1*} + w_\lambda^2 \times w_\lambda^{2*}$. The representation of the spectral family $E_T(\lambda)$ is, therefore,

$$E_{T\lambda}(x, x') = \frac{1}{4\pi} \int_0^\lambda d\lambda'\ \frac{e^{i\sqrt{\lambda'}(x-x')} + e^{i\sqrt{\lambda'}(x-x')}}{\sqrt{\lambda'}},$$

and its action on a test function,

$$(E_T(\lambda)\psi)(x) = \frac{1}{2\pi} \int_0^\lambda d\lambda' \int_{-\infty}^{\infty} dx'\ \frac{\cos\left[\sqrt{\lambda'}(x - x')\right]}{\sqrt{\lambda'}} \psi(x').$$

The reader may find it instructive to solve this problem by Fourier transformation.

2. Generalization to an arbitrary Hilbert space

2.1 *Generalization of the space of test functions.* – With appropriate modifications the results discussed in the previous section for the $L^2(-\infty,\infty)$-space are valid for any Hilbert space $\mathbb{H}$. The first thing needed is a subset of $\mathbb{H}$ that generalizes the space of test functions. Such a subset is formed by vectors ψ that "vanish sufficiently fast at infinity" with respect to a given basis on $\mathbb{H}$ in the sense that, if $\{b_k\}$, $k = 1, 2, \ldots$, is the given orthonormal basis and $\psi = \sum_k \alpha_k b_k$, then $\lim_{k\to\infty} k^m \alpha_k = 0$ for all $m \geq 0$. Clearly different bases may yield different subsets. We choose one such subset; it constitutes a vector space $\mathbb{H}^o$. For a reason similar to the one given in 1.1 for the space of test functions, $\mathbb{H}^o$ is not a subspace of the infinite-dimensional space $\mathbb{H}$. To make $\mathbb{H}^o$ complete one defines convergence of a sequence of vectors $\psi_n = \sum_k \alpha_{nk} b_k \in \mathbb{H}^o$ to the limit ψ by requiring that $k^m|\alpha_{nk} - \alpha_k|$ tend to zero for all k and $m \geq 0$.

2.1.1 That the space of test functions is a particular case of $\mathbb{H}^o$ can be seen by choosing as a basis in $L^2(-\infty,\infty)$ the harmonic oscillator eigenfunctions of VI.6.

$$\phi_k(x) = \frac{1}{\sqrt{2^k \pi^{\frac{1}{2}} k!}} e^{-\frac{1}{2}x^2} H_k(x),$$

where $H_k(x)$ is k-th Hermite polynomial. For a test function $\psi(x)$ one can verify that the components $\alpha_k = \int dx\ \psi^*(x)\phi_k(x)$ decrease at infinity faster than any negative power of k, as required.

2.2 *Generalization of the space of distributions.* – In the same way as the space $\mathbb{H}^o$ generalizes the space of test functions, its dual $\mathbb{H}^\ddagger = (\mathbb{H}^o)^\dagger$, that is, the vector space formed by the continuous linear functionals $\langle u; \psi \rangle$, $\psi \in \mathbb{H}^o$, generalizes the space of distributions. As before the elements of $\mathbb{H}^o$ can be regarded as elements of $\mathbb{H}$ or of $\mathbb{H}^\ddagger$ while the elements of $\mathbb{H}$ can be regarded as elements of $\mathbb{H}^\ddagger$. Although when $f \in \mathbb{H}$ is regarded as an element of $\mathbb{H}^\ddagger$ it should be written as $\langle f; \psi \rangle$, it is written simply as f if there is no danger of misunderstanding; note that in this case the functional $\langle f; \psi \rangle$ equals the scalar product $\langle f, \psi \rangle$ defined on $\mathbb{H}$. In general there are functionals $\langle u; \psi \rangle$ that cannot be regarded as vectors in $\mathbb{H}$. Even those functionals are usually written simply as u, omitting the independent variable.

2.2.1. In this appendix, when necessary for clarity, we call the elements of $\mathbb{H}^\ddagger$ "generalized vectors" and those of $\mathbb{H}^o$ "restricted vectors" to distinguish them from the vectors of $\mathbb{H}$. Also, as before, we refer to the operators on $\mathbb{H}$ that have a continuous restriction to $\mathbb{H}^o$ – the generalization of the operators introduced in 1.12 – as "good operators".

2.2.2. *If and only if* $\mathbb{H}$ *is finite-dimensional do the spaces* $\mathbb{H}^o$, $\mathbb{H}$ *and* $\mathbb{H}^\ddagger$ *coincide.* We thus have the relation

$$\mathbb{H}^o \subseteq \mathbb{H} \subseteq \mathbb{H}^\ddagger$$

with the equalities holding for unitary spaces and only for them.[2]

2.2.3. We shall not state explicitly the obvious generalization of all the pertinent results in 1.13–16. Suffice it to say that these results hold *verbatim* if the function spaces L^2, S and $S^\dagger$ are replaced, respectively, by the abstract vector spaces $\mathbb{H}$, $\mathbb{H}^o$ and $\mathbb{H}^\ddagger$. Nor shall we rewrite here the equations in a notation appropriate to an abstract vector space, for this will be done in the next section using Dirac's notation which is extremely well suited for such a purpose.

2.2.4. The question arises whether in the structure briefly presented in this section it is possible to omit all reference to the Hilbert space and deal exclusively with the space $\mathbb{H}^o$ and its dual $\mathbb{H}^\ddagger$. The answer is negative: the space $\mathbb{H}$ is necessary to define the hermiticity and unitarity of the operators and, moreover, the Hilbert space spectral theorem is needed in the proof of the generalized spectral results of 1.13–16.

2.3 *Different types of distributions.* – The generalization just sketched can be applied to the Hilbert space $L^2(a,b)$ of square integrable functions of the variable x with values on the interval (a,b) of the real axis. The space $S(a,b)$ of test functions is formed in this case by the infinitely differentiable functions that vanish outside (a,b). The continuous linear functionals on the test functions form the space $S^\dagger(a,b)$ of distributions. They are the ones originally defined by Schwartz and are usually referred to as distributions without any qualifier. In the space $S^\dagger(a,b)$, the δ-function can be defined as in Section 1 and so can be defined derivatives, integrals, limits and, *mutatis mutandis*, products of distributions. In this case, however, the Fourier transform theory does not apply.

2.3.1. More generally one can consider the space $L^2(\Delta)$ of functions defined over an arbitrary open region Δ of the real line. If Δ is bounded we have a situation almost identical to the one of the previous paragraph. If Δ is unbounded and coincides with the whole of the real line we have the case studied in Section 1. Otherwise the space $S(\Delta)$ of test functions is formed by the infinitely differentiable functions that vanish outside Δ and fall at infinity faster than any power of x. On the dual space $S^\dagger(\Delta)$, one can again define the δ-function as well as derivatives, limits and integral of distributions in the same manner as for tempered distributions.

[2] The set of the three spaces just introduced is often referred to in the literature as a *rigged Hilbert space*.

3. Dirac's formulation

As discussed in Chapter V, there are physical reasons which lead us to expect that pure states of physical systems are adequately represented in a Hilbert space by vectors that "vanish sufficiently fast at infinity" in the sense of 2.1. At the same time, as was also remarked there, it is necessary to consider idealized observables that correspond to unbounded operators with continuous spectra. These two circumstances point to the structure discussed in the preceding section as a well suited one to describe the states and the observables of quantum mechanics. In this section we present Dirac's formulation as an appropriate notation into which to fit that structure.

3.1 *Bras and kets.* – As discussed in Section 2 we work with a triplet of spaces $\mathbb{H}^o \subset \mathbb{H} \subset \mathbb{H}^\ddagger$ and represent the restricted vectors of $\mathbb{H}^o$ by ψ, ϕ, the vectors of $\mathbb{H}$ by f, g and the generalized vectors of $\mathbb{H}^\ddagger$ by u, v. The scalar product $\langle u; \psi\rangle$ is written in Dirac's notation as $\langle u|\psi\rangle$. Of course, as already remarked, if $u = f$ then $\langle f|\psi\rangle$ equals the scalar product $\langle f, \psi\rangle$ defined on $\mathbb{H}$. The bra $\langle u|$ represents a vector of $\mathbb{H}^\ddagger$ and the ket $|\psi\rangle$ a vector of $\mathbb{H}^o$. One could consistently abide by the convention of representing the generalized vectors by bras and the restricted vectors by kets but it is more convenient to define

$$\langle\psi|u\rangle = \langle u|\psi\rangle^*.$$

3.1.1. To give meaning to the symbol $\langle\psi|$ one can introduce the notion of the complex conjugate $\mathbb{V}^*$ of a vector space $\mathbb{V}$. To every element $f \in \mathbb{V}$ there corresponds a unique $f^* \in \mathbb{V}^*$ in such a way that to $\alpha f + \beta g$ there corresponds $\alpha^* f^* + \beta^* g^*$. Evidently $\mathbb{V}$ and $\mathbb{V}^*$ can be identified with each other but in the present context it is convenient not to do so.

3.1.2. We think of $\langle f|$ as the bra in the complex conjugate space $\mathbb{H}^*$ corresponding to the ket $|f\rangle \in \mathbb{H}$. The bras $\langle\psi|$ form the space $\mathbb{H}^{o*}$, the complex conjugate of $\mathbb{H}^o$, and the antilinear functionals $\langle\psi|u\rangle$ on $\mathbb{H}^{o*}$ constitute the space $\mathbb{H}^{\ddagger*}$, the complex conjugate of $\mathbb{H}^\ddagger$. To sum up: a ket is either a restricted vector or the complex conjugate of a generalized vector; a bra is either a generalized vector or the complex conjugate of a restricted vector.

3.1.3. The outcome of all this rather trivial proliferation of complex conjugate spaces is that the symbol $\langle\cdot|\cdot\rangle$ has a meaning whenever one of the arguments represent a vector of $\mathbb{H}^o$ and the other a vector of $\mathbb{H}^\ddagger$.

3.2 *Operators.* – The operators used in Dirac's formulation are hermitean or unitary "good" operators in the sense of 2.2.1 that have furthermore an absolutely continuous spectrum when regarded as operators on $\mathbb{H}$. Except when necessary for stress we shall omit the qualifiers and refer to them simply as operators. Let A be an operator on $\mathbb{H}$, A^o its continuous restriction to $\mathbb{H}^o$ and $A^\ddagger$ the extension of $A^\dagger$ to $\mathbb{H}^\ddagger$. No distinction in notation is made between A and A^o, or $A^\dagger$ and $A^\ddagger$; the restricted vector $A\psi$ is written as $A|\psi\rangle$ and the generalized vector $A^\ddagger u$ as $\langle u|A^\dagger$, while the respective complex conjugate vectors are represented by $\langle\psi|A$ and $A^\dagger|u\rangle$. Consequently the scalar products $\langle A^\ddagger u;\psi\rangle = \langle u; A^o\psi\rangle$ are both written as $\langle u|A|\psi\rangle$, while $\langle u|A^\dagger|\psi\rangle$ represents either of the corresponding products $\langle A^{\dagger\ddagger}u;\psi\rangle = \langle u; A^{\dagger o}\psi\rangle$ for the adjoint $A^\dagger$ (observe that $A^{\dagger\ddagger}$ is the extension to $\mathbb{H}^\ddagger$ of A). These products can also be written in terms of the complex conjugate vectors,

$$\langle\psi|A|u\rangle^* = \langle u|A^\dagger|\psi\rangle, \qquad \langle\psi|A^\dagger|u\rangle = \langle u|A|\psi\rangle^*,$$

expressions that are formally identical to the corresponding ones in the finite-dimensional case.

3.2.1. The generalization of the operator $u\times v^*$ of 1.12.1 is written as $|u\rangle\langle v|$. It is the operator from $\mathbb{H}^o$ to $\mathbb{H}^{\ddagger *}$ defined by $(|u\rangle\langle v|)\,|\psi\rangle = |u\rangle\langle v|\psi\rangle$. In the finite-dimensional case when $\mathbb{H}^o$ and $\mathbb{H}^{\ddagger *}$ coincide with $\mathbb{H}$ and $\mathbb{H}^*$, respectively, it becomes the operator introduced in A.2.14.2; as in that paragraph it is formally advantageous to define $|u\rangle\alpha\langle v| = \alpha|u\rangle\langle v|$. When $|f\rangle \in \mathbb{H}$, $|f\rangle\langle f|$ is just the one-dimensional projection on the subspace spanned by $|f\rangle$.

3.2.2. A most important consequence of restricting the formulation to good operators is the elimination of domain difficulties: the algebraic operations between operators are well defined. For example, if A and B are good operators then $A|\psi\rangle \in \mathbb{H}^o$, $\langle u|A \in \mathbb{H}^\ddagger$ and, therefore, $(BA)|\psi\rangle$ and $\langle u|(BA)$ are defined.

3.3 *The generalized eigenvalue problem.* – As in 1.13, the generalized eigenvalue problem for an operator A is said to have a solution $\langle\mu| \in \mathbb{H}^\ddagger$ if there exists a complex number μ such that,

$$\langle\mu|A|\psi\rangle = \mu^*\langle\mu|\psi\rangle,$$

for all $|\psi\rangle \in \mathbb{H}^o$. One writes the preceding equation in two equivalent abbreviated forms

$$\langle\mu|A^\dagger = \mu^*\langle\mu|, \qquad A|\mu\rangle = \mu|\mu\rangle.$$

The solutions $\langle\mu|$ or $|\mu\rangle$ are, respectively, the **eigenbras** or **eigenkets** of A belonging to the generalized eigenvalue μ. When the context permits, one omits often the adjective and refers to μ simply as an eigenvalue. As before, we discuss the eigenvalue problem and its relationship to the spectral problem for hermitean operators and point out later the modifications for unitary operators. Again, for the sake of clarity and because of its special interest, we present first the case of operators with a pure continuous spectrum that coincides with the whole of the real axis.

3.4 *Operators with a simple pure continuous spectrum.* – Let H be a hermitean operator on $\mathbb{H}$ with a pure continuous spectrum Σ. For each $\alpha \in \Sigma$ the eigenvalue problem

$$A|\alpha\rangle = \alpha|\alpha\rangle$$

has as a unique set of linearly dependent solutions of the form $c|\alpha\rangle$. The family of eigenkets $|\alpha\rangle$ can and will be chosen to form a complete family in the sense that for any pair of vectors $|\phi\rangle, |\psi\rangle \in \mathbb{H}^o$, we have

$$\langle\phi|\psi\rangle = \int_\Sigma d\alpha\ \langle\phi|\alpha\rangle\langle\alpha|\psi\rangle;$$

a relation that can be written as the operator equation

$$I = \int_\Sigma d\alpha\ |\alpha\rangle\langle\alpha|,$$

which, if applied to the vector $|\psi\rangle$, yields the equivalent relation

$$|\psi\rangle = \int_\Sigma d\alpha\ |\alpha\rangle\langle\alpha|\psi\rangle.$$

The three equivalent completeness relations are often expressed by the, symbolic but useful, "orthonormality" relation[3]

$$\langle\alpha|\alpha'\rangle = \delta(\alpha - \alpha').$$

In this case the generalization of the operator $u_\alpha \times u_\alpha^*$ of 1.14.1 is $|\alpha\rangle\langle\alpha|$ so that the spectral resolution of H is

$$H = \int_\Sigma d\alpha\ \alpha|\alpha\rangle\langle\alpha|,$$

[3] It is customary to distinguish the families of eigenbras and eigenkets belonging to different operators by using different letters, while different members of the family are distinguished by primes. Thus $|\alpha\rangle$, $|\alpha''\rangle$, represent two members of the family of eigenkets of H, while $\langle\beta|$, $\langle\beta'|$ represents two members of the family of eigenbras of some other operator.

and the formulae in 1.14.6 for the spectral family $E_H(\alpha)$ generalize to

$$|\alpha\rangle\langle\alpha| = \frac{dE_H(\alpha)}{d\alpha}$$

and

$$E_H(\alpha) = \int_{-\infty}^{\alpha} d\alpha' \; |\alpha'\rangle\langle\alpha'|.$$

3.4.1. If we write

$$\psi(\alpha) = \langle\alpha|\psi\rangle,$$

the first of the three forms of the completeness relation becomes

$$\langle\phi|\psi\rangle = \int_{\Sigma} d\alpha \phi^*(\alpha)\psi(\alpha)$$

which is identical to the expression of the scalar product in the function space $L^2(\Sigma)$. In fact the functions $\psi(\alpha) = \langle\alpha|\psi\rangle$, $\psi \in \mathbb{H}^o$, constitute the space $S(\Sigma)$ of test functions in $L^2(\Sigma)$. The function $\psi(\alpha)$ is the representation of ψ in the H-function space of B.3.2. The corresponding representation of a generalized vector $u \in \mathbb{H}^\ddagger$ is a continuous linear functional, i.e., a distribution in $S^\dagger(\Sigma)$,

$$\langle u|\psi\rangle = \int_{\Sigma} d\alpha \; u^*(\alpha)\psi(\alpha).$$

In Dirac's notation one writes $u(\alpha) = \langle\alpha|u\rangle = \langle u|\alpha\rangle^*$ so that the preceding equation becomes

$$\langle u|\psi\rangle = \int_{\Sigma} d\alpha \; \langle u|\alpha\rangle\langle\alpha|\psi\rangle.$$

The new meaning of the symbol $\langle\cdot|\cdot\rangle$ that we have just defined is consistent with the "orthonormality relation" in 3.4 which simply says that the representation of the vector $\langle\alpha'| \in \mathbb{H}^\ddagger$ is the distribution $\delta(\alpha - \alpha')$. With this definition the above equation can be obtained by substituting the expression for the identity operator in 3.4 into the equality $\langle u|\psi\rangle = \langle u|I|\psi\rangle$.

3.4.2. The H-function space representation of the vector $F(H)|\psi\rangle$ can be obtained from the spectral resolution of H in 3.4,

$$\langle\alpha|F(H)|\psi\rangle = \int_{\Sigma} d\alpha' \; F(\alpha')\langle\alpha|\alpha'\rangle\langle\alpha'|\psi\rangle = F(\alpha)\langle\alpha|\psi\rangle,$$

that is,

$$(F(H)\psi)(\alpha) = F(\alpha)\psi(\alpha).$$

The operator $F(H)$ is represented in the function space by multiplication by $F(\alpha)$ in agreement with B.3.2.

3.4.3. Let A be either a good operator on $\mathbb{H}$ or an operator mapping continuously $\mathbb{H}^o$ onto $\mathbb{H}^\ddagger$. Inserting the expression for the identity into the equality $A = IAI$ we obtain formally

$$A = \int_\Sigma d\alpha \int_\Sigma d\alpha' \; |\alpha\rangle\langle\alpha|A|\alpha'\rangle\langle\alpha'| = \int_\Sigma d\alpha \int_\Sigma d\alpha' \; \langle\alpha|A|\alpha'\rangle|\alpha\rangle\langle\alpha'|.$$

This expression is, in fact, valid in the sense that

$$\langle\phi|A|\psi\rangle = \int_\Sigma d\alpha \int_\Sigma d\alpha' \; \langle\alpha|A|\alpha'\rangle\langle\phi|\alpha\rangle\langle\alpha'|\psi\rangle$$

which, upon comparison with 1.12, shows that $\langle\alpha|A|\alpha'\rangle$ is the distribution of two variables that represents the operator A in the space of functions of α.

3.4.4. The results of the preceding paragraph constitute a generalization of the decomposition of an operator given in A.2.14.3 in terms of the vectors $|b_i\rangle$ of an orthonormal basis.

$$A = \sum_i \sum_j A_{ij}|b_i\rangle\langle b_j|, \qquad A_{ij} = \langle b_i|A|b_j\rangle.$$

In the present context, A_{ij} is the distribution of two discrete variables that represents A in the discrete function space induced by an operator with a simple spectrum and eigenvectors b_i.

3.5 *Operators with a general spectrum.* – Let H be a good hermitean operator on $\mathbb{H}$ with a mixed spectrum. As in 1.15, it will have a discrete spectrum Σ_d formed of eigenvalues α_{ki} of degeneracy k and a continuous spectrum $\Sigma_c = \bigcup_k \Sigma_k$ where Σ_k has degeneracy $k = 1, 2, \ldots$. The generalization of the results stated in 1.15 tells us that the scalar product $\langle\phi|\psi\rangle$ of any two restricted vectors can be written as

$$\langle\phi|\psi\rangle = \sum_k \sum_{r=1}^{k} \Big(\sum_i \langle\phi|\alpha_i, r, k\rangle\langle\alpha_i, r, k|\psi\rangle + \int_{\Sigma_k} d\alpha \; \phi|\alpha, r, k\rangle\langle\alpha, r, k|\psi\rangle \Big).$$

3.5.1. The orthonormal vectors $|\alpha_i, r, k\rangle \in \mathbb{H}$, $r = 1, \ldots, k$ are eigenvectors of H belonging to the k-degenerate eigenvalues $\alpha_{ki} \in \Sigma_d$ and the $|\alpha, r, k\rangle \in \mathbb{H}^{\ddagger *}$

are generalized eigenvectors belonging to the k-degenerate generalized eigenvalues $\alpha \in \Sigma_k$. They obey the eigenvalue equations

$$H|\alpha_i, r, k\rangle = \alpha_{ki}|\alpha_i, r, k\rangle, \qquad H|\alpha, r, k\rangle = \alpha|\alpha, r, k\rangle.$$

3.5.2. The preceding expansion of $\langle\phi|\psi\rangle$ can be written in abbreviated form as the resolution of the identity

$$\sum_k \sum_{r=1}^{k} \Big(\sum_i |\alpha_i, r, k\rangle\langle\alpha_i, r, k| + \int_{\Sigma_k} d\alpha \; |\alpha, r, k\rangle\langle\alpha, r, k| \Big) = I.$$

Or as an expansion of $|\psi\rangle$ in terms of the set of eigenkets

$$|\psi\rangle = \sum_k \sum_{r=1}^{k} \Big(\sum_i |\alpha_i, r, k\rangle\langle\alpha_i, r, k|\psi\rangle + \int_{\Sigma_k} d\alpha \; |\alpha, r, k\rangle\langle\alpha, r, k|\psi\rangle \Big).$$

The completeness is again expressed by "orthonormality" relations

$$\langle\alpha_i, r, k|\alpha_{i'}, r', k'\rangle = \delta_{kk'}\delta_{ii'}\delta_{rr'}, \qquad \langle\alpha, r, k|\alpha_i, r, k\rangle = 0,$$

$$\langle\alpha, r, k|\alpha', r', k'\rangle = \delta_{kk'}\delta_{rr'}\delta(\alpha - \alpha').$$

3.5.3. As in 1.16 we can use the expression of the identity to obtain the following, in general non-unique, expression for H

$$H = \sum_k \Big(\sum_i \alpha_{ki} \sum_{r=1}^{k} \sum_{r=1}^{k} |\alpha_i, r, k\rangle\langle\alpha_i, r, k| + \int_{\Sigma_k} d\alpha \; \alpha \sum_{r=1}^{k} |\alpha, r, k\rangle\langle\alpha_i, r, k| \Big).$$

3.5.4. The generalization of the operators $\Gamma_H(\alpha)$ of 1.16 is

$$\Gamma_H(\alpha) = \sum_{r=1}^{k} |\alpha, r, k\rangle\langle\alpha, r, k|, \qquad \alpha \in \Sigma_k.$$

Introducing also the projection operators

$$E_{Hki} = \sum_{r=1}^{k} |\alpha_i, r, k\rangle\langle\alpha_i, r, k|,$$

the spectral resolution of H can be written

$$H = H_d + H_c$$

with

$$H_d = \sum_k \sum_i \alpha_{ki} E_{Hki}, \qquad H_c = \int_{\Sigma_c} d\alpha\ \alpha \Gamma_H(\alpha).$$

The spectral family of H_c, the part of H with a pure continuous spectrum, is

$$E_c(\alpha) = \int_{-\infty}^{\alpha} d\alpha'\ \Gamma_H(\alpha'),$$

where $\Gamma_H(\alpha) = 0,\ \alpha \notin \Sigma_c$.

3.6 *Representation with respect to a family of eigenkets.* – We saw that the complete family of eigenkets $|\alpha\rangle$ of 3.4. induced a representation of $\mathbb{H}$ as a function space. The question arises whether this is true also of the general complete family of eigenkets of 3.5. The answer is in the affirmative but the function space is, as one can expect, somewhat more complicated.

3.6.1. To elucidate its structure it is useful to consider first the case of a complete set of eigenkets belonging to an operator H with a simple mixed spectrum. We call the discrete spectrum Λ and the continuous spectrum Σ and write its spectral resolution

$$H = H_d + H_c,$$

$$H_d = \sum_i \alpha_i |\alpha_i\rangle\langle\alpha_i|, \qquad H_c = \int_{\Sigma} d\alpha\ \alpha|\alpha\rangle\langle\alpha|,$$

where $\alpha_i \in \Lambda$ are the eigenvalues and $\alpha \in \Sigma$ the generalized eigenvalues. The completeness relation reads then

$$\langle\phi|\psi\rangle = \sum_i \langle\phi|\alpha_i\rangle\langle\alpha_i|\psi\rangle + \int_{\Sigma} d\alpha\ \langle\phi|\alpha\rangle\langle\alpha|\psi\rangle.$$

As stated in B.2.4.4, the Hilbert space $\mathbb{H}$ can be decomposed into a direct sum $\mathbb{H}_d \oplus \mathbb{H}_c$ in such a way that H_d can be regarded as an operator on $\mathbb{H}_d$ with a simple pure discrete spectrum and H_c as one on $\mathbb{H}_c$ with a simple pure continuous spectrum. The H_d-representation of $\mathbb{H}_d$ induced by the operator H_d is the l^2-space of functions $f_d(\alpha)$ of the discrete variable $\alpha \in \Lambda$ defined by $f_d(\alpha_i) = \langle\alpha_i|f_d\rangle$, $|f_d\rangle \in \mathbb{H}_d$ (or, equivalently, of the sequences $\{\langle\alpha_i|f_d\rangle\}$) on which test functions $\psi_d(\alpha)$ and distributions $u_d(\alpha)$ can also be defined. The H_c-representation of $\mathbb{H}_c$ is identical to the one studied in 3.4.2.

3.6.2. The representation of $\mathbb{H}$ is, therefore, the direct sum $l^2 \oplus L^2$ and the representation of the vector $|f\rangle = |f_d\rangle + |f_c\rangle$, is the function

$$f(\alpha) = f_d(\alpha) \oplus f_c(\alpha),$$

where

$$f_d(\alpha) = \langle\alpha|f_d\rangle = \langle\alpha|f\rangle,\ \alpha \in \Lambda; \qquad f_c(\alpha) = \langle\alpha|f_c\rangle = \langle\alpha|f\rangle,\ \alpha \in \Sigma.$$

An expression that applies, as in 3.4.1, even if $\langle f| \in \mathbb{H}^\ddagger$ in which case $f(\alpha)$ is a distribution. With this notation the scalar product $\langle\phi|\psi\rangle$ becomes

$$\langle\phi|\psi\rangle = \sum_{\alpha\in\Lambda} \phi_d^*(\alpha)\psi_d(\alpha) + \int_\Sigma d\alpha\ \phi_c(\alpha)\psi_c(\alpha).$$

3.6.3. Consider finally the general case of the complete family of eigenkets of the operator in 3.5. The Hilbert space can be decomposed into a direct sum $\mathbb{H} = \mathbb{H}_1 \oplus \mathbb{H}_2 \oplus \ldots$, and correspondingly the operator H can be written $H = H_1 + H_2 + \ldots$ such that H_k regarded as an operator on $\mathbb{H}_k$ has a mixed spectrum of degeneracy k; let Λ_k be its discrete spectrum and Σ_k its continuous one, $\mathbb{H}_k$ can further be decomposed into k orthogonal subspaces $\mathbb{H}_k = \mathbb{H}_k^1 \oplus \ldots \oplus \mathbb{H}_k^k$ and correspondingly $H_k = H_k^1 + \ldots + H_k^k$ where each of the operators H_k^r has a simple mixed spectrum consisting of Λ_k and Σ_k. According to 3.6.2, H_k^r induces a representation of $\mathbb{H}_k^r$ as a space of functions where the vector $|f_k^r\rangle = |f_{kd}^r\rangle + |f_{kc}^r\rangle$ is represented by the function

$$f_k^r(\alpha) = f_{kd}^r(\alpha) \oplus f_{kc}^r(\alpha),$$

where,

$$f_{kd}^r(\alpha) = \langle\alpha|f_{kd}^r\rangle = \langle\alpha|f_k^r\rangle, \qquad \alpha \in \Lambda_k,$$

$$f_{kc}^r(\alpha) = \langle\alpha|f_{kc}^r\rangle = \langle\alpha|f_k^r\rangle, \qquad \alpha \in \Sigma_k.$$

The function space representation of the vector $f_k = \bigoplus_{r=1}^k f_k^r \in \mathbb{H}_k$ is then

$$f_k(\alpha) = \bigoplus_{r=1}^k f_k^r(\alpha).$$

One final step gives the representation of $f = \bigoplus_k f_k \in \mathbb{H}$ as the function.

$$f(\alpha) = \bigoplus_k f_k(\alpha).$$

Again, if $\langle f| \in \mathbb{H}^{\ddagger}$, $f(\alpha)$ is a distribution. We shall not give the general explicit expressions for these representations, for their complication makes it much more useful to write them in the particular cases in which they are needed.

D A REMINDER FROM CLASSICAL MECHANICS

1. Canonical transformations

1.1 *Lagrangian formalism.* – A classical system is characterized by a **lagrangian** $L(q, \dot{q}, t)$ depending on generalized coordinates $q = \{q_i\}$, $i = 1, 2, \ldots, n$, their corresponding velocities $\dot{q} = \{\dot{q}_i\}$ and the time. **Lagrange's equations** of motion

$$\frac{d}{dt}\left(\frac{\partial L}{\partial \dot{q}_i}\right) - \frac{\partial L}{\partial q_i} = 0$$

are obtained from the requirement that **Hamilton's functional**

$$\mathbf{\Phi}[q(\cdot)] = \int_{t_1}^{t_2} dt \; L(q, \dot{q}, t)$$

be stationary with respect to functions $q(t)$ fixed at t_1 and t_2. The canonical momentum conjugate to the coordinate q_i is $p_i = \partial L / \partial \dot{q}_i$.

1.2 *Hamiltonian formalism.* – The transformation theory of classical mechanics is more conveniently studied from the hamiltonian point of view where the state of a system is a point in the phase space of the canonical momenta $p = \{p_i\}$, and coordinates $q = \{q_i\}$, $i = 1, 2, \ldots, n$, and the system is characterized by a **hamiltonian** $H(p, q, t) = \sum_i p_i \dot{q}_i - L$.

1.2.1. In the hamiltonian formalism the motion of a system is a one-to-one, time depending mapping of the phase space onto itself: $p \mapsto p(t)$, $q \mapsto q(t)$. This **dynamical mapping** is obtained from the stationary points of Hamilton's functional which, in this formalism, is the functional $\mathbf{\Phi} = \mathbf{\Phi}[p(\cdot), q(\cdot)]$ of the $2n$ functions

$p_i(t), q_i(t)$ defined by

$$\Phi = \int_{t_1}^{t_2} dt \left[\sum_i \dot{q}_i p_i - H(p,q,t)\right]$$

with $q(t_1)$ and $q(t_2)$ fixed. The corresponding Euler-Lagrange equations are the **canonical equations** of motion, or **Hamilton's equations**

$$\dot{p}_i = -\frac{\partial H(p,q,t)}{\partial q_i}, \qquad \dot{q}_i = \frac{\partial H(p,q,t)}{\partial p_i}.$$

1.3 *Poisson brackets.* – The **Poisson bracket** of two dynamical variables $f(p,q,t)$ and $g(p,q,t)$ is, defined by

$$(f,g) = \sum_i \left(\frac{\partial f}{\partial p_i}\frac{\partial g}{\partial q_i} - \frac{\partial f}{\partial q_i}\frac{\partial g}{\partial p_i}\right).$$

1.3.1. The Poisson brackets are skew-symmetric and bilinear in their arguments,

$$(f,g) = -(g,f), \qquad (f, \alpha g + \beta h) = \alpha(f,g) + \beta(f,h).$$

and they obey the **Jacobi identity**

$$((f,g),h) + ((g,h),f) + ((h,f),g) = 0.$$

These relations are identical to the relations satisfied by the commutators of operators on a Hilbert space. It follows that classical results that depend only on the algebra of Poisson brackets can be directly transcribed into quantum mechanics and vice versa.

1.3.2. In terms of Poisson brackets Hamilton's equations can be written

$$\dot{x}_m = (H, x_m), \qquad m = 1, 2, \ldots, 2n,$$

where we have put $x_i = p_i$, $x_{i+n} = q_i$, $i = 1, 2, \ldots, n$. We shall use this notation whenever convenient, and write $x = \{x_m\} = \{p, q\}$.

1.3.3. Taking the derivative along the motion of a dynamical variable $f(x,t)$ we obtain

$$\dot{f}(x,t) = \sum_i \left(\frac{\partial f}{\partial p_i}\dot{p}_i + \frac{\partial f}{\partial q_i}\dot{q}_i\right) + \frac{\partial f}{\partial t},$$

or, by Hamilton's equations,

$$\dot{f}(x,t) = (H,f) + \frac{\partial f}{\partial t}.$$

The equations in 1.3.2 are particular cases of this one. In case $\dot{f}(x,t) = 0$, the dynamical variable $f(x,t)$ is called a **constant of the motion**.

1.4 *Dynamical transformations.* – We will consider general one-to-one transformations of the phase space onto itself and of the time $\{x,t\} \mapsto \{\bar{x},\bar{t}\}$. We write $\bar{x} = \bar{\xi}(x,t)$, $\bar{t} = \bar{\tau}(x,t)$, with inverse $x = \xi(\bar{x},\bar{t})$, $t = \tau(\bar{x},\bar{t})$. The transform $\bar{f}(x,t)$ of the dynamical variable $f(x,t)$ is defined by requiring that the value of $\bar{f}$ at the image $\{\bar{x},\bar{t}\}$ of $\{x,t\}$ be equal to the value of f at $\{x,t\}$, that is,

$$\bar{f}(\bar{x},\bar{t}) = f\left(\xi(\bar{x},\bar{t}),\tau(\bar{x},\bar{t})\right)$$

[which is often abbreviated to read $\bar{f}(\bar{x},\bar{t}) = f(x,t)$], or, equivalently,

$$\bar{f}\left(\bar{\xi}(x,t),\bar{\tau}(x,t)\right) = f(x,t)$$

1.4.1. Such transformations do not, in general, preserve the functional form of f. If a particular one does, i.e., if $\bar{f}(x,t) = f(x,t)$ or,

$$f(x,t) = f(\bar{\xi}(x,t),\bar{\tau}(x,t)),$$

one says that the dynamical variable is *invariant* under that transformation.

1.4.2. Of special interest are transformations that preserve the form of Hamilton's equations. More explicitly, transformations under which the canonical equations of motion become

$$\dot{\bar{p}}_i = -\frac{\partial \bar{H}'(\bar{p},\bar{q},\bar{t})}{\partial \bar{q}_i}, \qquad \dot{\bar{q}}_i = \frac{\partial \bar{H}'(\bar{p},\bar{q},\bar{t})}{\partial \bar{p}_i},$$

where the dot denotes differentiation along the motion parameterized by $\bar{t}$. The hamiltonian $\bar{H}'$ need not be the transform of H.

1.4.3. Although the two sets of trajectories in phase space defined by the equations of motion of 1.2.1 and 1.4.2 represent the possible motions of the same physical system, these sets do not in general coincide. If they do, that is, if the dynamical mapping is preserved, the transformation is called a **dynamical symmetry** of the system.

1.5 *Canonical transformations.* – In classical mechanics a very important role is played by the **canonical transformations** which constitute a subset of the transformations that preserve the form of Hamilton's equations. They are defined as phase space transformations $p \mapsto \bar{p},\ q \mapsto \bar{q}$ such that

$$\sum_i p_i dq_i - \sum_i \bar{p}_i d\bar{q}_i + (H' - H)dt = d\Psi,$$

where $\Psi = \Psi(q, \bar{q}, t)$ is an arbitrary function. It follows that the corresponding Hamilton's functional $\bar{\mathbf{\Phi}} = \mathbf{\Phi}[\bar{p}(\cdot), \bar{q}(\cdot)]$ differs from $\mathbf{\Phi}$ by a constant so that their stationary points represent the same motion. In terms of p and q the motion is given by the system of canonical equations in 1.2.1. In terms of $\bar{p}$ and $\bar{q}$ by the Euler-Lagrange equations for $\bar{\mathbf{\Phi}}$, namely,

$$\dot{\bar{p}}_i = -\frac{\partial \bar{H}'}{\partial \bar{q}_i}, \qquad \dot{\bar{q}}_i = \frac{\partial \bar{H}'}{\partial \bar{p}_i}$$

which (as the equations of 1.4.2) are also of Hamilton's form but, in general, with a different hamiltonian in a sense to be explained shortly.

1.5.1. If the $2n$ variables $q,\ \bar{q}$, can be regarded as independent, the function $\Psi(q, \bar{q}, t)$ is called a **generating function** of the canonical transformation $p \mapsto \bar{p} = \bar{P}(p, q, t),\ q \mapsto \bar{q} = \bar{Q}(p, q, t)$, which can be obtained by solving for $\bar{p}$ and $\bar{q}$ the system of equations

$$p_i = \frac{\partial \Psi(q, \bar{q}, t)}{\partial q_i}, \qquad \bar{p}_i = -\frac{\partial \Psi(q, \bar{q}, t)}{\partial \bar{q}_i}.$$

In accordance with our previous notation we write the inverse transformation as $p = P(\bar{p}, \bar{q}, t),\ q = Q(\bar{p}, \bar{q}, t)$.

1.5.2. The hamiltonians H and H' are seen from 1.5 to be related by

$$H'(q, \bar{q}, t) = H(q, \bar{q}, t) + F(q, \bar{q}, t), \qquad F(q, \bar{q}, t) = \frac{\partial \Psi(q, \bar{q}, t)}{\partial t}.$$

By using the change of variable $\bar{q} = \bar{Q}(p, q, t)$ of the previous paragraph one can write the above relation as

$$H'(p, q, t) = H(p, q, t) + F(p, q, t),$$

where the shorthand notation $f(p, q, t) = f(q, \bar{Q}(p, q, t), t)$ has been used. Applying the canonical transformation to the above relation one gets

$$\bar{H}'(\bar{p}, \bar{q}, t) = \bar{H}(\bar{p}, \bar{q}, t) + \bar{F}(\bar{p}, \bar{q}, t).$$

1.5.3. If and only if the generating function is time independent, that is, $\Psi(q, \bar{q}, t) = \Psi(q, \bar{q})$ or, equivalently, $F(q, \bar{q}, t) = 0$, are the hamiltonians H' and H equal. An equality that, as seen from the previous paragraph can be written as $H' = H$ or $\bar{H}' = \bar{H}$ (in general, of course, $\bar{H} \neq H$ and $\bar{H}' \neq H'$).

1.5.4. A canonical transformation is a dynamical symmetry if and only if, up to an additive function of t, H and $\bar{H}'$ have the same functional form: $\bar{H}'(x,t) = H(x,t) + c(t)$, for only in such case is the set of trajectories preserved. In case $\bar{H}'(x,t) = H(x,t)$, we shall say that the transformation *preserves* the hamiltonian. Special cases of such transformations are the time independent canonical transformations under which the hamiltonian is invariant.

1.5.5. By applying appropriate Legendre transformations to $\Psi(q, \bar{q}, t)$ one obtains a variety of generating functions. Every canonical transformation can be obtained from at least one of them. We are interested here in transformations that are close to the identity; they are most conveniently studied with a generating function

$$\Phi(q, \bar{p}, t) = \Psi + \sum_i \bar{p}_i \bar{q}_i$$

which leads to the equations

$$p_i = \frac{\partial \Phi}{\partial q_i}, \qquad \bar{q}_i = \frac{\partial \Phi}{\partial \bar{p}_i}, \qquad H' = H + \frac{\partial \Phi}{\partial t}.$$

1.5.6. In particular the *space inversion* $p \mapsto -p$, $q \mapsto -q$, has generating function $\sum_i -\bar{p}_i q_i$, while the identity is generated by $\sum_i \bar{p}_i q_i$.

2. Infinitesimal canonical transformations.

2.1. *Definition.* – We now consider a one-parameter family of canonical transformations with generating functions of the form

$$\Phi_\epsilon(q, \bar{p}, t) = \sum_i \bar{p}_i q_i + \epsilon \Gamma(\bar{p}, q, t, \epsilon),$$

depending continuously on the parameter ϵ. We are interested in the dependence on ϵ of $\bar{p}$, $\bar{q}$ and other dynamical variables for transformations Φ_ϵ close to the identity, that is, for ϵ small. Denoting by p_ϵ, q_ϵ and H_ϵ the approximations of $\bar{p}$, $\bar{q}$ and H' to first order in ϵ, we put $\delta_\epsilon p = p_\epsilon - p$, $\delta_\epsilon q = q_\epsilon - q$, $\Delta_\epsilon H = H_\epsilon - H$, and $G(p, q, t) = \Gamma(p, q, t, 0)$. By 1.5.5 we obtain, then, the transformation equations

$$\delta_\epsilon p_i = -\epsilon \frac{\partial G}{\partial q_i}, \qquad \delta_\epsilon q_i = \epsilon \frac{\partial G}{\partial p_i}, \qquad \Delta_\epsilon H = \epsilon \frac{\partial G}{\partial t},$$

the first two can also be written

$$\delta_\epsilon x_m = \epsilon(G, x_m).$$

These equations define the **infinitesimal canonical transformation** which gives the approximation to first order in ϵ of the canonical transformation generated by Φ_ϵ. The quantity $G(p,q,t)$, which is a dynamical variable, is the **generator** of the (family of) infinitesimal canonical transformation(s).

2.2 *Constants of the motion.* – Let $p(t), q(t)$ be the motion of a system with hamiltonian $H(p,q,t)$. Taking $p = p(t)$, $q = q(t)$, $G(p,q,t) = H(p,q,t)$ and differentiating with respect to ϵ the expressions for $\delta_\epsilon p$ and $\delta_\epsilon q$ of the previous paragraph, one finds

$$\left.\frac{\partial p_{\epsilon i}}{\partial \epsilon}\right|_{\epsilon=0} = -\frac{\partial H(p,q,t)}{\partial q_i}, \qquad \left.\frac{\partial q_{\epsilon i}}{\partial \epsilon}\right|_{\epsilon=0} = \frac{\partial H(p,q,t)}{\partial p_i}.$$

Now, one also has

$$\left.\frac{\partial p_i(t+\epsilon)}{\partial \epsilon}\right|_{\epsilon=0} = \dot{p}_i(t) = -\frac{\partial H(p,q,t)}{\partial q_i},$$

$$\left.\frac{\partial q_i(t+\epsilon)}{\partial \epsilon}\right|_{\epsilon=0} = \dot{q}_i(t) = \frac{\partial H(p,q,t)}{\partial p_i},$$

which shows that, instantaneously, *the motion is an infinitesimal canonical transformation whose generator is the hamiltonian.*[1]

2.2.1. A similar procedure applied to the equation for $\Delta_\epsilon H$ gives $\dot{H} = \partial H/\partial t$, a corollary of which is: *if the hamiltonian does not depend explicitly on time, then it is a constant of the motion.*

2.2.2. Consider again an infinitesimal transformation with generator $G(p,q,t)$ and a dynamical variable $f(p,q,t)$. Let $\delta_\epsilon f$ be, to first order in ϵ, the difference $\bar{f}(p,q,t) - f(p,q,t)$. According to 1.4 this difference can be written

$$f\left(P(p,q,t), Q(p,q,t), t\right) - f(p,q,t),$$

and, to first order,

$$\delta_\epsilon f = f(p - \delta_\epsilon p, q - \delta_\epsilon q, t) - f(p,q,t) = -\sum_i \left(\frac{\partial f}{\partial p_i}\delta_\epsilon p_i + \frac{\partial f}{\partial q_i}\delta_\epsilon q_i\right).$$

[1] For arbitrary time intervals, the motion, that is, the transformation $q(t_1) \mapsto q(t_2)$, $p(t_1) \mapsto p(t_2)$, is a canonical transformation whose generating function is Hamilton's principal function, i.e., the stationary value of Hamilton's functional.

Using the transformation equations of 2.1 and the definition of Poisson bracket, we have

$$\delta_\epsilon f = \epsilon(f, G).$$

2.2.3. In particular if the dynamical variable is the hamiltonian $H(p, q, t)$,

$$\delta_\epsilon H = \epsilon(H, G).$$

By the definition of Δ_ϵH in 2.1 and the relation between $\bar{H}'$ and $\bar{H}$ in 1.5.2, we find that $\Delta_\epsilon H = \delta_\epsilon H + \epsilon \partial G/\partial t$. By the above paragraph, the right member becomes

$$\epsilon(H, G) + \epsilon \partial G/\partial t,$$

which, according to 1.3.3 equals the derivative along the motion of ϵG. Thus we have

$$\Delta_\epsilon H = \epsilon \dot{G}.$$

If G is a constant of the motion, then $\Delta_\epsilon H = 0$, and we can state: *The constants of the motion are the generators of infinitesimal canonical transformations that preserve the hamiltonian.* As we know, these transformations are dynamical symmetries.

In particular, the constants of the motion that do not depend explicitly on time are the generators of infinitesimal canonical transformations under which the hamiltonian is invariant.

2.2.4. As an important example consider a system of particles and take the canonical coordinates to be the cartesian coordinates of the particles. Using vector notation, we put $q = \{\mathbf{x}_k\}$, $p = \{\mathbf{p}_k\}$ where the index k runs over all particles and $\mathbf{p}_k$ is the momentum conjugate to $\mathbf{x}_k$. Take the generator to equal the component of the total momentum along the unit vector $\mathbf{e}$: $G(p, q) = \mathbf{e} \cdot \sum_k \mathbf{p}_k$, then the transformation equations of 2.1 give $\delta_\epsilon \mathbf{p}_k = 0$, $\delta_\epsilon \mathbf{x}_k = \epsilon \mathbf{e}$. The corresponding infinitesimal canonical transformation is the family of displacements $\epsilon \mathbf{e}$: *the momentum is the generator of displacements*. Similarly one can see that the *angular momentum is the generator of rotations* by taking $G(p, q) = \mathbf{e} \cdot \sum_k \mathbf{x}_k \times \mathbf{p}_k$, the component of the total angular momentum along $\mathbf{e}$. The resulting transformation is a family of rotations by an angle ϵ about the axis $\mathbf{e}$. From the conclusion of the previous paragraph we thus have the following conservation laws. If the hamiltonian is invariant under displacements along $\mathbf{e}$, the component of the total momentum along $\mathbf{e}$ is conserved. If the hamiltonian is invariant under rotations about an axis, the component along that axis of the total angular momentum is conserved.

2.3 *Dynamical symmetries.* – Canonical transformations under which the hamiltonian is invariant are dynamical symmetries, but not all dynamical symmetries are canonical transformations. An example of fundamental importance is the time reversal $p \mapsto p,\ q \mapsto -q,\ t \mapsto -t$ which is a dynamical symmetry for many systems.

2.3.1. If and only if the Poisson bracket (f, g) of any two dynamical variables is unchanged by the transformation $p, q \mapsto \bar{p}, \bar{q}$, is this transformation canonical. It follows that the relations

$$(p_i, p_j) = (q_i, q_j) = 0, \qquad (p_i, q_j) = \delta_{ij}$$

are characteristic of a set of $2n$ canonical variables. This property shows that the time reversal transformation just mentioned is indeed not canonical.

NOTES

A bibliographical list is given at the end of these notes. The works listed have been selected because they have been directly used in the preparation of this book, or because they may be useful to complement the material presented in it. References to books are given to the most current edition, in English if available. Those interested in the history of the subject may find an excellent account and bibliography in Jammer (1966).

Introductory textbooks that cover the necessary background material are Cohen-Tannoudji *et al.* (1977) and Griffiths (1995). Classic graduate textbooks such as Bohm (1951), Dirac (1958), Merzbacher (1970) or Messiah (1962), as well as more recent ones like Capri (1985) or Peebles (1992), can be used profitably in conjunction with the present book.

Section I.1. – No interpretation of quantum mechanics is being implied here. When we say that an experiment can be described in plain language we are simply stating an empirical fact. An account of the different interpretations with an extensive bibliography can be found in Hughes (1989), part II.

Section I.3. – Although the point of view introduced in this section is widely used implicitly, I have not found any reference in which it is explicitly stated. The molecular beam experiment with helium schematically described here was first performed by Estermann and Stern (1930).

Section II.2. – The use of the spin 1/2 as an example to present the basic quantum mechanical ideas was suggested to me by Kaempffer (1965). The idealized Stern-Gerlach apparatus used in the examples was used by Bohm (1951), Ch. 22, §11 and by Wigner (1963) in discussions of measurement, and by Feynman *et al.* (1965), Ch. 5 to introduce the matrix of amplitudes.

Section II.3. – The view here taken that observables are hermitean operators is almost universally held. A more general class of observables has been proposed by Ludwig (1983), Ch. IV; for a pedagogical exposition, see Kraus (1983). An account of the discovery of the electron spin is given by Jammer (1966), §3.4.

Section III.2. – The distinction between kinds of measurements was introduced by Pauli (1933).

Section III.3. – The term "commensurable", which I prefer to "compatible" or "simultaneously measurable" is used by Ludwig (1983), Ch. IV.

Section IV.3. – Because the algebra of observables is central to the concept of system, the material has been included here rather than in the appendices. In the infinite-dimensional Hilbert space the algebra generated by an irreducible set of hermitian operators consists of all bounded operators; for a discussion see Jordan (1969), Ch. IV. Readers interested in the mathematical details are referred to Dixmier (1981), Part I, Ch. 1 §§1,2. In the Dirac formalism the subset of operators that belong to the algebra consists of all "good" operators (in the sense of C.2.2.1) whose adjoints are also good operators; a precise mathematical characterization can be found in Dubin and Hennings (1990), §4.1.

Section V.1. – This formulation of the rules is essentially the one of von Neumann (1955), Ch. III. In §3 of this chapter, von Neumann stresses that an actual instrument cannot measure an observable with a continuous spectrum but only approximations to it whose spectra are discrete.

Section V.2. – For his own formulation of the rules see Dirac (1958), §12. The restriction of the formalism to "good" operators also implies that we can regard any observable as a function of the operators of an irreducible set without having to worry about boundedness (cf. notes to Section IV–3).

Section V.3. – It is possible to distill the algebraic relations from the linear operators on a Hilbert space, one arrives in this way to an algebraic structure, a "lattice," in which one can describe in its utmost purity the kinematical structure of quantum mechanics. Such approach is taken, for example, by Jauch (1968) and Ludwig (1983). A review is given by Hughes (1989), Ch.7. An introductory text is Cohen (1989).

Sections V.5–6. – Pontrjagin (1939), is the standard reference on topological groups of which the Lie groups are a particular case. A discussion of the latter can also be found in Hamermesh (1962), Ch. 8, and a short summary is given in Bogolioubov *et al.* (1975), Part Two, App. C. In Hamermesh (1962), §§12.5–7 the irreducible representations of the full Galilei group are obtained. In the applications, most authors study the galilean invariance of *isolated* systems only. Jauch (1968), Chs. 12 and 13, is an exception.

Section V.7. – The existence of superselection rules, which at a time played an important part in discussions on "hidden variables," was pointed out by Wick *et al.* (1952). A more detailed discussion of the material in this section can be found in Jordan (1969) §28 and in Jauch (1960).

Sections VI.1–3. – The approach used here is inspired in Jauch (1968), Ch. 12.

Section VI.2. – The operators P and Q are unbounded. In the Hilbert space in which they form an irreducible set, they generate the algebra of all bounded operators. In the von Neumann approach of Section V.1, these bounded operators constitute the observables of the system so that P and Q themselves are not observables. Physically this is reasonable, for the configurations of actual instruments are bounded. It would be pleasing, however, if the mathematical theory allowed unbounded operators such as P and Q as observables that represent idealized instruments. Such is the case in the Dirac formalism (see notes to Section IV.3).

Section VII.6. – Introduced by Wigner (1932), the Wigner distribution function is discussed extensively by Moyal (1949). The Weyl correspondence is given in Weyl (1931), Ch. IV, §14.

Section VIII.4. – For the connection between time reversal invariance and "microscopic reversibility" see Messiah (1962) Ch. XV, §20.

Section VIII.5. – See the references to Sections V.5–6 for the Galilei group. In Currie *et al.* (1963) both Lorentz and galilean invariance of the hamiltonian are studied. The galilean invariance is also presented in Jordan (1969), Ch. 7, and a more detailed study in Ludwig (1983), Ch. VII.

Section IX.2–3. – Hamermesh (1962), Ch. 9, and Inui *et al.* (1990) give group theoretical treatments of angular momentum. More elementary presentations can be found in Messiah (1962), Ch. XIII, and in Merzbacher (1970), Ch. 16.

Section IX.4. – See Messiah (1962), Ch XIX, §§29–34, for a derivation of the low energy limit of the relativistic electron theory.

Section X.1. – Readers interested in the mathematical aspects may find the relevant results on the algebra of operators on the tensor product of Hilbert spaces in Dixmier (1981), Part I, Ch. 2, §§3,4 (particularly Exercise 4). Corresponding results for the Dirac formalism are given by Dubin and Hennings (1990) §4.13.

Section X.2. – The need to supplement galilean invariance by additional conditions on the single-particle observables was also pointed out by Currie *et al.* (1963) in their discussion of an isolated system of particles. The form of the interaction for two particles with spin was given by Eisenbud and Wigner (1941).

Section X.4. – This problem was solved by von Neumann (1955), Ch. VI, §2. The use of the tensor product formalism of A.4.7 results in a more concise treatment, which is along the lines of Jauch (1968), §11–8.

Section XI.1. – For brevity, we are using the notation for operators with a discrete spectrum. It can be translated to the continuum case in the usual way.

Section XI.3. – Hamermesh (1962), Ch. 7, gives a thorough account of the symmetric group; the main results are given in Inui *et al.* (1990). Merzbacher (1970), Ch. 21, §4, gives examples of the use of the number representation in atomic problems.

Section XI.5–6. – Applications of the theory developed here to systems of weakly interacting particles and its connection with the notion of a quasi-particle are given in Kaempffer (1965), §30.

Sections A.1–3. – A good reference for unitary spaces is Halmos (1958). Most important results are also given in Kato (1966), Ch. I.

Section A.4. – Jauch (1968), §11–7, and Ludwig (1983), App. IV, §14, define the tensor product of two Hilbert spaces as a set of transformations.

Section A.6. – The theorem in this section was given by Wigner (1959), Ch. 20 (appendix). The proof sketched here is due to Bargmann (1964).

Appendix B. – A very readable reference on the Hilbert space is Riesz and Sz.-Nagy (1955); Kato (1966), Ch. V, §§1–3, contains the main results; Reed and Simon (1972) give an extensive presentation. Von Neumann (1955) and Prugovečki (1971) discuss the Hilbert space in the context of quantum mechanics.

Sections C.1–2. – For an introduction to distributions see Lighthill (1960). A more detailed account of the basic material used in these sections can be found in Bogolioubov *et al.* (1975), Part One, Chs. 1–3. A leisurely presentation is given in Schücker (1991).

Section C.3. – The mathematical interpretation of Dirac's formalism in terms of rigged Hilbert spaces is due to Roberts (1966) and Antoine (1969). For a short but detailed presentation based on an example see Böhm (1978). Dubin and Hennings (1990) give a mathematical formulation of the framework of quantum mechanics in the spirit of Dirac's.

Appendix D. – Two good references for the canonical transformations of classical mechanics are Arnold (1978), Ch. 9, and Goldstein (1980), Ch. 9.

BIBLIOGRAPHY

Antoine, J. P. (1969). Dirac Formalism and Symmetry Problems in Quantum Mechanics. I. General Dirac Formalism. *J. Math. Phys.* **10**, 53-69.

Arnold, V. I. (1978). *Mathematical Methods of Classical Mechanics.* Springer-Verlag, New York.

Bargmann, V. (1964). Note on Wigner's Theorem on Symmetry Operations. *J. Math. Phys.* **5**, 862-868.

Bogolioubov, N. N., Logunov, A. A. and Todorov, I. T. (1975). *Introduction to Axiomatic Field Theory.* Benjamin, Reading, Mass.

Bohm, D. (1951). *Quantum Theory.* Prentice-Hall, New York, Reprint: Dover, New York, 1989.

Böhm, A. (1978). *The Rigged Hilbert Space and Quantum Mechanics.* Springer-Verlag, Berlin.

Capri, A. Z. (1985). *Non Relativistic Quantum Mechanics.* Benjamin/Cummings, Menlo Park, Cal.

Cohen D. W., (1989). *An Introduction to Hilbert Space and Quantum Logic.* Springer-Verlag, New York.

Cohen-Tannoudji, C., Diu, B. and Laloë, F. (1977). *Quantum Mechanics.* Hermann, Wiley, New York.

Currie, D. G., Jordan, T. F. and Sudarshan, E. C. G. (1963). Relativistic Invariance and Hamiltonian Theories of Interacting Particles. *Rev. Mod. Phys.* **35**, 350-375.

Dirac, P. A. M. (1958). *The Principles of Quantum Mechanics.* Oxford University Press, London.

Dixmier, J. (1981). *Von Neumann Algebras.* North-Holland, Amsterdam.

Dubin D. A. and Hennings M. A. (1990). *Quantum Mechanics Algebras and Distributions.* Longman. Essex.

Eisenbud, L. and Wigner, E. P. (1941). Invariant Forms of Interaction between Nuclear Particles. *Proc. Nat. Acad. Sciences,* **27**, 281-289.

Estermann, I. and Stern, O. (1931). Beugung von Molekularstrahlen. *Zeits. f. Phys.* **61**, 95-125.

Feynman, R. P., Leighton, R. B. and Sands, M. (1965). *The Feynman Lectures in Physics.* Vol. III. Addison-Wesley, Reading, Mass.

Goldstein, H. (1980). *Classical Mechanics.* Addison-Wesley, Reading, Mass.

Griffiths, D. J. (1995). *Introduction to Quantum Mechanics.* Prentice Hall, Englewood Cliffs, New Jersey.

Halmos, P. R. (1958). *Finite-Dimensional Vector Spaces.* Van Nostrand, Princeton.

Hamermesh, M. (1962). *Group Theory and its Application to Physical Problems.* Addison Wesley, Reading, Mass. Reprint: Dover, New York, 1989.

Hughes, R. I. G. (1989). *The Structure and Interpretation of Quantum Mechanics.* Harvard University Press, Cambridge, Mass.

Inui, T., Tanabe, Y., and Onodera, Y, (1990). *Group Theory and its Applications in Physics.* Springer-Verlag, Berlin.

Jammer, M. (1966). *The Conceptual Development of Quantum Mechanics.* McGraw-Hill, New York.

Jauch, J. M. (1960). Systems of Observables in Quantum Mechanics. *Helv. Phys. Acta,* **33**, 711-726.

Jauch, J. M. (1968). *Foundations of Quantum mechanics.* Addison- Wesley. Reading, Mass.

Jordan, T. F. (1969). *Linear Operators for Quantum Mechanics.* Wiley, New York.

Kaempffer, F. A. (1965). *Concepts in Quantum Mechanics.* Academic Press, New York.

Kato, T. (1966). *Perturbation Theory for Linear Operators.* Springer-Verlag, New York.

Kraus, K. (1983). *States, Effects and Operations.* Springer-Verlag, Berlin.

Lighthill, M. J. (1960). *Fourier Analysis and Generalised Functions.* Cambridge University Press, Cambridge.

Ludwig, G. (1983). *Foundations of Quantum Mechanics.* Springer-Verlag, New York.

Merzbacher, E. (1970). *Quantum Mechanics.* Wiley, New York.

Messiah, A. (1962). *Quantum Mechanics.* Vols. I and II. North Holland, Amsterdam. Wiley, New York.

Moyal, J. E. (1949). Quantum Mechanics as a Statistical Theory. *Proc. Cambridge Philos. Soc.* **45**, 99-124.

Peebles, P. J. E. (1992). *Quantum Mechanics.* Princeton University Press. Princeton.

Pauli, W. (1958). Die allgemeinen Prinzipien der Wellenmechanik. *Encyclopedia of Physics.* Vol. 5, part 1, p. 73.

Pontrjagin, L. (1939). *Topological Groups.* Princeton University Press, Princeton.

Prugovečki, E. (1971). *Quantum Mechanics in Hilbert Space.* Academic Press, New York.

Reed, M. and Simon, B. (1972). *Methods of Modern Mathematical Physics.* Vol I. Academic Press, New York.

Riesz, F. and Sz.-Nagy B. (1955). *Functional Analysis.* Ungar, New York.

Roberts, J. E. (1966). The Dirac Bra and Ket Formalism. *J. Math. Phys.* **7**, 1097-1104.

Schücker, T. (1991). *Distributions, Fourier Transforms and Some of their Applications to Physics.* World Scientific, Singapore.

von Neumann, J. (1955). *Mathematical Foundations of Quantum Mechanics.* Princeton University Press, Princeton.

Weyl, H. (1931). *The Theory of Groups and Quantum Mechanics.* Methuen, London. Reprint: Dover, New York, 1950.

Wick, G. C., Wightman, A. S., and Wigner, E. P. (1952). The Intrinsic Parity of Elementary Particles. *Phys. Rev.* **88**, 101-105.

Wigner, E. P. (1932). On the Quantum Correction for Thermodynamic Equilibrium. *Phys. Rev.* **40**, 749-759.

Wigner, E. P. (1959). *Group Theory.* Academic Press, New York.

Wigner, E. P. (1963). The Problem of Measurement. *Am. J. Phys.* **31**, 6-15.

INDEX

Page numbers in italics denote principal references.